Wastewater Stabilization Lagoon Design, Performance and Upgrading

Wastewater Stabilization Lagoon Design, Performance and Upgrading

E. Joe Middlebrooks
Charlotte H. Middlebrooks
James H. Reynolds
Gary Z. Watters
Sherwood C. Reed
Dennis B. George

Macmillan Publishing Co., Inc.
NEW YORK

Collier Macmillan Publishers
LONDON

Copyright © 1982 by Macmillan Publishing Co., Inc.

All rights reserved. No part of this book may be reproduced or transmitted in any form or by any means, electronic or mechanical, including photocopying, recording, or by any information storage and retrieval system, without permission in writing from the Publisher.

Scientific and Technical Books
A Division of Macmillan Publishing Co., Inc.
866 Third Avenue, New York, N. Y. 10022

Collier Macmillan Canada, Inc.

Library of Congress Catalog Card Number: 81-68476

Printed in the United States of America

printing number

1 2 3 4 5 6 7 8 9 10

Library of Congress Cataloging in Publication Data
Main entry under title:

Wastewater stabilization lagoon design, performance,
    and upgrading.

Includes index.
    1. Sewage lagoons.   I. Middlebrooks, E. Joe.
TD746.5.W363   628.3′51   81-68476
ISBN 0-02-949500-8   AACR2

Contents

Preface

APPROXIMATELY 90 PERCENT of the wastewater lagoons in the United States are located in small communities of 5000 people or fewer. These communities, many with an average daily wastewater flow of only 200,000 gal (757,000 l) or less, do not have the resources to keep workers at the lagoon sites throughout the day. A high degree of technical know-how is usually lacking in these communities. Often, only periodic inspection or maintenance is carried out by the general municipal work force. Therefore, the development of a relatively inexpensive method that does not require sophisticated and constant operation or extensive maintenance is needed to upgrade these effluents. Such a need exists throughout the world.

Most regulatory agencies are adopting more stringent water quality standards, enforcement of which will necessitate changes in both present treatment methods and the philosophy of wastewater treatment. Small communities that use stabilization ponds will be affected most drastically by the new standards. Because most communities using ponds are relatively small and install ponds primarily to lower operating costs, it is unlikely that modifications requiring significant increases in operation will be acceptable. Therefore, the more sophisticated alternatives for upgrading treatment must be excluded from a practicable solution to solids removal from stabilization pond effluent.

The removal of algae from stabilization pond effluent may be accomplished by many methods, and under certain conditions each process may be shown to be economical and operational. It is important that the most promising procedures available as a means of polishing stabilization pond effluents be discussed and proper design procedures be presented. That is the principal purpose of this book. An attempt has been made to present the overall picture of the application and design of wastewater stabilization ponds so that they meet new standards without excessive costs being incurred. Low maintenance and ease of operation must be designed into a system to allow small communities to continue to satisfactorily treat wastewater.

Wastewater Stabilization Lagoon Design, Performance and Upgrading

CHAPTER

1

Introduction to Lagoon Systems

THERE ARE MORE THAN 5000 publicly owned wastewater treatment lagoons in the United States. Generally, these are located in small communities and are designed for flows of less than 2 million gallons per day (mgd). Lagoons have been used where land is available, because operation is simple and operating costs are small. The low energy requirements of these systems are particularly attractive. The great majority of the existing systems are continuous flow facultative or oxidation lagoons.

There is a wide variation in the design of these systems, and until recently comprehensive performance data were lacking. An extensive evaluation supported by the EPA of four facultative and five aerated lagoons has provided considerable information on the capabilities of these systems, and a brief summary of these studies can be found in Chapter 2.

The results of these studies indicate that:

1. The biochemical oxygen demand (BOD) concentration in the effluent, in general, meets discharge requirements, and this can be achieved in all cases by improved design. Natural oxygen sources (i.e., algae, surface turbulence) are reduced or eliminated during the winter in ice-covered ponds. The BOD levels would not be met under these conditions for systems with a short detention time.
2. The suspended solids, (SS) concentration equivalent to "secondary" effluent quality is generally not achieved because of the algae in the effluent.
3. The pH value of the effluent varies markedly, depending on

1

alkalinity-CO_2 relationships. The variation is, however, rarely sufficient to require pH adjustment.

Aerated lagoons with properly designed SS separation can meet the adjusted effluent quality requirements. Because partial-mix systems permit some algae growth, SS can be high at times.

Controlled discharge ponds have been used in northern climates and can meet the federal requirements if properly operated. Such lagoons and other long-retention, multiple cell systems achieve significant reductions in fecal coliform (FC) concentrations. A FC concentration of approximately 200/100 ml can be achieved without disinfection if adequate hydraulic residence times are provided.

As indicated above, algae are generally responsible for the high SS concentrations found in lagoon effluents. Algae are naturally present in wastewater treatment lagoons, and the nonaerated types are specifically designed to rely on photosynthetic oxygenation. Algae cells are an integral part of the treatment system, do not settle readily, and may be carried out of the lagoon as SS in the effluent. Methods for removing algae from lagoon effluents have been developed and are described in Chapter 3.

Terminology

Many often conflicting terms are used to describe wastewater treatment lagoons or ponds. In this book a wastewater treatment (stabilization) pond or lagoon is defined as a basin within which natural stabilization processes occur. The terms pond and lagoon are frequently used interchangeably. The oxygen necessary to sustain some of these processes can come from photosynthetic and/or mechanical sources. Lagoons are also characterized by the hydraulic flow pattern in use. The following terminology is used for the lagoons discussed herein.

1. *Oxidation Lagoon*—A lagoon that is aerobic throughout. The depth is shallow to permit light penetration to support the photosynthetic activity of contained algae. The oxygen sources to support aerobic stabilization are from algae activity and wind action at the liquid surface.

2. *Facultative Lagoon*—A lagoon having an aerobic zone near the surface with a gradient to anaerobic conditions near the bottom. Oxygen sources are the same as described for oxidation ponds, but the oxygen provided cannot maintain total aerobic conditions in the deeper facultative pond. These are the most common types of domestic waste stabilization ponds in the United States.

3. *Aerated/Partial Mix Lagoon*—A lagoon designed for mechanical

aeration as the oxygen source. The mixing intensity is not sufficient to keep all solids in suspension. As a result, there will be some sludge deposition and related anaerobic zones at or near the lagoon bottom. The incomplete mixing also permits light penetration and can result in significant algae growth at times. Algae and turbulence at the liquid surface will provide some dissolved oxygen, but the design is usually based on mechanical aeration as the sole oxygen source.

4. *Aerated/Complete Mix Lagoon*—A lagoon designed for mechanical aeration as the oxygen source, and also with sufficient mixing intensity to keep solids in suspension. Algae will generally not be a factor, because of the turbulent conditions and lack of light penetration. Many complete mix lagoons are actually designed as a variation of the activated sludge process, including clarification and sludge recycle. Such complete mix systems must continue to satisfy the basic secondary treatment requirements of 40 CFR, Part 133.[1] However, lagoon systems that might have a complete mix aeration cell (for odor control or partial oxidation of strong wastes) followed by partial mix or facultative cells could be subject to the SS criteria discussed in this chapter.

5. *Continuous Discharge Lagoon*—A lagoon designed without imposed constraints on discharge. The actual discharge may be intermittent because of low seasonal flow, seasonal evaporation, or the like, but the design would permit continuous unrestricted discharge. All the lagoons described above (items 1–4) can be designed for this mode of operation.

6. *Controlled Discharge Lagoon*—A lagoon designed to retain the wastewater without discharge for a significant period of time (6 months–1 year). Discharge is then planned for a relatively short period (1–3 weeks) when lagoon characteristics are compatible with receiving water conditions. It might be possible to design any of the ponds defined above (1, 2, 3, 4) for this mode of operation; however, the long detention period reduces the need for mechanical aeration, so this mode of operation is most commonly found with facultative ponds.

7. *Complete Retention Lagoon*—A lagoon designed for evaporation and/or seepage as the hydraulic pathway so there is no discharge to surface waters. The method may be acceptable in locations having suitable climatic conditions and with proper regard for groundwater protection, odor control, and water rights (Figure 1–1). These lagoons are similar in configuration to the previously described oxidation type having a shallow depth and large surface area to provide maximum potential for evaporation. Another type of "no discharge" lagoon is a component in a land treatment system. These are discussed in Chapter 3.

Typical design factors for these different types of ponds are summarized in Table 1–1.

FIGURE 1-1
Aerial photograph of a complete retention lagoon located in Wellsville, Utah.

Performance Requirements

The Federal Water Pollution Control Act Amendments of 1972 (PL 92–500) established the minimum performance requirements for publicly owned treatment works. Section 301(b)(1)(B) of that act requires that such treatment works must, as a minimum, meet effluent limitations based on secondary treatment as defined by the EPA administrator. The EPA published information on secondary treatment in August 1973 in 40 CFR Part 133.[1] This

Table 1-1
Design Parameters for Wastewater Lagoons

Type of Pond	Detention Time (Days)	Depth (FT)	BOD Loading (LB/ACRE•DAY)	Number of Cells	Size of Cell (ACRE)
Oxidation	10–40	1.5–3	60–120	3+	2–10
Facultative					
Winter average					
air temperature					
Above 60°F	25–40	3.5	40–80	3–4	2–10
32–60°F	40–60	4–6	20–40	3–4	2–10
Below 32° F	80–180	5–7	10–20	3–4+	2–10+
Partial mix aerated	7–20	8–10	30–100	3+	2–10

Source: From Refs. 2 and 3.

contained criteria for BOD, SS, FC bacteria, and pH. Subsequently, the requirements for FC were deleted from 40 CFR Part 133 on July 25, 1976, leaving BOD, SS, and pH as originally defined.

Wastewater treatment lagoons have historically been accepted as a secondary treatment process and have been particularly advantageous for smaller communities. Treatment performance with respect to BOD removal has been generally acceptable if lagoons have been conservatively designed and properly operated. Treatment performance with respect to SS is often complicated by the presence of algae cells in the pond effluent. After considerable study and discussion, the EPA published revised SS limitations for wastewater treatment ponds on October 7, 1977.

The effluent criteria currently applicable to wastewater treatment ponds are as follows.

1. BOD_5—The arithmetic mean for 30 consecutive days shall not exceed 30 mg/l or 85% removal, whichever results in the lesser effluent concentration. The arithmetic mean of the values for 7 consecutive days shall not exceed 45 mg/l.

2. SS—Wastewater treatment lagoons that are the sole process for secondary treatment and with maximum facility design capacity of 2 mgd or less and meet the BOD_5 limitations as prescribed by 40 CFR 133.102(a) are required to meet an effluent limitation for SS in accordance with values set by the state or appropriate EPA regional office. The current values set by the states and regional offices are listed in Table 1-2. These values correspond to a 30-day consecutive average or an average over the period of discharge when the duration of the discharge is less than 30 days. In some cases states have developed additional values, such as weekly averages or daily maximums for compliance monitoring purposes. These additional values are not indicated in Table 1-2.

3. pH Values—The effluent pH values shall remain within the range of 6.0–9.0, unless variations are due to natural causes (i.e., natural pH of wastewater and/or phenomena resulting from biological activity in the lagoon). Adjustment of the pH value is required only when inorganic chemicals are added for treatment or when contributions from industrial sources cause the pH to be outside the range of 6.0–9.0.

More stringent performance requirements than defined above may be necessary to meet other criteria such as state water quality standards, including disinfection requirements.

Design Features

Design features common to most lagoon systems include earthen dikes and inlet and outlet works. Designs can usually be based on a balanced cut and

Table 1-2
Effluent Limitations for SS Concentrations in Lagoon Effluents from Systems Treating 2 mgd or Less[a]

LOCATION	SUSPENDED SOLIDS LIMIT (MG/L)
Alabama	90
Alaska	70
Arizona	90
Arkansas	90
California	95
Colorado	
Aerated Ponds	75
All others	105
Connecticut	N.C.[b]
Delaware	N.C.
District of Columbia	N.C.
Florida	N.C.
Georgia	90
Guam	N.C.
Hawaii	N.C
Idaho	N.C.
Illinois	37
Indiana	70
Iowa[c]	
Controlled Discharge, 3 cell	Case-by-case but not greater than 80
All others	80
Kansas	80
Kentucky	N.C.
Louisiana	90
Maine	45
Maryland	90
Massachusetts	N.C.
Michigan	
Controlled seasonal discharge	
Summer	70
Winter	40
Minnesota	N.C.
Mississippi	90
Missouri	80
Montana	100
Nebraska	80
North Carolina	90
North Dakota	
North and east of Missouri River	60
South and west of Missouri River	100
Nevada	90
New Hampshire	45
New Jersey	N.C.
New Mexico	90

(continued)

Table 1 - 2 (*continued*)

LOCATION	SUSPENDED SOLIDS LIMIT (MG/L)
New York	70
Ohio	65
Oklahoma	90
Oregon	
East of Cascade Mountains	85
West of Cascade Mountains	50
Pennsylvania	N.C.
Puerto Rico	N.C.
Rhode Island	45
South Carolina	90
South Dakota	110
Tennessee	100
Texas	90
Utah	N.C.
Vermont	55
Virginia[c]	
East of Blue Ridge Mountains	60
West of Blue Ridge Mountains	78
Eastern slope counties	case-by-case
Loudoun, Fauquier, Rappa-	application of 60/78
hannock, Madison, Green,	limits
Albermarle, Nelson, Amherst,	
Bedford, Franklin, Patrick	
Virgin Islands	N.C.
Washington	75
West Virginia	80
Wisconsin	60
Wyoming	100
Trust Territories and N. Marianas	N.C.

SOURCE: From Ref. 4.

[a] The values tabulated above can be considered as interim. Further revision and refinement will be possible as additional data become available. Wastewater treatment ponds with a maximum facility design capacity in excess of 2 mgd will have to satisfy the basic SS requirements of 40 CFR Part 133 (i.e., 30 mg/l or 85% removal on a 30-day average and, 45 mg/l on a 7 day average). Other lagoons that are not eligible for an adjustment in SS limitations include: basins or lagoons used as a final polishing step for other secondary treatment systems and ponds which include complete-mix aeration and sludge recycle, since these systems are in essence a variation of the activated sludge process.

[b] N.C.—no change from existing criteria.

[c] The values set for Iowa and Virginia incorporate a specific case-by-case provision; however, in accordance with 40 CFR 133.103(c), adjustments of the SS limitations for individual ponds in all states are to be authorized on a case-by-case basis.

fill so that most of the excavated material can be used in dike construction. Outside slopes of dikes are usually 3:1 or flatter to permit grass mowing. Inside slopes are steeper, ranging from 2:1 to 3:1. When the size of pond cells are much greater than 10 acres or if the system is in a particularly windy location, the inside slopes of the dikes should be protected from wave erosion. Membrane liners are easily punctured, and some are sensitive to solar radiation; therefore, it is common practice to overlay at least the above-water portion with soil and rip rap.

The inlet structure for small ponds is generally at the center. For large lagoons the use of inlet diffusers with multiple outlet ports is desirable to distribute the SS over a larger area. Transfer and outlet structures should permit lowering the water level at a rate of less than 1 ft/week when the facility is receiving its normal load. Manhole type structures are commonly used with either valved piping or adjustable "stop log" type overflow gates used to control depth.

Typical design parameters for continuous flow lagoons are given in Table 1-1. Most existing wastewater lagoons were designed within the range of parameters listed in Table 1-1. There are many local modifications employed. It is not uncommon for state agencies and EPA regions to have specific criteria for lagoon design. Several attempts have been made to develop a more rational basis for facultative lagoon design and are presented in Chapter 2. None of these can be used to predict consistently the performance of actual lagoons in a variety of settings, probably because of the variations in hydraulic residence times that occur in actual systems. Until methods are available to define hydraulic residence time more accurately, lagoon design will probably remain an empirical procedure based on successful past experience.

Controlled Discharge Lagoons. The unique features of controlled discharge lagoons are long-term retention and periodic, controlled discharge, usually once or twice a year. Lagoons of this type have operated satisfactorily in the North Central United States, using the following design criteria:

> Overall organic loading: 20–25 lb BOD_5/acre.
> Liquid depth: Not more than 6 ft for the first cell, not more than 8 ft for subsequent cells.
> Hydraulic detention: At least 6 months above the 2-ft liquid level (including precipitation), but not less than the period of ice cover.
> Number of cells: At least three for reliability, with piping flexibility for parallel or series operation.

The design of the controlled discharge lagoon must include an analysis showing that receiving stream water quality standards will be maintained during discharge intervals and that the receiving watercourses can accommodate the discharge rate from the lagoon. The design must also include a recommended discharge schedule.

Selecting the optimum day and hour for release of the lagoon contents is critical to the success of this method. The operation and maintenance manual must include instructions on how to correlate lagoon discharge with effluent and stream quality. The lagoon contents and stream must be carefully examined, before and during the release of the lagoon contents.

In a typical program, discharge of effluents follows a consistent pattern for all lagoons. The following steps are usually taken:

1. Isolate the cell to be discharged, usually the final one in the series, by valving-off the inlet line from the preceding cell.
2. Arrange to analyze samples for BOD, SS, volatile SS, pH, and other items which may be required for a particular location.
3. Plan to work so as to spend full time on control of the discharge throughout the period.
4. Sample contents of the cell to be discharged for dissolved oxygen, noting turbidity, color, and any unusual conditions.
5. Note conditions in the stream to receive the effluent.
6. Notify the state regulatory agency of results of these observations and plans for discharge; obtain approval.
7. If discharge is approved, commence discharge and continue so long as weather is favorable, dissolved oxygen is near or above saturation values, and turbidity is not excessive, following the prearranged discharge flow pattern among the cells. Usually this consists of drawing down the last two cells in the series (if there are three or more) to about 18–24 in. after isolation; interrupting the discharge for a week or more to divert raw waste to a cell that has been drawn down and resting the initial cell before its discharge. When this first cell is drawn down to about a 24-in. depth, the usual series flow pattern, without discharge, is resumed. During discharge to the receiving waters, samples are taken at least three times each day near the discharge pipe for immediate dissolved oxygen analysis. Additional testing may be required for suspended solids.

Experience with these lagoons is limited to northern states with seasonal and climatic influences on algae growth. The concept will be quite effective for BOD removal in any location, and if SS are within the limits given in Table 1–2, it should be effective in warmer climates. The process will also work with a discharge cycle that is more frequent than semiannually, depending on receiving water conditions and requirements. Operating the isolation cell on a fill and draw batch basis is similar to the "phase isolation" technique discussed in Chapter 3.

Partial Mix Aerated Lagoons. The design of partial mix aerated lagoons is commonly based on first-order reaction rate equations for complete mixed flow, even though by definition the ponds are not completely mixed. As indicated in Table 1–1, at least three cells are usually provided. The aeration is usually tapered with higher intensity near the inlet of the first cell and a quiescent zone near the end of the final cell.

The detention time or lagoon volume is based on the low-temperature wintertime reaction rates. The oxygen requirements are based on the higher-temperature summer reaction rates. No allowance is made for photosynthetic oxygenation, even though algae will be present. An allowance must be made for sludge accumulation and for winter ice formation in northern climates. In Alaska, which might represent the worst case, an allowance of 5 % is made for sludge storage and 15 % for ice formation in calculating total pond volume. The total depth of the lagoon is then based on the requirements for the type of aeration equipment chosen. Special attention is required for design of surface aerators in cold climates because of ice problems.

Partial mix lagoons may have high SS on an infrequent basis because of algae. If these values exceed the limits given in Table 1–2, it may be possible to operate the final cell in an intermittent discharge mode during algae blooms.

Complete Retention Lagoons.   Complete retention lagoons may be feasible in locations with low-cost land and high evaporation rates. Many existing complete retention systems probably depend to a greater degree than is desirable on seepage. Many states are adopting increasingly stringent seepage requirements for wastewater lagoon systems.[5]

General Considerations

The following criteria apply to the waste treatment lagoons described above.

Pathogen Control.   Natural die-off of pathogens is very effective in long retention time facultative and controlled discharge lagoons. The FC concentrations of approximately 200/100 ml can be achieved without disinfection if adequate hydraulic residence times are provided. A positive disinfection technique may be necessary for ponds to comply with site-specific discharge requirements. Chlorination can achieve the required FC reductions. A mathematical model designed to be applicable to most lagoon systems, and nomographs that can be used to calculate chlorine dosages to yield adequate residuals without lysing algae cells have been presented by Johnson et al.[6] and are summarized in Chapter 6.

Control of Short-Circuiting.   Short-circuiting of flow occurs to varying degrees in most existing ponds. In a dye study of a multiple cell system, it was found that the measured mean detention time (centroid of the area beneath the dispersion curve) in the cells varied from 25 to 89 % of the theoretical design detention time.[7]

The use of multiple cells operated in series and multiple port inlet structures is effective in reducing short-circuiting. In-basin baffles can also be effective, but special attention is required in northern climates to avoid problems with ice. Multiple cells are probably the most effective approach. No fewer than three cells should be provided for a lagoon.

Seepage Control. Lining of the lagoon bottom and inner dike surfaces may be necessary if compaction of the *in situ* soils does not produce an acceptable level of impermeability. In general, all the states require protection of the beneficial use of groundwater beneath a lagoon. Only a few states define a specific seepage limitation. Most states do not have a specific value but decide on a case-by-case basis for protection of groundwater. Lining materials include locally available clays, bentonite, asphalt, concrete, soil cement, and various membranes. Some of the low seepage rates required would be difficult to achieve with solid stabilization techniques; therefore, constructed liners or membranes might be necessary. Middlebrooks et al.[5] describe detail techniques for pond lining (see Chapter 5).

Sludge Accumulation. Sludge will accumulate to varying degrees on the bottom of all the ponds. Most of the accumulation will occur at or near the inlet structures. Decomposition of these benthic sludge deposits is via anaerobic processes. This sludge can in time exert a significant oxygen demand on the system. The problem is particularly critical in northern temperate climates, where a temperature-induced "turnover" of lagoon contents can occur in the spring and fall of each year. This "turnover" can resuspend some of the benthic material and result in odor as well as temporary effluent quality problems.

Sludge will accumulate at faster rates in lagoons in cold climates, since the low winter temperatures inhibit the anaerobic reactions. In Alaska, which is probably the worst case, it is common practice to reserve up to 5% of the design volume for sludge accumulation.

References

1. Federal Register. 1977. *Secondary Treatment Information*, 40 CFR Part 133, 38 FR 22298, August 17, 1973; amended 41 FR 30786, July 26, 1976; amended 42 FR 54664, October 7, 1977.

2. Reed, S. C., and A. B. Hais. 1978. Cost Effective Use of Municipal Wastewater Treatment Ponds. *Proc. Amer Soc. Civil Eng. Ann. Conv.*

3. Hill, D. O., and A. Shindala. 1977. *Performance Evaluation of Kilmichael Lagoon.* EPA-600/2-77-109. Municipal Environmental Research Laboratory, U.S. Environmental Protection Agency, Cincinnati, Ohio.

4. Environmental Protection Agency. 1980. *Wastewater Treatment Ponds, MCD-14.* Office of Water Program Operations, Washington, D.C.

5. Middlebrooks, E. J., C. D. Perman, and I. S. Dunn. 1979. *Wastewater Stabilization Pond Linings.* U.S. Army, Cold Regions Research and Engineering Laboratory, Special Report 18–3.

6. Johnson, B. A., J. L. Wight, D. S. Bowles, J. H. Reynolds, and E. J. Mid-

dlebrooks. 1979. *Waste Stabilization Lagoon Microorganism Removal Efficiency and Effluent Disinfection with Chlorine.* EPA-600/2-79-018. Minicipal Environmental Research Laboratory, U.S. Environmental Protection Agency, Cincinnati, Ohio.

7. Middlebrooks, E. J., D. H. Falkenborg, and R. F. Lewis (Editors). 1979. *Performance and Upgrading of Wastewater Stabilization Ponds.* Proceedings of a Conference at Utah State University, EPA-600/9-79-011.

CHAPTER 2

Performance and Design of Lagoons

Facultative Wastewater Stabilization Lagoons

Performance results from existing lagoon systems and an evaluation of several facultative waste stabilization pond design equations are presented in this chapter. To satisfy the need for reliable lagoon performance data, in 1974 the U.S. Environmental Protection Agency sponsored four intensive facultative lagoon performance studies. These studies were located at Peterborough, New Hampshire,[1] Kilmichael, Mississippi,[2] Eudora, Kansas,[3] and Corinne, Utah.[4] These studies encompassed 12 full months of data collection, including four separate 30-consecutive-day sample periods, once each season.

Several equations have been proposed for use in design of facultative wastewater stabilization lagoons. Design engineers must choose between these often contradictory methods when designing a facultative pond system. These techniques include simple design criteria based on organic loading and hydraulic detention time, empirical design equations, and rational design equations. Examples of each technique are used in conjunction with the data collected at the four sites mentioned above.

Site Description

Peterborough, New Hampshire. The Peterborough facultative waste stabilization lagoon system consists of three cells operated in series with a total surface area of 8.5 ha (21 acres) followed by chlorination. A schematic drawing of the facility is shown in Figure 2–1. An effluent

FIGURE 2-1
Schematic flow diagram and aerial photograph of the facultative lagoon system at Peterborough, New Hampshire (Ref. 1).

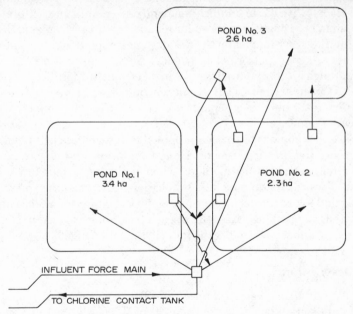

POND No. 3
2.6 ha

POND No. 1
3.4 ha

POND No. 2
2.3 ha

INFLUENT FORCE MAIN

TO CHLORINE CONTACT TANK

chlorine residual of 2.0 mg/l is maintained at all times. The facility was designed in 1968 on an areal loading basis of 19.6 kg BOD_5/day·ha (17.5 lb BOD_5/day·acre) with an initial average hydraulic flow of 1893 m^3/day (0.5 mgd). At the design depth of 1.2 m (4 ft), the theoretical hydraulic detention time would be 57 days. The results of the study conducted during 1974–1975 indicated an actual mean areal loading of 15.6 kg BOD_5/day·ha (13.9 lb BOD_5/day·acre) and a mean hydraulic flow of 1011 m^3/day (0.267 mgd). Thus the actual theoretical hydraulic detention time was 107 days.

Kilmichael, Mississippi. The Kilmichael facultative waste stabilization lagoon system consists of three cells operated in series with a total surface area of 3.3 ha (8.1 acres). The effluent is not chlorinated. A schematic drawing of the facility is shown in Figure 2–2.

The design load for the first cell in the series was 67.2 kg BOD_5/day·ha (60 lb BOD_5/day·acre). The second cell was designed with a surface area equivalent to 40% of the surface area of the first cell. The third cell was designed with a surface area equivalent to 16% of the first cell. The system was designed for a hydraulic flow of 693 m^3/day (0.183 mgd). The average depth of the lagoons is approximately 2 m (6.6 ft). This provides for a theoretical hydraulic detention time of 79 days. The result of the study indicated that the actual average organic load on the first cell averaged 27.2 kg BOD_5/day·ha (24.3 lb BOD_5/day·acre) and that the average hydraulic inflow to the system was 281 m^3/day (0.074 mgd). Thus the actual theoretical hydraulic detention time in the system was 214 days.

Eudora, Kansas. The Eudora facultative waste stabilization lagoon system consists of three cells operated in series with a total surface area of 7.8 ha (19.3 acres). A schematic diagram of the system is shown in Figure 2–3. The effluent is not chlorinated.

The facility was designed on an areal loading basis of 38 kg BOD_5/day·ha (34 lb BOD_5/day·acre) with a hydraulic flow of 1514 m^3/day (0.4 mgd). At the design operating depth of 1.5 m (5 ft), the theoretical hydraulic detention time would be 47 days. The results of the study indicated that the actual average organic load on the system was 19.0 kg BOD_5/day·ha (16.7 lb BOD_5/day·acre) and the actual average hydraulic flow to the system was 506 m^3/day (0.13 mgd). Thus the actual theoretical hydraulic detention time in the system was 231 days.

Corinne, Utah. The Corinne facultative waste stabilization lagoon system consists of seven cells operated in series with a total surface area of 3.86 ha (9.53 acres). A schematic drawing of the system is shown in Figure 2–4. The effluent is not chlorinated.

The facility was designed on an areal loading basis of 36.2 kg BOD_5/day·ha (32.2 lb BOD_5/day·acre) with a hydraulic flow of 265 m^3/day (0.07 mgd). With a design depth of 1.2 m (4 ft), the system has a theoretical hydraulic detention time of 180 days. The results of the study indicated that the actual average organic load on the system was 14.1 kg BOD_5/day·ha

FIGURE 2-2
Schematic flow diagram and aerial photograph of the facultative lagoon system at Kilmichael, Mississippi (Ref. 2).

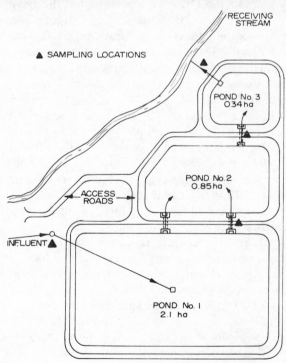

FIGURE 2-3

Schematic flow diagram and aerial photograph of the facultative lagoon system at Eudora, Kansas (Ref. 3).

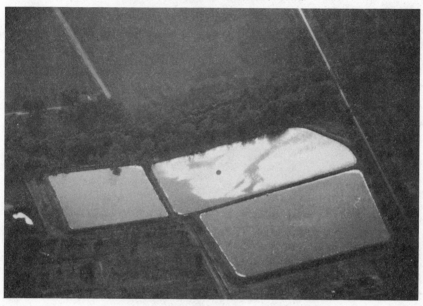

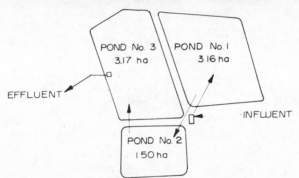

(12.6 lb BOD$_5$/day·acre) and the actual average hydraulic flow to the system was 694 m^3/day (0.18 mgd). Thus the actual theoretical hydraulic detention time in the system was 70 days.

Performance

Biochemical Oxygen Demand (BOD$_5$) Performance. The monthly average effluent BOD$_5$ concentrations for the four previously described facultative lagoon systems are compared with the Federal Secondary Treatment Standard of 30 mg/l in Figure 2–5.

FIGURE 2-4
Schematic flow diagram and aerial photograph of the facultative lagoon
system at Corinne, Utah (Ref. 4).

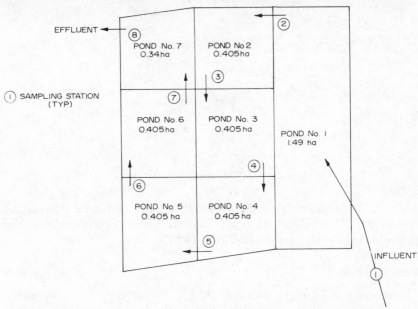

FIGURE 2-5
Monthly average effluent (BOD₅).

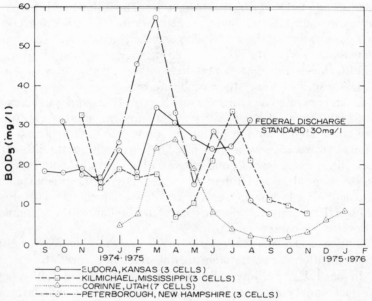

In general, all the systems were capable of providing a monthly average effluent BOD₅ concentration of less than 30 mg/l during the major portion of the year. Monthly average effluent BOD₅ concentrations ranged from 1.4 mg/l during September 1975 at the Corinne, Utah site to 57 mg/l during March 1975 at the Peterborough, New Hampshire site. Monthly average effluent BOD₅ concentrations tended to be higher during the winter months (January, February, March, and April) at all the sites. This was especially evident at the Peterborough site, when the ponds were covered over by ice due to freezing winter temperatures. The ice cover caused the ponds to become anaerobic. However, even when the ponds at the Corinne site were covered over with ice, the monthly average effluent BOD₅ concentration did not exceed 30 mg/l.

None of the systems studied was significantly affected by the fall overturn; however, the spring overturn did cause significant increases in effluent BOD₅ concentrations at two of the sites. At the Corinne site two different spring overturns occurred. The first occurred in March 1975, with a peak daily BOD₅ concentration of 36 mg/l. The second occurred during April 1975, with a peak daily effluent BOD₅ concentration of 39 mg/l. At the Eudora site the peak daily effluent BOD₅ concentration of 57 mg/l occurred during April 1975. The Kilmichael and Peterborough sites were not severely affected by the spring overturn period.

The monthly average effluent BOD₅ concentration of the Corinne lagoon system never exceeded 30 mg/l throughout the entire study. The Eudora lagoon system monthly average effluent BOD_5 concentration exceeded 30 mg/l twice during the entire study. The Kilmichael lagoon system monthly average effluent BOD_5 concentration exceeded 30 mg/l on only two occasions during the study. The Federal Secondary Treatment of 30 mg/l was exceeded by the Peterborough lagoon system monthly average effluent BOD_5 concentration 4 of the 12 months studied.

The results of these studies indicate that properly designed, maintained, and operated facultative waste stabilization pond systems can produce a high-quality effluent. Although these systems are subject to seasonal upsets, they are capable of producing a low BOD_5 effluent that can be discharged to a waterway in many cases. When a high-quality effluent is required, the effluent from a pond system can be upgraded with inexpensive processes to produce an effluent that will satisfy most effluent standards. Since facultative lagoon effluents exceed 30 mg/l during a relatively small portion of the year, it is possible to control the discharge in such a manner as not to exceed discharge standards.

Suspended Solids (SS) Performance.   The monthly average effluent SS concentrations for each system are illustrated in Figure 2–6.

In general, the effluent SS concentrations of the facultative lagoons follow a seasonal pattern. Effluent SS concentrations are high during summer months when algal growth is intensive and also during the spring and

FIGURE 2-6
Monthly average effluent SS for typical facultative lagoons.

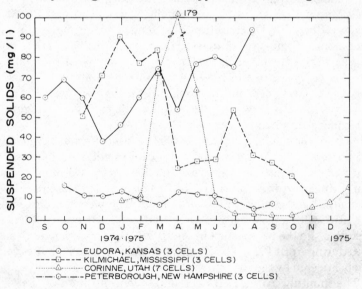

fall overturn periods when settled SS are resuspended from bottom sediments due to mixing. The monthly average SS concentrations ranged from 2.5 mg/l during September 1975 at the Corinne site to 179 mg/l during April 1975, also at the Corinne site. The high monthly average effluent concentration of 179 mg/l at the Corinne site occurred during the spring overturn period which caused a resuspension of settled solids.

The Eudora and Kilmichael sites illustrate the increase in effluent SS concentrations due to algal growth during the warm summer months. However, the Peterborough and Corinne sites were not significantly affected by algal growth during the summer months. In general, the Corinne and Peterborough sites produced monthly average effluent suspended solids concentrations of less than 20 mg/l. During 10 of the 13 months studied, the monthly average effluent SS concentration at the Corinne site never exceed 20 mg/l. However, the monthly average effluent SS concentration at the Eudora site was never less than 39 mg/l throughout the entire study.

The results of the studies indicate that facultative lagoons can produce an effluent that has a low SS concentration; however, effluent SS concentrations will be high at various times thoughout the year. In general, these SS are composed of algal cells which may not be particularly harmful to receiving streams. In areas in which effluent SS standards are stringent, some type of polishing device will be necessary to reduce facultative lagoon effluent SS concentrations to acceptable levels.

Fecal Coliform (FC) Removal Performance. The monthly geometric mean effluent coliform concentrations for the four facultative lagoon systems are compared with a concentration of 200/100 ml in Figure 2–7.

Only the Peterborough, New Hampshire facultative lagoon system provides disinfection. As illustrated in Figure 2–7, the chlorinated Peterborough lagoon effluent never exceeds a concentration of 10 FC organisms/100 ml. This clearly indicates that facultative lagoon effluent may be satisfactorily disinfected by the chlorination process.

For the three facultative lagoon systems without disinfection processes, the geometric mean monthly effluent FC concentration ranged from 0.1 organisms/100 ml in June and September 1975 at the Corinne, Utah lagoon system to 13,527 organisms/100 ml in January 1975 at the Kilmichael, Mississippi lagoon system. In general, geometric mean effluent fecal coliform concentrations tend to be higher during the colder periods. Periods of ice cover during winter months would seriously affect FC die-off due to sunlight effects. The Eudora, Kansas and the Kilmichael, Mississippi geometric mean monthly effluent FC concentrations consistently exceeded 200 organisms/100 ml during winter operation.

The FC concentration in the Corinne, Utah lagoon system effluent never exceeded 200 organisms/100 ml, even though this system did not include any form of disinfection. This system is composed of seven cells in series. Analysis of the FC concentrations between the seven cells indicated

FIGURE 2-7

Effluent monthly geometric mean fecal coliform concentrations from typical facultative lagoons.

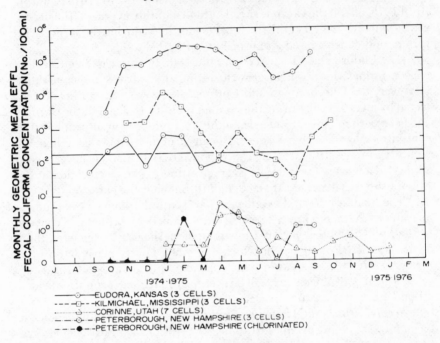

that FC were essentially absent after the fourth cell in the series.[4] The other two facultative lagoon systems without disinfection only utilize three cells in series; however, FC die-off is primarily a function of hydraulic residence time rather than the absolute number of cells in series.

The results of these studies indicate that facultative lagoon effluent can be chlorinated sufficiently to produce fecal coliform concentration less than 10 organisms/100 ml. Two of the systems studied could not produce an effluent containing less than 200 FC/100 ml. This was probably due to hydraulic short-circuiting; however, the Corinne, Utah system study clearly indicated that properly designed facultative lagoon systems can significantly reduce FC concentrations.

Evaluation of Design Methods

A summary of the facultative lagoon performance data used to evaluate the various design methods is presented in Table 2–1. These data were collected at the four facultative lagoon systems described in the first sections of this chapter. Only the characteristics of the influent wastewater and the effluent from the primary (first) cell of the systems are presented in Table 2–1 for the four systems.

TABLE 2-1
Performance Data for Four Facultative Lagoons

INF BOD MG/L	CELL #1 BOD	INF SOL BOD	CELL #1 SBOD	INF COD MG/L	CELL #1 COD	INF SOL COD	CELL #1 SCOD	DET TIME DAYS	TEMP DEG C	LIGHT LAN	TSS MG/L	VSS MG/L	LOCATION
122.	31.	40.	5.	173.	118.	81.	47.	44.43	2.	190	52.	45.	CORINNE,UTAH
107.	38.	28.	5.	140.	125.	64.	48.	22.72	1.	265	61.	54.	CORINNE,UTAH
58.	57.	19.	10.	135.	126.	47.	39.	19.56	5.	385	55.	52.	CORINNE,UTAH
49.	33.	16.	6.	114.	113.	46.	37.	23.23	9.	495	69.	59.	CORINNE,UTAH
62.	33.	15.	5.	105.	117.	40.	38.	28.47	12.	590	74.	61.	CORINNE,UTAH
52.	30.	17.	6.	78.	95.	37.	45.	27.29	18.	630	56.	43.	CORINNE,UTAH
40.	29.	9.	5.	75.	131.	35.	45.	20.73	22.	640	65.	53.	CORINNE,UTAH
40.	36.	11.	5.	81.	162.	38.	44.	18.97	19.	550	87.	76.	CORINNE,UTAH
92.	35.	25.	5.	141.	168.	54.	38.	21.07	16.	480	95.	81.	CORINNE,UTAH
87.	34.	22.	4.	178.	146.	50.	39.	17.64	9.	335	85.	71.	CORINNE,UTAH
85.	30.	24.	5.	132.	114.	58.	42.	37.14	4.	210	66.	57.	CORINNE,UTAH
99.	21.	32.	6.	189.	89.	64.	46.	63.66	2.	145	27.	25.	CORINNE,UTAH
140.	21.	50.	9.	192.	68.	88.	54.	55.76	2.	190	18.	15.	CORINNE,UTAH
332.	41.	125.	11.	633.	183.	277.	83.	83.82	22.	430	89.	76.	EUDORA,KANSAS
258.	49.	116.	13.	552.	225.	225.	107.	87.86	17.	285	124.	95.	EUDORA,KANSAS
303.	41.	184.	11.	576.	156.	265.	63.	89.00	10.	230	79.	68.	EUDORA,KANSAS
400.	53.	182.	22.	614.	163.	260.	71.	101.34	4.	180	71.	65.	EUDORA,KANSAS
326.	56.	169.	18.	573.	174.	224.	74.	102.99	4.	190	84.	78.	EUDORA,KANSAS
303.	35.	123.	10.	631.	180.	234.	.76.	66.08	3.	280	74.	70.	EUDORA,KANSAS
373.	49.	181.	14.	635.	186.	236.	60.	70.36	7.	345	112.	97.	EUDORA,KANSAS
284.	44.	129.	15.	580.	172.	173.	71.	80.47	14.	440	85.	72.	EUDORA,KANSAS
209.	57.	95.	13.	375.	284.	117.	76.	95.46	21.	530	130.	102.	EUDORA,KANSAS
179.	42.	78.	11.	458.	204.	128.	52.	101.12	24.	560	121.	109.	EUDORA,KANSAS
270.	55.	140.	19.	533.	265.	201.	83.	116.83	26.	580	172.	151.	EUDORA,KANSAS
298.	69.	178.	15.	544.	246.	224.	91.	109.41	25.	435	137.	120.	EUDORA,KANSAS
197.	43.	70.	11.	334.	203.	136.	128.	48.61	9.	240	70.	0.	PETERBOROUGH,N.H.
170.	46.	49.	10.	303.	207.	102.	82.	50.89	6.	165	74.	0.	PETERBOROUGH,N.H.
144.	48.	49.	22.	245.	155.	101.	84.	50.78	4.	115	44.	0.	PETERBOROUGH,N.H.
123.	65.	40.	51.	204.	158.	94.	106.	48.29	5.	130	27.	0.	PETERBOROUGH,N.H.
131.	70.	37.	53.	201.	154.	73.	100.	45.74	3.	230	26.	0.	PETERBOROUGH,N.H.
128.	68.	43.	46.	263.	151.	94.	93.	37.90	4.	280	21.	0.	PETERBOROUGH,N.H.
101.	50.	33.	36.	181.	128.	78.	79.	35.87	7.	400	23.	0.	PETERBOROUGH,N.H.
157.	36.	73.	31.	425.	110.	250.	77.	42.09	17.	450	37.	0.	PETERBOROUGH,N.H.
133.	38.	56.	21.	249.	174.	116.	101.	44.61	21.	525	47.	0.	PETERBOROUGH,N.H.
113.	37.	49.	22.	223.	213.	114.	126.	43.06	24.	500	63.	0.	PETERBOROUGH,N.H.
123.	30.	56.	12.	315.	230.	142.	98.	43.21	22.	450	76.	0.	PETERBOROUGH,N.H.
137.	25.	50.	9.	313.	161.	125.	88.	41.30	17.	340	49.	0.	PETERBOROUGH,N.H.
200.	23.	57.	5.	254.	138.	114.	68.	189.80	14.	260	57.	0.	KILMICHAEL,MISS.
172.	24.	52.	4.	333.	123.	101.	52.	189.80	9.	200	82.	0.	KILMICHAEL,MISS.
106.	21.	36.	3.	204.	120.	81.	42.	189.80	10.	205	86.	0.	KILMICHAEL,MISS.
135.	20.	50.	3.	232.	128.	81.	35.	182.50	11.	270	74.	0.	KILMICHAEL,MISS.
107.	27.	39.	3.	225.	164.	78.	30.	54.31	12.	340	107.	0.	KILMICHAEL,MISS.
187.	16.	55.	3.	470.	103.	108.	38.	115.98	18.	450	65.	0.	KILMICHAEL,MISS.
140.	17.	37.	5.	312.	90.	81.	49.	82.38	23.	550	47.	0.	KILMICHAEL,MISS.
278.	26.	80.	5.	628.	89.	154.	50.	185.37	27.	530	52.	0.	KILMICHAEL,MISS.
278.	25.	96.	6.	626.	121.	192.	71.	191.48	29.	450	43.	0.	KILMICHAEL,MISS.
321.	31.	96.	7.	746.	140.	235.	94.	456.34	29.	470	55.	0.	KILMICHAEL,MISS.
301.	15.	74.	3.	572.	52.	168.	60.	171.11	20.	370	46.	0.	KILMICHAEL,MISS.
247.	17.	75.	4.	535.	108.	164.	73.	111.93	20.	340	33.	0.	KILMICHAEL,MISS.
200.	14.	79.	3.	458.	102.	191.	67.	165.37	18.	250	28.	0.	KILMICHAEL,MISS.

Most of the kinetic analyses of the systems were limited to the performance obtained in the primary cell because the parameters used to measure pollution (BOD_5 and COD) appear to impact the performance of the primary cells of the systems far more than the following cells. Algae succession, changes in nutrient concentration, and the buffering capacity of the system appear to exert more influence on the cells following the primary cell. Since all lagoon system designs are based on some form of organic loading expressed in terms of BOD_5 or COD, most attempts to evaluate the design methods were limited to the primary cells of the four systems. The commonly used design methods are discussed individually in the following sections.

Design Criteria for Organic Loading and Hydraulic Detention Time.
Canter and Englande[5] reported that most states have design criteria for
organic loading and/or hydraulic detention time for facultative waste
stabilization ponds. Design criteria are used by the states to ensure that
pond effluent water quality meets state and federal discharge standards.
Repeated violations of effluent quality standards by pond systems that meet
state design criteria indicate the inadequacy of the criteria. Reported
organic loading design criteria averaged 21.2 kg BOD_5/ha·day (26.0 lb
BOD_5/acre·day) in the north region (above 42° latitude), 49.4 kg
BOD_5/ha·day (44 lb BOD_5/acre·day) in the southern region (below 37°
latitude) and 37.0 kg BOD_5/ha·day (33 lb BOD_5/acre·day) in the central
region. Reported design criteria for detention time averaged 117 days in the
north, 82 days in the central, and 31 days in the south region.

Design criteria for organic loading in New Hampshire was 39.3 kg
BOD_5/ha·day (35 lb BOD_5/acre·day). The Peterborough treatment system
was designed for a loading of 19.6 kg BOD_5/ha·day (17.5 lb BOD_5/acre
·day) in 1968 to be increased as population increased to 39.9 kg BOD_5/
ha·day (35 lb BOD_5/acre·day) in the year 2000. Actual loading during
1974–1975 averaged 16.2 kg BOD_5/ha·day (14.4 lb BOD_5/acre·day) with
the highest loading being 21.2 kg BOD_5/ha·day (18.9 lb BOD_5/acre·day).
Although the organic loading was substantially below the state design limit,
the effluent exceeded the federal standard of 30 mg BOD_5/l during the
months of October 1974 and February, March, and April 1975.

Mississippi's design criteria for organic loading was 56.2 kg BOD_5/
ha·day (50 lb BOD_5/acre·day). The Kilmichael treatment system was
designed for a loading of 43 kg BOD_5/ha·day (38 lb BOD_5/acre·day). Actual
loading during 1974–1975 averaged 17.5 kg BOD_5/ha·day (15.6 lb
BOD_5/acre·day) with a maximum of 24.7 kg BOD_5/ha·day (22 lb
BOD_5/acre·day), and yet the federal BOD_5 effluent standard was exceeded
twice during the sample year (November and July).

The design load for the Eudora, Kansas, system was the same as the
state design limit, 38.1 kg BOD_5/ha·day (34 lb BOD_5/acre·day). Actual
loading during 1974–1975 averaged only 18.8 kg BOD_5/ha·day (16.7 lb
BOD_5/acre·day) with maximum of 31.5 kg BOD_5/ha·day (28 lb
BOD_5/acre·day). The federal BOD_5 effluent standard was exceeded three
months during the sample year (March, April, and August).

Utah has both an organic loading design limit, 45 kg BOD_5/ha·day (40
lb BOD_5/acre·day) on the primary cell and a winter detention time design
criteria of 180 days. Design loading for the Corinne system was 36.2 kg
BOD_5/ha·day (32.2 lb BOD_5/acre·day) and design detention time was 180
days. Although the organic loading averaged 33.6 kg BOD_5/ha·day (29.8 lb
BOD_5/acre·day) on the primary cell, during two months of the sample year
it exceeded 56.2 kg BOD_5/ha·day (50 lb BOD_5/acre·day). Average organic
loading on the total system was 13.0 kg BOD_5/ha·day (14.6 lb BOD_5/acre

·day) and the hydraulic detention time was estimated to be 88 days during the winter. Regardless of the deviations from the state design criteria, the monthly BOD_5 average never exceeded the federal effluent standard.

A summary of the state design criteria for each location and actual design values for organic loading and hydraulic detention time are shown in Table 2–2. Also included is a list of the months the federal effluent standard for BOD_5 was exceeded. Note that the actual organic loading for all four systems are nearly equal, yet as the monthly effluent BOD_5 averages shown in Figure 2–5 indicate, the Corinne system consistently produced a higher-quality effluent. This may be a function of the larger number of cells in the Corinne system; seven as compared to three for the rest of the systems. Hydraulic short-circuiting may be occurring in the three cell systems, resulting in a shorter actual detention time than exists in the Corinne system. Detention time may also be affected by the location of pond cell inlet and outlet structures. As shown in Figure 2–4, the outlet structures are at the furthest point possible from the inlet structures in the Corinne, Utah system. At the Eudora, Kansas system shown in Figure 2–3, large "dead zones" undoubtedly occur in each cell because of the unnecessarily short distance between inlet and outlet structures. These dead zones result in decreased hydraulic detention and increased effective organic loading rate.

Empirical Design Equations. In a survey of primary facultative ponds in tropical and temperate zones, McGarry and Pescod[6] showed that areal BOD_5 removal (L_r, lb/acre·day) may be estimated through knowledge of areal BOD_5 loading (L_o, lb/acre·day) using

$$L_r = 9.23 + 0.725 L_o \qquad (2–1)$$

The regression equation had a correlation coefficient of .995 and a 95% confidence interval of ± 29.3 lb BOD_5/acre·day removal (Figure 2–8). The equation was reported to be valid for any loading between 30 and 500 lb BOD_5/acre·day. McGarry and Pescod[6] also found that under normal operating ranges hydraulic detention time and pond depth have little influence on percentage or areal BOD_5 removal. With such a large 95% confidence interval, it is impractical to apply the equation to the design of lagoon systems loaded at rates of 50 lb BOD_5/acre·day or less, as was the situation with the majority of the months of operation for the four facultative lagoon systems described above.

Relationships between organic removal per acre·day and organic loading per acre·day for the lower rates observed at the four facultative lagoon systems were developed using the BOD_5, soluble biochemical oxygen demand ($SBOD_5$), chemical oxygen demand (COD), and soluble chemical oxygen demand (SCOD) removal and application rates. Statistically significant relationships were observed for all four carbon-estimating parameters, but the best relationships were observed when the organic

Table 2-2
Summary of Design and Performance Data

LOCATION	ORGANIC LOADING (KG BOD$_5$/HA·DAY)			THEORETICAL HYDRAULIC detention time (DAYS)			MONTHS EFFLUENT EXCEEDED 30 MG/L BOD$_5$
	State Design Standard	Design	Actual (1974–1975)	State Design Standard	Design	Actual	
Peterborough, NH	39.3	19.6	16.2	None	57	107	Oct., Feb., Mar., Apr.,
Kilmichael, MS	56.2	43.0	17.5	None	79	214	Nov., July
Eudora, KS	38.1	38.1	18.8	None	47	231	Mar., Apr., Aug.
Corinne, UT	45.0[a]	36.2[a]	29.7[a]	180	180	70	None
			14.6[b]			88[c]	

(kg/ha·day) × 0.889 = (lb/acre·day)

[a] Primary cell.
[b] Entire system.
[c] Estimated from dye study.

26

FIGURE 2-8
McGarry and Pescod equation for areal BOD5 removal as a function of BOD5 loading (Ref. 9).

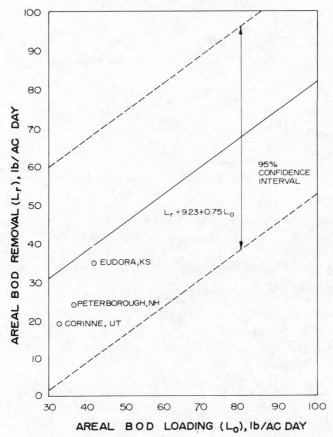

removals were calculated using the influent BOD_5 and the effluent $SBOD_5$ ($ITBOD_5$ and $ESBOD_5$) and the influent COD and the effluent SCOD (IT-COD and ESCOD). The BOD_5 and $SBOD_5$ relationship is shown in Figure 2–9, and the COD and SCOD relationship is shown in Figure 2–10. The 95% confidence intervals for the BOD_5–$SBOD_5$ and the COD-SCOD relationships are much smaller than the value reported by McGarry and Pescod[6] and are shown in Figures 2–9 and 2–10.

Larsen[7] proposed an empirical design equation, developed by using data from a 1–year study at the Inhalation Toxicology Research Institute, Kirtland Air Force Base, New Mexico. The institute's facultative pond system consists of one 0.66 ha (1.62 acre) cell receiving waste from 151 staff members, 1300 beagle dogs, and several thousand small animals. Larsen

FIGURE 2-9
Relationship of organic loading and removal rates (facultative wastewater stabilization ponds, BOD$_5$).

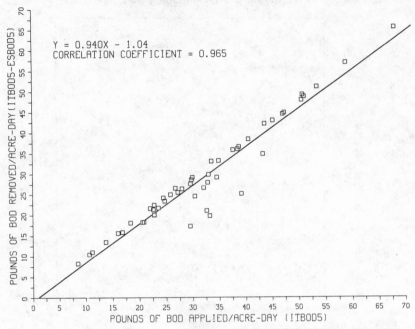

found that the required pond surface area could be estimated by use of the following equation.

$$MOT = (2.468^{RED} + 2.468^{TTC} + 23.9/TEMPR + 150.0/DRY)*10^6 \quad (2\text{-}2)$$

where the dimensionless products

$$MOT = 1.0783 \times 10^{-7} \frac{(\text{surface area, ft}^2)(\text{solar radiation, Btu/ft}^2\text{–day})^{1/3}}{(\text{influent flow rate, gal/day})(\text{influent BOD}_5\text{, mg/l})^{1/3}}$$

$$RED = \frac{(\text{influent BOD}_5\text{, mg/l}) - (\text{effluent BOD}_5\text{, mg/l})}{(\text{influent BOD, mg/l})}$$

$$TTC = 0.0879 \frac{(\text{wind speed, miles/hr})(\text{influent BOD}_5\text{, mg/l})^{1/3}}{(\text{solar radiation, Btu/ft}^2\text{-day})^{1/3}}$$

$$TEMPR = \frac{(\text{lagoon liquid temperature, }^\circ F)}{(\text{air temperature, }^\circ F)}$$

$$DRY = \text{relative humidity, } \%$$

FIGURE 2-10
Relationship of organic loading and removal rates (facultative wastewater stabilization ponds, COD).

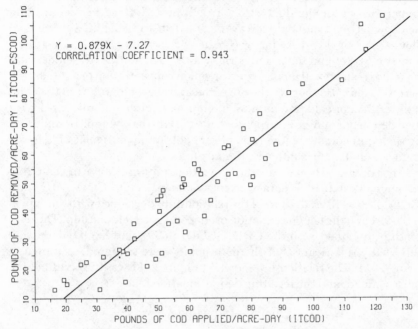

$Y = 0.879X - 7.27$
CORRELATION COEFFICIENT = 0.943

POUNDS OF COD REMOVED/ACRE-DAY (ITCOD-ESCOD)

POUNDS OF COD APPLIED/ACRE-DAY (ITCOD)

To determine the effect of using the Larsen equation on multicell facultative ponds, it was applied to both the entire system and the primary cell for each of the four locations. In each case the Larsen equation underestimated the pond surface area required for a particular BOD$_5$ removal. Prediction errors for multiple-cell lagoons ranges from 190 to 248%. Prediction errors for the single-cell surface areas range from 18 to 98%. Use of the Larsen equation is not recommended.

Gloyna[8] proposed the following empirical equation for the design of facultative wastewater stabilization ponds:

$$V = 3.5 \times 10^{-5} QL_a [\theta^{(35-T)}] ff' \qquad (2-3)$$

where

V = pond volume (m^3)
Q = influent flow rate (l/day)
L_a = ultimate influent BOD$_u$ or COD (mg/l)
θ = temperature coefficient
T = pond temperature ($^\circ$C)
f = algal toxicity factor
f' = sulfide oxygen demand

The BOD$_5$ removal efficiency can be expected to be 80–90% based on unfiltered influent samples and filtered effluent samples. A pond depth of 1.5 m is suggested for systems with significant seasonal variations in temperature and major fluctuations in daily flow. Surface area design using the Gloyna equation should always be based on a 1 m depth. According to Gloyna, the algal toxicity factor (f) can be assumed to be equal to 1.0 for domestic wastes and many industrial wastes. The sulfide oxygen demand (f') is also equal to 1.0 for SO$_4$ equivalent ion concentration of less than 500 mg/l. Gloyna[8] also suggests the use of the average temperature of the pond in the critical or coldest month. Sunlight is not considered to be critical in pond design but may be incorporated into the Gloyna equation by multiplying the pond volume by the ratio of sunlight in the particular area to the average found in the Southwest United States.

The data used to evaluate the Gloyna equation are shown in Table 2–1. Although ultimate BOD data were not available, COD, SCOD, BOD$_5$, and SBOD$_5$ data were used. None of the relationships developed with the data in Table 2–1 and the Gloyna equation produced a good relationship. The relationships obtained with the COD, SCOD, BOD$_5$, and SBOD$_5$ data were statistically significant, but the data points were scattered, as shown in Figure 2–11. The relationship shown in Figure 2–11 was the best fit of the data obtained, and the resulting design equation follows:

$$\frac{V}{Q} = 0.035\,(\text{BOD}_5,\,\text{mg/l})\,(1.099)^{\frac{\text{LIGHT}\,(35\text{-T})}{250}} \qquad (2\text{-}4)$$

where LIGHT equals solar radiation, in langleys, and the other terms are as defined above.

The validity of the above expression is questionable because of the scattered data, but the relationship is statistically significant. The use of this relationship is left to the discretion of the design engineer.

Although not directly comparable, the results obtained with the Gloyna equation and results obtained with the relationship shown in Figure 2–9 are presented below to show the variation between the two approaches.

Assuming a design flow rate of 3785 m³/day (1 mgd), a solar radiation intensity of 250 langleys, an influent BOD$_5$ concentration of 300 mg/l, and a temperature of 10°C, the Gloyna equation (Equation 2–4) yields a surface area of 420,918 m² (assuming a water depth of 1 m), and the organic loading rate relationship (Figure 2–9) yielded a loading rate of 27 kg/ha·day (24 lb/acre·day). At an organic loading rate of 27 kg/ha·day (24 lb/acre·day), the organic removal rate will be 24.1 kg/ha·day (21.5 lb/acre·day) (Figure 2–9) or an 89.7% reduction in BOD$_5$. The percentage of reduction is within the range of 80–90% expected with a design using the Gloyna equation. The results obtained with the Gloyna equation appear to be conservative when compared with the organic loading-removal relationship (Figure 2–9),

FIGURE 2-11
Gloyna equation (facultative wastewater stabilization ponds, BOD).

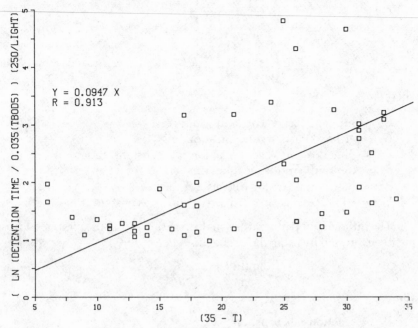

principally because of the temperature correction factor in the Gloyna equation. Although it is logical to expect temperature to exert an influence on BOD_5 removal, the plot shown in Figure 2–9 indicates essentially no influence by the temperature. The most logical explanation for this phenomenon is that the systems are so large that the temperature influence is masked in the process.

Rational Design Equations. Kinetic models based on plug flow and complete mix hydraulics and first-order reaction rates have been proposed by many authors to describe the performance of wastewater stabilization ponds. The basic models are modified to reflect the influence of temperature. The basic models are presented below:

Plug Flow:

$$\frac{C_e}{C_o} = e^{-k_p t} \tag{2–5}$$

or:

$$\ln \left(\frac{C_e}{C_o}\right) = -k_p t \tag{2–6}$$

Complete Mix:

$$\frac{C_e}{C_0} = \frac{1}{1 + k_c t} \qquad (2\text{-}7)$$

or:

$$\left(\frac{C_o}{C_e} - 1\right) = k_c t \qquad (2\text{-}8)$$

where

C_o = influent carbon concentration, mg/l
C_e = effluent carbon concentration, mg/l
k_p = plug flow first-order reaction rate constant, time^{-1}
t = hydraulic residence time, time
e = base of natural logarithms, 2.7183
k_c = complete mix first-order reaction rate constant, time^{-1}

The influence of temperature on the reaction rate constants is most frequently expressed by using the Arrhenius relationship:

$$\frac{d (\ln K)}{dt} = \frac{E_a}{RT^2} \qquad (2\text{-}9)$$

where

K = reaction rate constant
E_a = activation energy, calories/mole
R = ideal gas constant, 1.98 calories/mole-degree
T = reaction temperature, °kelvin

Integrating Equation 2–9 yields the following expression:

$$\ln K = - \frac{E_a}{RT} + \ln B \qquad (2\text{-}10)$$

where B is a constant. Experimental data can be plotted as shown in Figure 2–12 to determine the value of E_a. Equation 2–9 can be integrated between the limits of T_1 and T_2 to obtain Equation 2–11:

$$\ln \left(\frac{K_2}{K_1}\right) = \frac{E_a}{R} \frac{T_2 - T_1}{T_2 T_1} \qquad (2\text{-}11)$$

In most biological wastewater treatment processes it is assumed that E_a/RT_2T_1 is a constant, C, and Equation 2–11 reduces to

$$\ln \left(\frac{K_2}{K_1}\right) = C (T_2 - T_1) \qquad (2\text{-}12)$$

or

$$\frac{K_2}{K_1} = e^{C(T_2 - T_1)} \qquad (2\text{-}13)$$

FIGURE 2-12
An Arrhenius plot to determine the
activation energy.

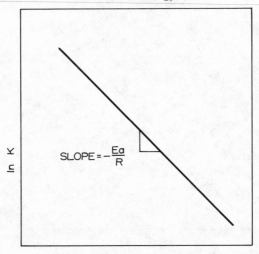

or

$$\frac{K_2}{K_1} = \theta^{(T_2 - T_1)} \qquad (2\text{-}14)$$

where θ = temperature factor. By plotting experimental values of the natural logarithms of K_2/K_1 versus $(T_2 - T_1)$, as shown in Figure 2–13, the value of θ can be determined.

The plug flow and complete mix models, along with modifications suggested by various investigators, were evaluated using the data shown in Table 2–1.

The influence of temperature on the calculated reaction rates was evaluated. As shown in Figures 2–14 and 2–15, the reaction rates calculated with the plug flow and complete mix equations were essentially independent of the temperature. This lack of influence by the liquid temperature was also observed in the section on empirical design equations (Figures 2–9 and 2–10). The logical explanation for the lack of influence by temperature is that the lagoon systems are so large that the temperature effect is masked by other factors. There is no doubt that temperature influences biological activity, but for the systems listed in Table 2–1, the influence was overshadowed by other parameters that may include dispersion, detention time, light, and species of organisms.

After observing the lack of influence by the water temperature, the data in Table 2–1 were fitted to the plug flow (Equation 2–5) and the complete mix (Equation 2–7) models to determine if the systems could be de-

FIGURE 2-13
Plot of reaction rate constants and temperature to determine the temperature factor.

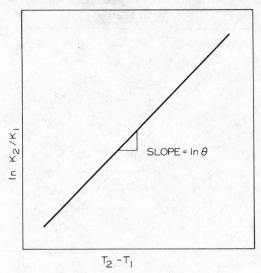

FIGURE 2-14
Relationship of decay rate and temperature (facultative wastewater stabilization pond, BOD).

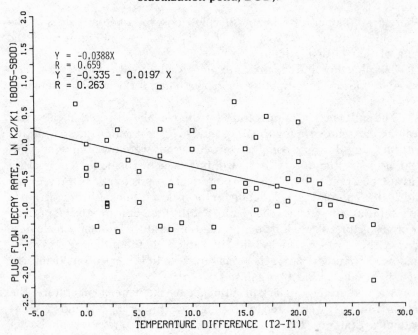

FIGURE 2-15
Relationship of decay rate and temperature (facultative wastewater stabilization pond, COD).

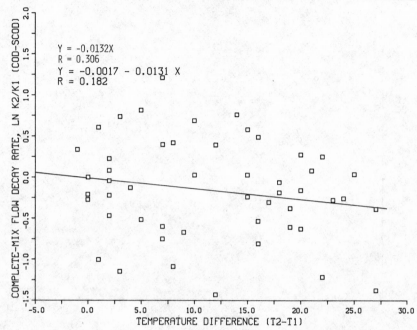

fined by these simple relationships. As shown in Figures 2–16 to 2–19, the fit of the data is less than ideal but is statistically highly significant (1 % level). Attempts to incorporate various types of temperature and light intensity relationship into the plug flow and complete mix models were not successful. The best statistical relationships obtained with the data are shown in Figures 2–16 to 2–19.

Thirumurthi[9] stated that a kinetic model based on plug flow or complete mix hydraulics should not be used for the rational design of stabilization ponds. Thirumurthi found that facultative ponds exhibit nonideal flow patterns and recommended the use of the following chemical reactor equation developed by Wehner and Wilhelm[10] for pond design:

$$\frac{C_e}{C_o} = \frac{4ae^{\frac{1}{2}d}}{(1 + a)^2 e^{a/2d} - (1 - a)^2 e^{-a/2d}} \qquad (2\text{–}15)$$

where

C_e = effluent BOD_5 (mg/l)
C_o = influent BOD_5 (mg/l)
K = first-order BOD_5 removal coefficient (l/day)
a = $\sqrt{1 + 4Ktd}$

FIGURE 2-16
Plug flow model [facultative wastewater stabilization ponds, ln (pond #1 TBOD$_5$/influent TBOD$_5$) versus detention time].

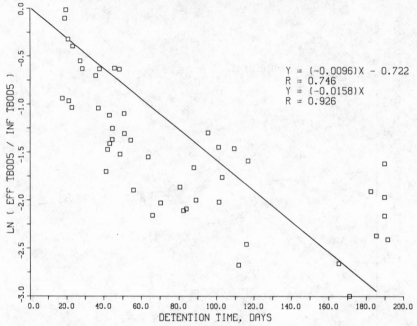

t = mean detention time (days)
d = dimensionless dispersion number

Thirumurthi[9] prepared the chart shown in Figure 2–20 to facilitate the use of Equation 2–15. The dimensionless term Kt is plotted versus the percentage of BOD$_5$ remaining for dispersion numbers varying from zero for an ideal plug flow unit to infinity for a complete mix unit. Dispersion numbers measured in wastewater stabilization ponds range from 0.1 to 2.0, with most values less than 1.0.

The data in Table 2–1 were used to calculate values of K for three different dispersion numbers (0.1, 0.25, and 1.0) by an iterative process. The values of K were normalized and plotted versus temperature, as illustrated in Figure 2–21. Less than 12% of the regression was explained by the linear relationship, indicating that temperature exerts little influence on the performance of the lagoons. Values of K calculated using the influent total BOD$_5$ and the effluent soluble BOD$_5$ concentrations with a dispersion number of 0.25 ranged from 0.0110 to 0.282/day, with a mean value of 0.0733/day and a median value of 0.0552/day. The mean temperature was 13.5°C, and the median temperature was 12°C. Using the design data presented with the Gloyna equation, a K of 0.0552/day adjusted linearly for

FIGURE 2-17

Plug flow model [facultative wastewater stabilization ponds, ln (pond #1 SBOD$_5$/influent TBOD$_5$) versus detention time].

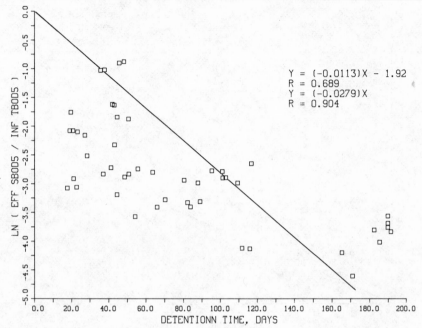

temperature to yield a value of 0.0460/day (10°C/12°C × 0.0552 = 0.0460), a dispersion number of 0.25, and assuming 90% removal of the BOD$_5$ (TBOD$_5$–SBOD$_5$); a detention time of 74 days is obtained. This detention time is less than the value obtained with the Gloyna equation (100 days), but considering the differences in approach, agreement is reasonable.

Ammonia Nitrogen Removal in Facultative Wastewater Stabilization Lagoons

Theoretical Considerations

Differences between influent and effluent ammonia-N concentrations in facultative wastewater stabilization ponds can occur through three processes:

1. Gaseous ammonia stripping to the atmosphere.
2. Ammonia assimilation in algal biomass.
3. Biological nitrification-denitrification.

FIGURE 2-18
Complete mix model [facultative wastewater stabilization ponds, (influent TBOD$_5$/pond #1 SBOD$_5$) -1 versus detention time].

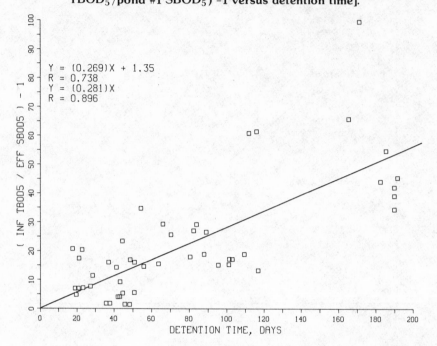

Ammonia-N assimilation in algal biomass depends on the biological activity in the system and is affected by several factors such as temperature, organic load, detention time, and wastewater characteristics. The rate of gaseous ammonia losses to the atmosphere depends mainly on the pH value, surface-to-volume ratio, temperature, and the mixing conditions in the pond. Alkaline pH shifts the equilibrium equation $NH_3^0 + H_2O \leftrightarrow NH_4^+ + OH^-$ toward gaseous ammonia production, while the mixing conditions affect the magnitude of the mass transfer coefficient. Temperature affects both the equilibrium constant and mass transfer coefficient.

At low temperatures, when biological activity decreases and the pond contents are generally well mixed because of wind effects, ammonia stripping will be the major process for ammonia-N removal in facultative wastewater stabilization ponds. The ammonia stripping in lagoons can be expressed by assuming a first-order reaction.[11,12] The mass balance equation will be:

$$V \frac{dC}{dt} = Q(C_o - C_e) - k_N A (NH_3^0) \qquad (2\text{--}16)$$

FIGURE 2-19
Complete mix model [facultative wastewater stabilization ponds, (influent TBOD$_5$/pond #1 TBOD$_5$) -1 versus detention time].

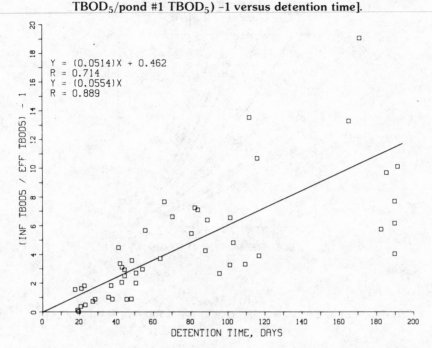

where

Q = flow rate, m^3/d
C_o = influent concentration of (NH$_4$$^+$ + NH$_3$°), mg/l as N
C_e = effluent concentration of (NH$_4$$^+$ + NH$_3$°), mg/l as N
C = average pond contents concentration of (NH$_4$$^+$ + NH$_3$°), mg/l as N
V = volume of the pond, m^3
k_N = mass transfer coefficient, m/d
A = surface area of the pond, m^2
t = time, days

The equilibrium equation for ammonia dissociation can be expressed as:

$$K_b = \frac{[NH_4^+][OH^-]}{[NH_3^°]} \qquad (2\text{-}17)$$

K_b = ammonia dissociation constant.

By modifying Equation 2–17, gaseous ammonia concentration can be

FIGURE 2-20
Wehner and Wilhelm equation chart. Courtesy of American Society of Civil Engineers (Ref. 9).

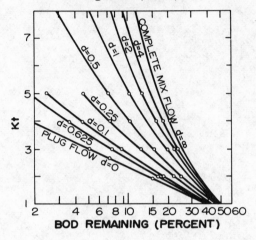

FIGURE 2-21
Relationship of decay rate and temperature (facultative wastewater stabilization pond, COD).

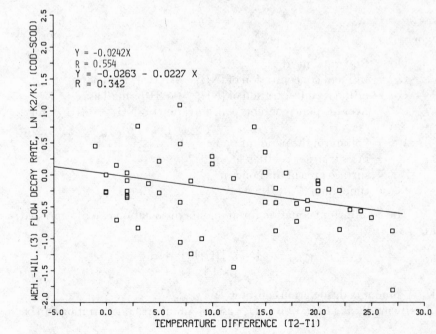

expressed as a function of the pH value and total ammonia concentration ($NH_4^+ + NH_3^\circ$) as follows:

$$[H^+] = \frac{K_W}{[OH^-]} \qquad (2\text{--}18)$$

K_w = water dissociation constant.

$$C = NH_4^+ + NH_3^\circ \qquad (2\text{--}19)$$

$$NH_3^\circ = \frac{C}{1 + 10^{pK_w - pK_b - pH}} \qquad (2\text{--}20)$$

where

$pk_w = -\log K_w$
$pK_b = -\log K_b$

Assuming steady-state conditions and a completely mixed pond where $C_e = C$, Equations 2–16 and 2–20 will yield the following relationship:

$$\frac{C_e}{C_o} = \frac{1}{1 + \frac{A}{Q} k_N \left[\dfrac{1}{1 + 10^{pK_w - pK_b - pH}} \right]} \qquad (2\text{--}21)$$

where

Q = flow rate, m^3/d
C_o = influent concentration of ($NH_4^+ + NH_3^\circ$), mg/l as N
C_e = effluent concentration of ($NH_4^+ + NH_3^\circ$), mg/l as N
C = average pond contents concentration of ($NH_4^+ + NH_3^\circ$), mg/l as N
k_N = mass transfer coefficient, *m/d*
A = surface area of the pond, m^2
V = volume of the pond, m^3
pK_w = $-\log K_w$
pK_b = $-\log K_b$
t = time, days

This relationship emphasizes the effect of pH, temperature (pK_w and pK_b are functions of temperature), and hydraulic loading rate on ammonia-N removal.

Experiments on ammonia stripping conducted by Stratton[11,12] showed that the ammonia loss rate constant was dependent on the pH value and temperature, as shown in the following relationships:

Ammonia loss rate constant $\propto e^{1.57(pH-8.5)}$ $(2\text{--}22)$
Ammonia loss rate constant $\propto e^{0.13(T-20)}$ $(2\text{--}23)$

King[13] reported that only 4% nitrogen removal was achieved by harvesting floating *Cladophora fracta* from the first pond in a series of four receiving secondary effluent. The major nitrogen removal in the ponds was attributable to ammonia gas stripping. The removal of total nitrogen was described by first-order kinetics using a plug flow model ($N_t = N_o e^{-0.03t}$ where N_t = total nitrogen concentration, mg/l at time t; N_o = initial total nitrogen concentration, mg/l; t = time, days).

Large-scale facultative wastewater stabilization pond systems only approach steady-state conditions, and only during windy seasons will well-designed ponds approach completely mixed conditions. Moreover, when ammonia removal through biological activity becomes significant or ammonia is released into the contents of the pond from anaerobic activity at the bottom of the pond, the expressions for ammonia removal in the system must include these factors, along with the theoretical consideration of ammonia stripping, as shown in Equation 2–21.

Pano and Middlebrooks[14] developed mathematical relationships for ammonia-N removal based on the performance of three full-scale facultative wastewater stabilization ponds considering the above theoretical approach and incorporating temperature, pH value, and hydraulic loading rate as variables. Rather than using the theoretical expression for ammonia-N stripping (Equation 2–21), the following equation was considered for ammonia-N removal in facultative ponds:

$$\frac{C_e}{C_o} = \frac{1}{1 + \frac{A}{Q}K_N \cdot f(pH)} \tag{2–24}$$

where:

K_N = removal rate coefficient, L/t
$f(pH)$ = function of pH
Other terms as defined above.

The K_N values are considered to be a function of temperature and mixing conditions. For a similar pond configuration and climatic region, the K_N values can be expressed as a function of temperature only. The function of pH is considered to be dependent on temperature, which affects the pK_w and pK_b values as well as the biological activity in the pond. To incorporate the effect of the pH function on ammonia-N stripping (Equation 2–21), the pH function was found be an exponential relationship. The selection of an exponential function to describe the pH function was based on statistical analyses that indicated an exponential relationship best described the data. Also, most reaction rate and temperature relationships are described by exponential functions such as the Van't Hoff–Arrhenius equation; therefore, it is logical to assume that such a relationship would apply in the application of the theoretical equation to a practical problem.

The Pond Systems

The stabilization pond systems located in Peterborough, New Hampshire, Eudora, Kansas, and Corinne, Utah described earlier, are exposed to similar climatic conditions. During the winter the water temperatures in these ponds range between 1 and 5°C (34–41°F) and ice cover is experienced during the winter. During the summertime the water temperature is approximately 20°C (68°F) and is generally less than 25°C (77°F).

The characteristics of the wastewater entering these systems were significantly different, as shown in Table 2–3. The Eudora wastewater is a typical medium-strength domestic wastewater, the Corinne wastewater is alkaline, and Peterborough has an acid wastewater when compared with typical domestic wastewater. The pH value in the ponds, and as a result the ammonia-N removal, is affected by these differences. In the Corinne ponds the pH values were above 9. In the Peterborough system, because of low buffer capacity of the wastewater, anaerobic conditions existed during the winter months, and the pH value during the summer was 6.7–7.4 which is very low when compared with properly operated ponds. It should be emphasized that other problems in the Peterborough system added to the poor conditions in the ponds. Accumulated solids in the chlorine contact chamber were pumped to the primary pond, and this resulted in anaerobic conditions in the system. Several times the river flooded into the chlorine contact chamber.

Ammonia-N Removal

The annual percentage ammonia-N removals in the Eudora and Corinne facultative pond systems were about 95%, and in the Peterborough ponds it was about 53%. The major removal occurred in the primary ponds. Table 2–4 summarizes these results.

Table 2-3
Characteristics of the Wastewaters at Peterborough, N.H., Eudora, Kansas, and Corinne, Utah

Parameter	System		
	Peterborough	*Eudora*	*Corinne*
BOD mg/l	138	270	74
COD mg/l	271	559	128
NH_4 mg/l	21.5	25.5	7.5
pH	7.0	7.7	8.4
Alkalinity mg/l as $CaCO_3$	107	428	576

Source: From Ref. 14.

Table 2-4
Ammonia-N Removal in Facultative Ponds

PARAMETER	SYSTEM		
	Peterborough	Eudora	Corinne
Influent NH$_4$-N, mg/l	21.47	25.48	7.53
#1 pond effluent NH$_4$-N, mg/l	16.47	9.20	1.44
Final effluent NH$_4$-N, mg/l	11.49	1.14	0.23
% removal in #1	23.3	63.9	80.9
Overall % removal	52.8	96.1	94.5
Theoretical detention time in			
Pond #1—days	44	92	29
In the overall system—days	108	227	74

During the months of June through September, ammonia-N removal in the Peterborough system reached 67%, while in Eudora ammonia-N removal was almost 99%. In the winter period of December to March in the primary pond of the Peterborough system there was no ammonia-N removal, and in the total system there was approximately 19% ammonia-N removal. In the Eudora and Corinne systems there was about 90% ammonia-N removal, with 53-62% removal occurring in the primary ponds. Tables 2-5 and 2-6 summarize these results.

Model Development

The theoretical consideration for ammonia stripping, as expressed in Equation 2-24, can be modified as follows:

$$\frac{C_e}{C_o} = \frac{1}{1 + \frac{A}{Q} K_N e^{apH}} \tag{2-25}$$

Table 2-5
Ammonia Removal during the Winter Months of December-March

PARAMETER	SYSTEM		
	Peterborough	Eudora	Corinne
Water temperature (°C)	2-5	3-7	1-5
#1 pond pH value	6.6-7.0	7.8-8.5	8.6-9.4
Influent NH$_4$-N, mg/l	20.75	29.56	9.82
#1 pond effluent NH$_4$-N, mg/l	21.54	13.81	3.77
Final effluent NH$_4$-N, mg/l	16.74	2.38	0.25
% removal in #1 pond	0.0	53.3	61.6
Overall removal (%)	19.3	92.0	97.5

SOURCE: From Ref. 14.

Table 2-6
Ammonia Removal during the Summer Months of June-September

PARAMETER	SYSTEM		
	Peterborough	*Eudora*	*Corinne*
Water temperature (°C)	17-25	22-27	16-22
#1 pond pH value	6.9-7.4	8.1-8.5	9.4-9.5
Influent NH_4-N, mg/l	21.04	21.97	5.21
#1 pond effluent NH_4-N, mg/l	13.0	4.48	0.26
Final effluent NH_4-N, mg/l	6.85	0.28	0.40
% removal in #1 pond	38.2	79.6	95.0
Overall removal (%)	67.4	98.7	92.3

SOURCE: From Ref. 14.

where K_N and a are assumed to be a function of the temperature and other factors as described above. Because the major part of ammonia-N removal takes place in the primary ponds, the analysis of ammonia removal has been focused on the primary ponds.

To determine the coefficients a and K_N in Equation 2–25 as a function of the temperature, the monthly mean concentrations of ammonia-N in the primary pond have been separated into four temperature ranges: (a) 1–5°C, (b) 6–10°C, (c) 16–20°C, (d) 21–25°C. For each temperature range a least-squares fit of the data to Equation 2–25 was performed. Table 2–7 presents a summary of the results of the regression analyses. Tables 2–8 through 2–11 contain the ammonia-N removal data used to obtain the regression analyses results presented in Table 2–7.

As discussed earlier, low pH values result in negligible gaseous ammonia concentrations. The ammonia-N removal performance in the three systems show that for a pH value of 6.6 or less, there is no ammonia-N removal. Therefore, Equation 2–25 can be written in a form that eliminates the effect of stripping below some minimum pH value (Equation 2–26):

$$\frac{C_e}{C_o} = \frac{1}{1 + \frac{A}{Q} K_N' \; e^{a(\text{pH}-6.6)}} \qquad (2\text{-}26)$$

Table 2-7
Summary of the Regression Analyses for Ammonia Removal versus pH Value

PARAMETER	TEMPERATURE (°C)			
	1-5	*6-10*	*16-20*	*21-25*
Number of data values	10	6	6	8
Correlation coefficient	0.853	0.905	0.938	0.874
Slope a in Equation 2-25	1.092	1.511	1.789	1.540
K-m/d in Equation 2-25	3.13×10^{-6}	2.27×10^{-7}	4.64×10^{-8}	1.94×10^{-7}

SOURCE: From Ref. 14.

Table 2-8
Ammonia-N Removal Data and pH Values at Corinne,
Eudora, and Peterborough Wastewater
Lagoon Systems for a Temperature Range of 1-5°C

LOCATION	pH	$\dfrac{C_o}{C_e} - 1$	t (DAYS)	$\dfrac{C_o}{C_e} - 1 \Big/ t$ (DAY^{-1})	$\dfrac{C_o}{C_e} - 1 \Big/ A/Q$ (M/DAY)
Peterborough	7.0	0.2203	50.8	0.00434	0.00566
Eudora	7.8	0.9241	101.4	0.00911	0.01313
Eudora	7.8	0.8947	103	0.00869	0.01251
Eudora	7.8	1.3678	66.1	0.02069	0.02981
Corinne	8.59	1.2475	19.6	0.06365	0.07766
Corinne	8.86	0.7964	22.7	0.03509	0.04268
Corinne	8.96	1.3136	55.8	0.02354	0.02869
Corinne	9.30	1.8196	44.4	0.04098	0.04986
Corinne	9.41	2.8586	63.7	0.04488	0.05467
Corinne	9.60	9.2421	37.2	0.24844	0.30291

SOURCE: From Ref. 14.

From Equation 2–26 the values of K_N' (m/d) were calculated to be 4.223×10^{-3} for $T = 1$–$5°C$, 4.865×10^{-3} for $T = 6$–$10°C$, 6.229×10^{-3} for $T = 16$–$20°C$, and 5.035×10^{-3} for $T = 21$–$25°$ C. The K_N' values calculated with Equation 2–26 are probably closer to the actual mass transfer coefficients than the K_N values obtained with Equation 2–25. The decrease in the

Table 2-9
Ammonia-N Removal Data and pH Values at Corinne,
Eudora, and Peterborough Wastewater
Lagoon Systems for a Temperature Range
of 6-10°C

LOCATION	pH	$\dfrac{C_o}{C_e} - 1$	t (DAYS)	$\dfrac{C_o}{C_e} - 1 \Big/ t$ (DAY^{-1})	$\dfrac{C_o}{C_e} - 1 \Big/ A/Q$ (M/DAY)
Peterborough	7.2	0.5270	50.9	0.01035	0.01350
Peterborough	7.4	1.2570	48.6	0.02589	0.03371
Eudora	7.8	1.1062	89.0	0.01243	0.01789
Eudora	8.5	1.5627	70.4	0.02220	0.03198
Corinne	9.19	4.5364	23.2	0.19553	0.23773
Corinne	9.44	9.8873	17.6	0.56178	0.68217

SOURCE: From Ref. 14.

C_o = Influent ammonia concentration in mg/l as N.
C_e = Primary pond effluent—ammonia concentration in mg/l as N.
t = Hydraulic detention time in primary pond—days
Q/A = Hydraulic loading rate—m/day

Table 2-10
Ammonia-N Removal Data and pH Values at Corinne,
Eudora, and Peterborough Wastewater
Lagoon Systems for a Temperature Range of 16-20°C

LOCATION	pH	$\frac{C_o}{C_e} - 1$	t (DAYS)	$\frac{C_o}{C_e} - 1 \; / t$ (DAY^{-1})	$\frac{C_o}{C_e} - 1 \; /A/Q$ (M/DAY)
Eudora	8.0	2.6069	87.9	0.02966	0.04273
Corinne	9.42	58.1250	21.1	2.75474	3.35877
Corinne	9.48	18.1739	19.0	0.95652	1.16637
Corinne	9.48	8.0000	27.3	0.29304	0.35683
Peterborough	6.7	0.2478	42.1	0.00589	0.00767
Peterborough	7.4	1.1632	41.3	0.02816	0.03672

SOURCE: From Ref. 14.

slope a and K_N' values above 20°C are probably due to decreased mixing intensity and stratification during the summer which decreases the actual residence time. Linear regression analyses of the slopes and K_N' values up to 20°C versus the mean temperature values resulted in correlation coefficients of .953 and .999, respectively. These correlation coefficients are significant at the .20 and .05 level, indicating that a good approximation is provided.

Table 2-11
Ammonia-N Removal Data and pH Values at Corinne,
Eudora, and Peterborough Wastewater
Lagoon Systems for a Temperature Range of 21-25°C

LOCATION	pH	$\frac{C_o}{C_e} - 1$	t (DAYS)	$\frac{C_o}{C_e} - 1 \; / t$ (DAY^{-1})	$\frac{C_o}{C_e} - 1 \; /A/Q$ (M/DAY)
Peterborough	6.9	0.2945	44.6	0.00660	0.00860
Peterborough	7.1	0.2428	43.1	0.00563	0.00735
Peterborough	7.1	1.2636	43.2	0.02925	0.03813
Corinne	9.4	13.6667	20.7	0.66022	0.80261
Eudora	7.9	0.9650	95.5	0.01010	0.01456
Eudora	8.1	3.4554	83.9	0.04118	0.05936
Eudora	8.1	2.7194	109.4	0.02486	0.03579
Eudora	8.5	3.7625	101.1	0.03722	0.05357

SOURCE: From Ref. 14

C_o	= Influent ammonia concentration, mg/l as N.
C_e	= Primary pond effluent, ammonia concentration in mg/l as N.
t	= Hydraulic detention time in primary pond, days.
Q/A	= Hydraulic loading rate, m/day.

The final form of the ammonia removal equation incorporating temperature up to 20°C is as follows:

$$\frac{C_e}{C_o} = \frac{1}{1 + \dfrac{A}{Q}(0.0038 + 0.000134T)\exp[(1.041 + 0.044T)(pH - 6.6)]}$$

$$(2\text{--}27)$$

where

C_e, C_o = ammonia-nitrogen concentrations, mg/l as N
Q/A = m/day
T = °C

Figure 2–22 shows a plot of the measured values of ammonia-N removal versus the predicted values for a range of temperatures of 1–20°C using Equation 2–27. The application of Equation 2–27 should be limited to ponds with an average depth of 1.22–1.52 m (4–5 ft) and located in a region with a climate similar to the three systems studied. For a temperature range of 21–25°C, it is recommended that Equation 2–28 be used:

$$\frac{C_e}{C_o} = \frac{1}{1 + 5.035 \times 10^{-3}\dfrac{A}{Q}\, e^{1.540(pH-6.6)}}$$

$$(2\text{--}28)$$

FIGURE 2-22
The measured values of ammonia-N removal plotted versus the predicted values for a temperature range of 0-20°C (Ref. 14).

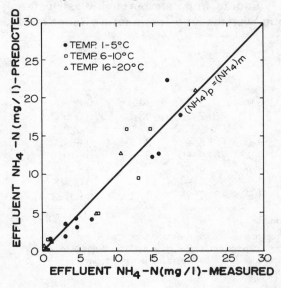

Model Verification

Ammonia-N removal data for the second, third, and fourth ponds of the Corinne, Eudora, and Peterborough systems and data collected at other pond systems were used to verify the model (Equations 2–27 and 2–28). Only influent ammonia-N concentrations greater than 1 mg/l and water temperatures within a range of 1–25°C were used. Because of water losses (exfiltration and evaporation) in the pond systems, the influent flow rates to the ponds following the primary ponds were estimated using the following equation:

$$Q_i = Q_{i-1} - \frac{\sum\limits_{j=1}^{i-1} A_j}{\sum\limits_{j=1}^{n} A_j} (Q_o - Q_e) \qquad (2\text{--}29)$$

where

Q_o, Q_e = influent and effluent flow rates, m³/d
n = number of ponds in series
A_j = area of j pond, m²
i = 2, n

Figure 2–23 shows a plot of the observed ammonia-N concentrations versus predicted values for a total of 41 months of performance data from the Corinne, Eudora, and Peterborough systems. If perfect agreement were obtained, all of the points would lie on the 45-degree line shown on Figure 2–23. Excellent agreement was obtained, and it is evident that the model can be used to predict ammonia-N removal in facultative wastewater stabilization ponds. Data collected from wastewater stabilization ponds with similar configurations and climatic conditions to those employed in the model development were also used to verify the model. The data are summarized in Table 2–12. Figure 2–24 shows the observed values and the predicted values. The results shown in Figure 2–24 further illustrate the applicability of the model toward predicting the ammonia-N removal in facultative wastewater stabilization ponds.

Summary

Annual ammonia-N removals above 90 % were observed for two facultative wastewater stabilization ponds. The major part of ammonia-N removal occurred in the primary ponds, and the process appeared to follow first-order kinetics.

Analyses of ammonia-N removal in the primary ponds showed a statistically significant relationship with the pH value, water temperature,

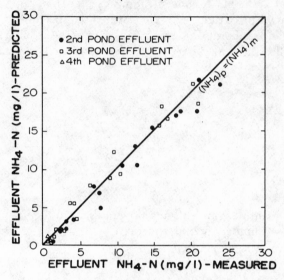

FIGURE 2-23
The predicted ammonia-N concentrations plotted versus the ammonia-N concentrations for the second, third, and fourth ponds in series at Corinne, Eudora, and Peterborough (Ref. 14).

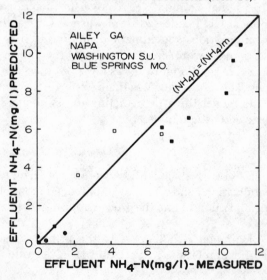

FIGURE 2-24
The predicted ammonia-N concentrations plotted versus the measured ammonia-N concentrations at other locations (Ref. 14).

Table 2 - 12
Characteristics of Wastewater Stabilization
Ponds Used for Model Verification

System	Type	Temperature (°C)	pH	Influent Ammonia-N Concentration MG/L	Ref.
Blue Springs, MO	Facultative				15
1st cell		4.5, 18.2	7.39, 7.51	10.67, 11.03	
2nd cell		4.1, 18.7	7.34, 7.55	10.25, 6.78	
3rd cell		3.7, 20.1	7.27, 7.62	10.63, 7.30	
Ailey, GA[a]	Facultative and polishing	19.9, 26.1, 9.5	8.7, 9.5, 8.6	5.26, 7.37, 3.86	16
Washington State University	Dairy manure Anaerobic lagoons				17
4th cell		23	9.8	57.6	
Napa System[b]	Polishing				18
2nd cell		11.9, 19.1	7.5, 8.0	6.71, 1.76	
3rd cell		11.8	7.7	6.75	
4th cell		11.6	7.8	4.22	

Source: From Ref. 14.

[a] Effluent from the polishing ponds.

[b] Flow rates were not reported. Hydraulic loading rates were estimated using the reported detention time and mean depths. Only the data for winter and spring were used when the influent concentrations were higher than 1 mg/l.

and hydraulic loading rate. A relationship similar to the theoretical ammonia stripping model, as a function of pH, temperature, and hydraulic loading rate, was developed. The relationship shows that ammonia-N removal increases as the pH value, detention time, and temperature (up to 20°C) increase. At 20°C with the pH value of 8.0 in the primary pond (40% of total area) and 8.4 in the other ponds and a 100-day detention time [average depth of 1.22 m (4ft)], the predicted ammonia removal will be about 98%. By increasing the temperature from 20–25°C, while the other factors remain constant, the predicted ammonia removal will be 92%. The decrease in the model coefficients with temperature above 20°C is probably due to thermal stratification in the pond system and poor mixing conditions.

The equations should be used under conditions similar to those at the three sites evaluated.

Aerated Wastewater Stabilization Lagoons

General

To satisfy the need for reliable lagoon performance data, in 1974 the U.S. Environmental Protection Agency sponsored five intensive aerated lagoon performance studies. The aerated lagoons are located at Bixby, Oklahoma,[19] Pawnee, Illinois,[20] Gulfport, Mississippi,[21] Lake Koshkonong, Wisconsin,[22] and Windber, Pennsylvania.[23] Data collection encompassed 12 months, with four separate 30-consecutive-day sampling periods once each season.

Aerated waste stabilization ponds are medium depth, man-made basins designed for the biological treatment of wastewater. A mechanical aeration device is used to supply supplemental oxygen to the system. In general, an aerated lagoon is aerobic throughout its entire depth. The mechanical aeration device may cause turbulent mixing (i.e., surface aerator) or may produce laminar flow conditions (diffused air systems). The five aerated lagoons were considered to be partial mix lagoons by the investigators although data describing the mixing characteristics were not collected.

Although the development, history, and design of aerated wastewater stabilization ponds have been reported by several investigators[24,25,26] very little reliable year-round performance data were available until the U.S. Environmental Protection Agency funded the evaluation of five aerated lagoon systems.[27] A portion of the data from these studies is presented in the following paragraphs.

Site Descriptions

Bixby, Oklahoma. A diagram of the Bixby, Oklahoma aerated lagoon system is shown in Figure 2–25.[19] The system consists of two aerated cells with a total surface area of 2.3 ha (5.8 acres). It was designed to treat 336 kg BOD_5/day (740 lb BOD_5/day) with a hydraulic loading rate of 1551 m^3/day (0.4 mgd). There is no chlorination facility at the site. The hydraulic retention time is 67.5 days.

Pawnee, Illinois. A diagram of the Pawnee, Illinois aerated lagoon system is shown in Figure 2–26.[20] The system consists of three aerated cells in series with a total surface area of 4.45 ha (11.0 acres). The design flow rate was 1893 m^3/day (0.5 mgd) with an organic load of 386 kg BOD_5/day (850 lb BOD_5/day) and a theoretical hydraulic retention time of 60.1 days. The facility is equipped with chlorination disinfection and a slow sand filter for polishing the effluent. Data reported below were collected prior to the filters and represent only lagoon performance.

Gulfport, Mississippi. A diagram of the Gulfport, Mississippi aerated lagoon system is shown in Figure 2–27.[21] The system consists of two aerated lagoons in series with a total surface area of 2.5 ha (6.3 acres). The system was designed to treat 1893 m^3/day (0.5 mgd) with a total theoretical hydraulic retention time of 26.2 days. The organic load on the first cell in the series was 374 kg BOD_5/day·ha (334 lb BOD_5/day·acre) and 86 kg BOD_5/day·ha (77 lb BOD_5/day·acre) on the second cell in the series. The system is equipped with a chlorination facility.

FIGURE 2-25
**Schematic flow diagram and aerial photograph of the aerated lagoon
system at Bixby, Oklahoma (Ref. 19).**

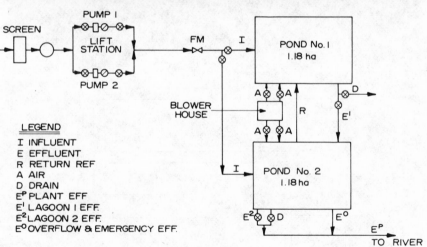

Lake Koshkonong, Wisconsin. A diagram of the Lake Koshkonong,
Wisconsin aerated lagoon system is shown in Figure 2–28.[22] The system consists of three aerated cells with a total surface area of 2.8 ha (6.9 acres)
followed by chlorination. The design flow was 2271 m³/day (0.6 mgd) with
a design organic load of 467 kg BOD$_5$/day (1028 lb BOD$_5$/day). The current

FIGURE 2-26
Schematic flow diagram and aerial photograph of the aerated lagoon system at Pawnee, Illinois (Ref. 20).

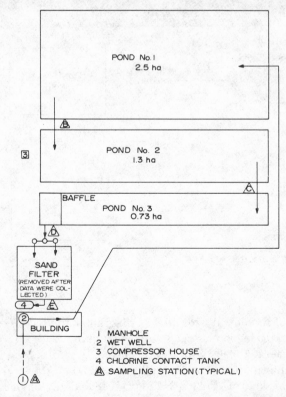

1 MANHOLE
2 WET WELL
3 COMPRESSOR HOUSE
4 CHLORINE CONTACT TANK
⚠ SAMPLING STATION(TYPICAL)

FIGURE 2-27
Schematic flow diagram and aerial photograph of the aerated lagoon
system located at North Gulfport, Mississippi (Ref. 21).

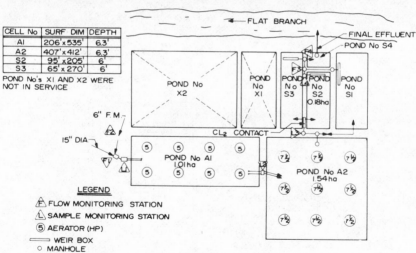

CELL No	SURF DIM	DEPTH
AI	206'x535'	6.3'
A2	407'x412'	6.3'
S2	95' x205'	6'
S3	65' x 270'	6'

POND No's XI AND X2 WERE
NOT IN SERVICE

LEGEND
F FLOW MONITORING STATION
△ SAMPLE MONITORING STATION
⑤ AERATOR (HP)
⟹ WEIR BOX
○ MANHOLE

organic loading rate is 248 kg BOD$_5$/day (545 lb BOD$_5$/day) with a
theoretical hydraulic retention time of 57 days.

Windber, Pennsylvania. The Windber, Pennsylvania aerated
lagoon system consists of three cells in series with a total surface area of 8.4
ha (20.7 acres) followed by chlorination (Figure 2–29).[23] The design flow

FIGURE 2-28
Schematic flow diagram and aerial photograph of the aerated lagoon
system at Lake Koshkonong, Wisconsin (Ref. 22).

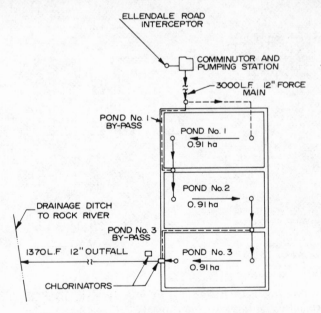

FIGURE 2-29
Schematic flow diagram and aerial photograph of the aerated lagoon
system located at Windber, Pennsylvania (Ref. 23).

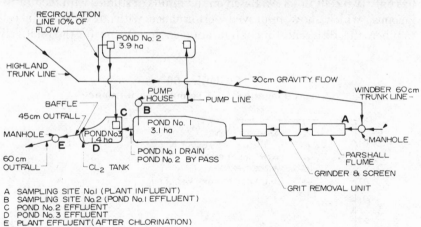

A   SAMPLING SITE No.I (PLANT INFLUENT)
B   SAMPLING SITE No.2 (POND No.I EFFLUENT)
C   POND No.2 EFFLUENT
D   POND No.3 EFFLUENT
E   PLANT EFFLUENT (AFTER CHLORINATION)

rate was 7576 m³/day (2.0 mgd) with a design organic loading rate of approximately 1369 kg BOD$_5$/day (3000 lb BOD$_5$/day). The design mean
hydraulic residence time was 30 days for the three cells operating in series.
Actual influent flow rates varied from 3000 to 5300 m³/day (0.8–1.4 mgd),
the actual organic loading rate was approximately 924 kg BOD$_5$/day (2030

lb BOD$_5$/day), and the theoretical mean hydraulic residence time was approximately 55 days.

Performance

Biochemical Oxygen Demand (BOD$_5$).   The monthly mean effluent BOD$_5$ concentrations are compared with the Federal Secondary Treatment Standard of 30 mg/l in Figure 2–30.

In general, all the systems studied, except the Bixby, Oklahoma system, were capable of producing a final effluent BOD$_5$ concentration of 30 mg/l. Mean monthly effluent BOD$_5$ concentrations appear to be independent of influent BOD$_5$ concentration fluctuations and are also not significantly affected by seasonal variations in temperature.

Mean monthly influent BOD$_5$ concentrations at Bixby, Oklahoma ranged from 212 mg/l to 504 mg/l, with a mean of 388 mg/l during the study period reported. The design influent BOD$_5$ concentration was 240 mg/l, or only 62% of the actual influent concentration. The mean flow rate during the period of study was 523 m^3/day (0.123 mgd) which is less than one-third of the design flow rate. The Bixby system was designed to treat 336 kg BOD$_5$/day (740 lb BOD$_5$/day), and apparently a load of only 203 kg BOD$_5$/day (446 lb BOD$_5$/day) was entering the lagoon. The only major difference between the Bixby and other aerated lagoons is the number of cells. Bixby has only two cells in series. Based on the results of studies with facultative lagoons, which show improved performance with an increase in cell number, this difference in configuration could account for the relatively

FIGURE 2-30
Aerated lagoon BOD$_5$ removal performance.

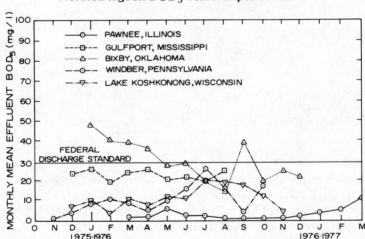

poor performance by the Bixby system. However, there are many other possible explanations, such as operating procedures or short-circuiting (related to number of cells in series).

The results of these studies indicate that aerated lagoons that are properly designed, operated, and maintained can consistently produce an effluent BOD_5 concentration of less than 30 mg/l. In addition, effluent quality is not seriously affected by seasonal climate variations.

Suspended Solids (SS) Removal Performance.   The monthly mean effluent SS concentrations for each system is illustrated in Figure 2–31. In general, the effluent SS concentration from three of the aerated lagoon systems tend to increase significantly during the warm summer months. However, two of the aerated lagoon systems (Windber, Pennsylvania, and Gulfport, Mississippi) produce a relatively constant effluent SS concentration throughout the entire year.

Mean monthly effluent SS concentrations ranged from 2 mg/l at Windber, Pennsylvania in November 1975 to 96 mg/l at Bixby, Oklahoma in June 1976. The mean monthly effluent SS concentration for the Windber, Pennsylvania site never exceeded 30 mg/l throughout the entire study period. In addition, the mean monthly effluent SS concentrations of the Pawnee, Illinois and the Lake Koshkonong, Wisconsin sites only exceeded 30 mg/l during one of the months reported.

The results of these studies indicate that aerated lagoon effluent SS concentrations are variable. However, a well-designed, operated, and maintained aerated lagoon can produce final effluents with low SS concentrations.

FIGURE 2-31
Aerated lagoon SS removal performance.

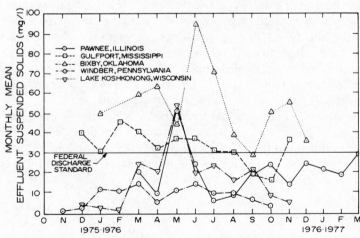

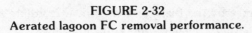

FIGURE 2-32
Aerated lagoon FC removal performance.

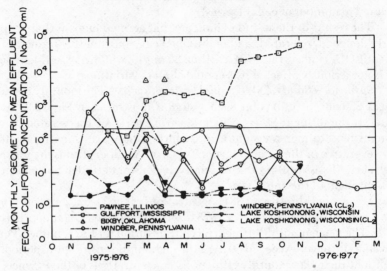

Fecal Coliform (FC) Removal Performance. FC data were not available for the Lake Koshkonong, Wisconsin site and only one monthly geometric mean FC value was available for Bixby, Oklahoma. The monthly geometric mean effluent FC concentration compared to a concentration of 200 organisms/100 ml is illustrated in Figure 2–32.

All the aerated lagoon systems, except the Bixby, Oklahoma site, have chlorine disinfection. Therefore, the data actually indicate the susceptibility of aerated lagoon effluent to chlorination. In general, the Windber, Pennsylvania and the Pawnee, Illinois systems produced final effluent monthly geometric mean FC concentrations of less than 200 organisms/100 ml. The nonchlorinated Bixby, Oklahoma, data indicate a high effluent FC concentration. The Gulfport, Mississippi system produced an effluent containing more than 200 FC/100 ml most of the time, but the FC were measured in effluent samples from a holding pond with a long detention time following the addition of the chlorine. Therefore, aftergrowth of the FC probably accounted for the high concentrations.

Summary. From the aerated lagoon performance data currently available, it appears that (a) aerated lagoons can produce an effluent BOD$_5$ concentration of less than 30.0 mg/l, (b) aerated lagoon SS concentrations are affected by seasonal variations, and (c) aerated lagoon effluent can be satisfactorily disinfected with chlorination.

Evaluation of Design Methods

A summary of the aerated lagoon performance data used to evaluate the various design methods is presented in Table 2–13. These data were col-

TABLE 2-13
Performance Data for Five Aerated Lagoons

INF BOD MG/L	CELL #1 BOD	INF SOL BOD	CELL #1 SBOD	INF COD MG/L	CELL #1 COD	INF SOL COD	CELL #1 SCOD	DET TIME DAYS	TEMP DEG C	LIGHT LAN	TSS MG/L	VSS MG/L	LOCATION
368.	68.	222.	55.	664.	152.	321.	68.	66.59	5.	190	93.	64.	BIXBY,OKLAHOMA
422.	150.	244.	87.	524.	186.	368.	133.	68.26	10.	250	63.	40.	BIXBY,OKLAHOMA
414.	60.	227.	40.	671.	192.	319.	101.	60.26	16.	350	101.	63.	BIXBY,OKLAHOMA
392.	98.	127.	23.	610.	212.	215.	79.	45.87	19.	435	74.	59.	BIXBY,OKLAHOMA
379.	81.	136.	9.	552.	202.	240.	61.	52.49	20.	590	74.	74.	BIXBY,OKLAHOMA
413.	87.	122.	13.	606.	175.	267.	104.	66.94	26.	600	61.	53.	BIXBY,OKLAHOMA
355.	58.	140.	14.	594.	138.	254.	54.	59.92	29.	560	67.	50.	BIXBY,OKLAHOMA
313.	64.	105.	5.	589.	180.	223.	62.	33.32	28.	550	89.	76.	BIXBY,OKLAHOMA
330.	87.	142.	5.	646.	285.	259.	68.	52.89	24.	470	109.	78.	BIXBY,OKLAHOMA
388.	106.	136.	8.	773.	268.	278.	57.	59.58	13.	350	166.	140.	BIXBY,OKLAHOMA
364.	117.	147.	7.	774.	254.	249.	71.	55.28	9.	265	127.	103.	BIXBY,OKLAHOMA
283.	61.	119.	9.	619.	178.	224.	67.	49.85	7.	200	59.	44.	BIXBY,OKLAHOMA
277.	13.	58.	7.	440.	86.	83.	61.	85.15	5.	520	56.	20.	PAWNEE,ILLINOIS
468.	33.	75.	17.	753.	131.	146.	89.	77.53	12.	485	51.	33.	PAWNEE,ILLINOIS
470.	20.	89.	10.	1147.	113.	231.	63.	82.36	19.	415	41.	29.	PAWNEE,ILLINOIS
452.	16.	130.	10.	1208.	82.	252.	60.	86.40	23.	555	33.	24.	PAWNEE,ILLINOIS
602.	20.	136.	11.	1921.	143.	361.	99.	83.69	26.	510	43.	30.	PAWNEE,ILLINOIS
578.	28.	143.	10.	1208.	132.	298.	68.	88.44	19.	395	58.	41.	PAWNEE,ILLINOIS
799.	9.	127.	5.	1572.	74.	290.	62.	86.61	11.	255	31.	17.	PAWNEE,ILLINOIS
548.	9.	129.	5.	1156.	77.	294.	58.	96.76	5.	220	53.	27.	PAWNEE,ILLINOIS
554.	9.	126.	4.	1115.	78.	246.	70.	87.09	3.	165	23.	16.	PAWNEE,ILLINOIS
395.	10.	97.	5.	779.	68.	213.	60.	115.11	3.	245	17.	8.	PAWNEE,ILLINOIS
296.	18.	83.	11.	649.	75.	170.	59.	76.36	8.	320	37.	21.	PAWNEE,ILLINOIS
233.	14.	39.	7.	367.	89.	53.	61.	66.57	6.	345	51.	18.	PAWNEE,ILLINOIS
177.	19.	74.	15.	267.	37.	58.	33.	29.73	19.	140	19.	12.	WINDBER,PENNSYLVANIA
203.	43.	76.	42.	376.	48.	97.	36.	29.53	14.	110	21.	13.	WINDBER,PENNSYLVANIA
152.	30.	61.	25.	211.	57.	74.	47.	14.81	12.	135	38.	28.	WINDBER,PENNSYLVANIA
220.	68.	110.	65.	192.	52.	64.	35.	14.47	12.	225	32.	22.	WINDBER,PENNSYLVANIA
186.	38.	81.	34.	206.	56.	75.	38.	15.71	13.	275	25.	18.	WINDBER,PENNSYLVANIA
155.	46.	58.	21.	270.	95.	135.	60.	14.52	14.	470	52.	37.	WINDBER,PENNSYLVANIA
202.	66.	108.	58.	435.	207.	211.	97.	20.00	20.	460	104.	72.	WINDBER,PENNSYLVANIA
182.	69.	103.	46.	431.	190.	190.	101.	22.92	26.	500	123.	81.	WINDBER,PENNSYLVANIA
145.	64.	91.	44.	356.	137.	210.	77.	16.24	25.	510	51.	31.	WINDBER,PENNSYLVANIA
172.	48.	82.	37.	749.	109.	199.	65.	15.71	24.	470	60.	41.	WINDBER,PENNSYLVANIA
173.	39.	69.	14.	1062.	131.	242.	52.	16.48	22.	345	88.	73.	WINDBER,PENNSYLVANIA
106.	45.	66.	34.	537.	146.	191.	52.	15.94	17.	225	26.	21.	WINDBER,PENNSYLVANIA
89.	11.	34.	22.	209.	34.	34.	22.	26.15	4.	85	7.	6.	LAKE KOSHKONONG,WISC.
96.	12.	46.	44.	229.	52.	46.	44.	28.91	1.	150	5.	4.	LAKE KOSHKONONG,WISC.
75.	10.	41.	37.	244.	45.	41.	37.	25.74	1.	225	7.	6.	LAKE KOSHKONONG,WISC.
37.	14.	29.	30.	116.	52.	29.	30.	18.82	4.	290	22.	20.	LAKE KOSHKONONG,WISC.
55.	12.	31.	28.	127.	48.	31.	28.	20.30	11.	415	16.	15.	LAKE KOSHKONONG,WISC.
71.	18.	26.	24.	160.	54.	26.	24.	22.41	12.	510	19.	18.	LAKE KOSHKONONG,WISC.
86.	14.	37.	29.	186.	56.	37.	29.	22.65	21.	590	20.	19.	LAKE KOSHKONONG,WISC.
93.	23.	34.	36.	229.	74.	34.	36.	22.86	23.	575	30.	27.	LAKE KOSHKONONG,WISC.
118.	25.	18.	25.	299.	44.	18.	25.	22.77	22.	530	12.	10.	LAKE KOSHKONONG,WISC.
113.	34.	21.	26.	194.	48.	21.	26.	23.40	16.	400	20.	13.	LAKE KOSHKONONG,WISC.
102.	25.	18.	21.	200.	36.	18.	21.	24.23	12.	255	6.	5.	LAKE KOSHKONONG,WISC.
87.	17.	38.	27.	168.	32.	38.	27.	48.95	7.	180	6.	5.	LAKE KOSHKONONG,WISC.
178.	49.	94.	32.	510.	105.	144.	60.	7.06	12.	240	38.	33.	NORTH GULFPORT,MISS.
214.	63.	88.	41.	457.	96.	120.	56.	6.04	14.	250	37.	33.	NORTH GULFPORT,MISS.
199.	68.	86.	46.	454.	105.	115.	55.	8.22	19.	365	50.	44.	NORTH GULFPORT,MISS.
192.	76.	117.	35.	369.	93.	138.	52.	8.21	16.	280	53.	41.	NORTH GULFPORT,MISS.
178.	67.	79.	36.	431.	105.	112.	55.	7.64	22.	505	52.	43.	NORTH GULFPORT,MISS.
175.	58.	76.	31.	291.	120.	102.	56.	7.03	23.	490	51.	35.	NORTH GULFPORT,MISS.
171.	66.	71.	37.	359.	101.	92.	46.	6.73	28.	550	98.	63.	NORTH GULFPORT,MISS.
134.	53.	55.	30.	203.	80.	69.	46.	5.75	29.	480	56.	33.	NORTH GULFPORT,MISS.
151.	56.	68.	36.	201.	86.	86.	51.	5.84	29.	475	61.	42.	NORTH GULFPORT,MISS.
170.	62.	74.	38.	230.	82.	89.	49.	7.19	27.	400	42.	30.	NORTH GULFPORT,MISS.
171.	82.	74.	41.	221.	101.	86.	57.	9.30	24.	345	75.	55.	NORTH GULFPORT,MISS.
206.	73.	119.	42.	324.	108.	151.	69.	7.84	18.	255	40.	27.	NORTH GULFPORT,MISS.

lected at the five aerated lagoon systems described earlier. Only the characteristics of the influent wastewater and the effluent from the primary (first) cell of the systems are presented in Table 2–13.

The majority of the kinetic analyses of the systems were limited to the performance in the primary cells, because the BOD_5 and COD appear to have a greater impact on the performance of the primary cells of the systems. Species succession, nutrient concentrations, and the buffering capacity of the systems appear to have more of an effect on the cells following the primary cell. The more commonly used design methods for aerated lagoons are discussed individually in the following paragraphs.

Design Equations.   The most commonly used aerated lagoon design equation is presented in the Ten State Standards (see Appendix B).

$$t = \frac{E}{2.3\,K_1\,(100-E)} \tag{2-30}$$

where

t = detention time, days
E = percent BOD$_5$ to be removed in an aerated pond
K_1 = reaction coefficient, aerated lagoon, base 10. For normal domestic wastewater, the K_1 value may be assumed to be 0.12/day at 20°C and 0.06/day at 1°C.

Equation 2–30 is equivalent to Equation 2–8 presented in the section on Rational Design Equations for facultative wastewater stabilization pond design. By manipulating Equation 2–30 as shown below, it can be shown that the two equations are equal.

$$2.3\,K_1 t = \frac{E}{100 - E} \tag{2-31}$$

$$E = \frac{C_o - C_e}{C_o} \times 100 \tag{2-32}$$

$$2.3\,K_1 t = \frac{\left(\dfrac{C_o - C_e}{C_o}\right) \times 100}{100 - \left[\left(\dfrac{C_o - C_e}{C_o}\right) \times 100\right]} \tag{2-33}$$

$$2.3\,K_1 t = \frac{100 - 100\,C_e/C_o}{100 - 100 + 100\,C_e/C_o} \tag{2-34}$$

$$2.3\,K_1 t = \frac{C_o}{C_e} - 1 \tag{2-35}$$

The only difference between Equation 2–30 and Equation 2–8 is that the constant in Equation 2–8 is expressed in terms of base e, and the other is expressed in terms of base 10.

Equation 2–8 is the most commonly used equation to design aerated lagoons. Practically every aerated lagoon in the United States has been designed with this simple complete mix model. In the design process, the reaction rate coefficient is adjusted to reflect the influence of temperature on the biological reactions in the aerated ponds by using Equation 2–14.

The plug flow (Equation 2–5,), complete mix (Ten State Equation and Equation 2–8) and Wehner–Wilhelm (Equation 2–15) models were evaluated using the data shown in Table 2–13. The effect of water temperature on the reaction rates calculated with each of the models using the influent total BOD_5 and the effluent $SBOD_5$ was evaluated, and the results are shown in Figures 2–33 through 2–35. The Wehner–Wilhelm equation was solved using a dispersion number of 0.25.

Although the relationships between decay rates and temperature differences are statistically significant at the 1 or 5% level, the data points are scattered, and less than 30% of the variation is explained by the regression analyses. Using the regression analyses for the regression of the data through the origin, temperature factors (v) of 1.07, 1.09, and 1.07 were obtained for the plug flow, complete mix and Wehner–Wilhelm models using the influent total BOD_5 and effluent soluble BOD_5 to estimate the substrate strength, respectively. The temperature factors are in agreement with values reported in the literature.[8,28,29] The above values are recommended for use when adjusting the reaction rates to compensate for temperature effects in aerated lagoons.

The relationship shown in Figures 2–33 through 2–35 were the best ob-

FIGURE 2-33
Relationship of decay rate and temperature (aerated wastewater stabilization pond, BOD).

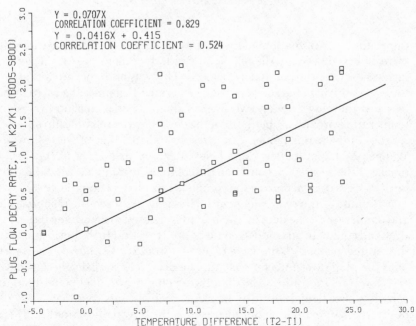

FIGURE 2-34
**Relationship of decay rate and temperature (aerated wastewater
stabilization pond, BOD).**

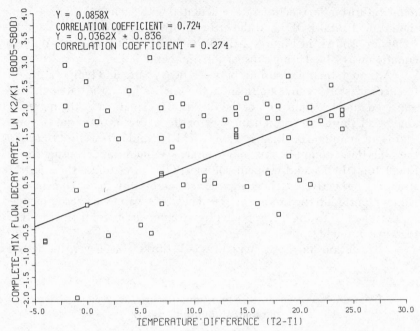

tained for all combinations of the BOD and COD data. Better relationships
were observed between the reaction rates and temperature for the aerated
lagoons than for the facultative lagoons. This was probably attributable to
the better mixing characteristics in the aerated lagoons. The relationship
between the plug flow and Wehner–Wilhelm models reaction rates and
temperature provided a better fit of the data then the relationship observed
for the complete mix model. These results indicate that the plug flow and
Wehner–Wilhelm models provide a better approximation of the performance of the aerated lagoons than the complete mix model. All the aerated
lagoons were designed using the complete mix model (Equation 2–8).

Because of the relatively poor relationship between the reaction rates
and temperature, the data in Table 2–13 were used to determine if the plug
flow and complete mix models could be used to estimate the performance of
the aerated lagoons. As shown in Figures 2–36 and 2–37, the fit of the data to
both the plug flow and complete mix models yield statistically significant
(1% level) relationships. The data fit the plug flow model better than the
complete mix model, even though the complete mix model was used to
design the systems. This is not surprising if the flow patterns in the diffused
air aeration systems is considered. Lagoons using the Hinde Engineering

FIGURE 2-35
Relationship of decay rate and temperature (aerated wastewater stabilization pond, BOD).

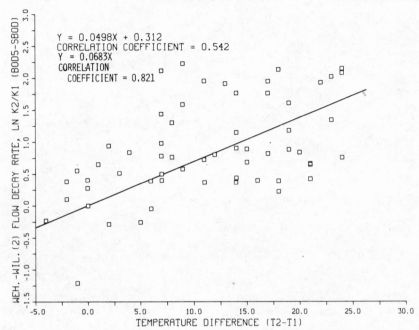

Company aeration system (diffused air) are operated essentially as a plug flow, tapered aeration, activated sludge system without cellular recycle. Therefore, the plug flow model would be expected to more closely describe the performance of these systems. Surface aeration systems could be designed so that either flow model would describe the performance of a lagoon. For example, an aeration basin with a high length-to-width (L/W) ratio with diffused aerators would approximate a plug flow pattern, but a circular or square basin with surface aerators would be expected to approach complete mix conditions. In practice, the flow patterns in lagoons are imperfect and vary with each system. Logically, a model such as the Wehner-Wilhelm equation (Equation 2–15) would be expected to provide the best estimate of the performance of lagoon systems. Unfortunately, none of the simple models is obviously superior to the others. Based on the analysis of the data in Table 2–13, aerated wastewater lagoons can be satisfactorily designed with any of the three flow models if the reaction rates are selected to describe the environmental conditions and if the hydraulic design of the basin is accurate.

Oxygen Requirements. There are no rational design equations to predict the mass rate of transfer of dissolved oxygen and the required mix-

FIGURE 2-36

Plug flow model [aerated wastewater lagoons, ln (pond #1 SBOD$_5$/influent TBOD$_5$) versus detention time].

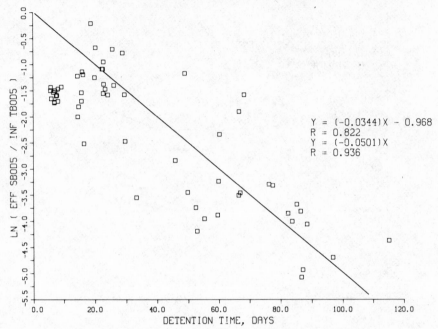

ing to keep the solids suspended in an aerated lagoon. Using the mass of BOD$_5$ entering the system as a basis to calculate the biological oxygen requirements is simple and as effective as other approaches. To predict the aeration needed for mixing, it is necessary to rely on empirical methods developed by equipment manufacturers. Oxygen requirements do not control the design of aeration equipment in aerated lagoons unless the detention time in the aeration tank is approximately 1 day. Such a short detention time is not recommended for an aerated lagoon; therefore, mixing is the major concern in the design.

Catalogs from equipment manufacturers must be consulted to ensure that adequate pumping or mixing is provided. Graphs or tables such as the relationships shown in Figure 2–38 are available from all aeration equipment manufacturers, and all types of equipment must be evaluated to ensure that the most economical and efficient system is selected. The oxygen demand can be estimated using the procedure outlined in Table 2–14. An examination of Figure 2–38 shows that the suspension of the solids (complete mix system) will require approximately 10 times as much power if only oxygen is to be supplied by the aerator. Therefore, an economic analysis along with engineering judgment must be used to select the proper aeration equipment.

FIGURE 2-37
Complete mix model [aerated wastewater lagoons, (influent TBOD$_5$/pond #1 SBOD$_5$) -1 versus detention time].

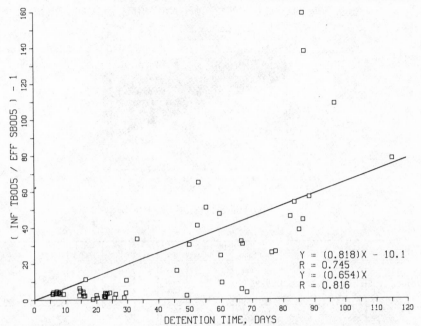

The system should be divided into a minimum of three basins and preferably four basins to improve the hydraulic characteristics and improve mixing conditions. The division of the basins can be accomplished by using separate basins or baffles. There are simple plastic baffles commercially available. Aerators should be selected and spaced through the basins to provide overlapping zones of mixing, and spaced in proportion to the expected oxygen demand. The oxygen demand will decrease as the liquid flows through the system.

Ammonia and Total Kjeldahl Nitrogen Removal in Aerated Lagoons

Theoretical Considerations

Ammonia nitrogen exists in aqueous solutions as ammonia or ammonium ions. At a pH value of 8.0, approximately 95% of the ammonia nitrogen is in the form of ammonium ion. Therefore, in biological systems such as aerated lagoons where the pH values are usually less than 8.0, the majority of the ammonia nitrogen is in the form of ammonium ion.

FIGURE 2-38
Aeration equipment information provided by equipment manufacturer.
Courtesy of Aqua-Aerobic Systems, Inc., Rockford, Illinois.

GUIDE FOR SELECTION OF DRAFT TUBE AND ANTI-EROSION EQUIPMENT

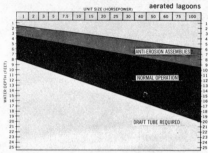

NOTE: These charts are intended for approximation purposes only. Requirements are dependent on basin geometry, etc., and the factory should be contacted for specific applications:

GUIDE FOR SELECTION OF AERATORS

| MODEL SERIES | | | | UNIT SIZE | | | Zcm | D | ZOD | Q |
FSS	CFSS	TFNI	STAINLESS	HP	RPM	Nc	(FEET)	(FEET)	(FEET)	(GPM)
—	—	—	3900111	1	1800	3.4	20		65	1,450
—	—	—	3900211	2	1800		28		90	1,740
4200311	4200317	—	3900311°	3	1800	3.8	40	6	145	2,750
4200331	4200337	—	3900310	1.2	1200		27		87	1,180
4200511	4200517	—	3900511	5	1800		45		150	3,390
4200531	4200537	—	3900510	2.1	1200		29		97	1,500
4200711	4200717	—	3900711	7.5	1800	3.6	50	8	160	3,780
4200731	4200737	—	3900710	3.1	1200		32		104	1,670
4201011	4201017	—	3901011	10	1800	3.4	51		172	5,060
4201031	4201037	—	3901010	4.2	1200		33		92	2,280
4201511	4201517	—	3901511	15	1800	3.5	62		200	6,140
4201531	4201537	—	3901510	6.2	1200		39		129	2,720
4202011	4202017	—	3902011	20	1200	3.2	72	10	230	8,320
4202031	4202037	—	3902010	8.3	900		46		149	3,700
4202511	4202517	—	3902511	25	1200	3.4	80		255	9,830
4202531	4202517	—	3902510	10.4	900		52		165	4,380
—	—	3703011	3903011	30	1200	3.5	88		280	12,570
—	—	3703010	3903010	13.3	900		59		181	5,960
—	—	3704011	3904011	40	1200	3.8	102		325	14,000
—	—	3704010	3904010	17.7	900		68		216	6,640
—	—	3705011	3905011	50	1200		105		330	18,560
—	—	3705010	3905010	22.2	900		70		220	8,830
—	—	3706011	3906011	60	1200	3.5	115	12	350	20,560
—	—	3706010	3906010	26.6	900		76		233	9,720
—	—	3707511	3907511	75	1200	3.0	130		380	22,550
—	—	3707510	3907510	33.2	900		86		253	10,660

Nc = Transfer Rate: lb. of oxygen/ brake hp/ hour @ standard conditions   Zcm = Zone of complete mix
D   = Nominal operating depth in which ZOD, Zcm and Q hold true   ZOD = Zone of complete oxygen dispersion
Q   = Pumping rate thru unit   DUAL SPEED MODEL AVAILABLE
       °STAINLESS STEEL MODELS FROM 3 to 25 HP ARE AVAILABLE ON SPECIAL ORDER.

Total Kjeldahl nitrogen is composed of the ammonia nitrogen and the organic nitrogen. Organic nitrogen is a potential source of ammonia nitrogen because of the deamination reactions during the metabolism of organic matter in wastewater.

Table 2-14
Calculation of the Oxygen Demand and Surface Aerator Size

Design Conditions

Design flow rate = 3785 m^3/day (1 MGD)

Influent BOD$_5$ = 300 mg/l

Pond temperature = 10°C

Barometric pressure = 760 mm of mercury

Relative oxygen solubility (β) = 0.9

C_L = dissolved oxygen concentration to be maintained during treatment = 2.0 mg/l

Relative oxygen transfer (α) = 0.9

Oxygen Requirements

BOD$_5$ in wastewater = 300 mg/l (3785 m^3/day) (1000 l/m^3)
                     = 1136 kg/day

Assume that O$_2$ demand of the solids at peak flows will be 1.5 times the mean value of 1136 kg/day

Oxygen requirements (N$_a$) = 1136 × 1.5 = 1703 kg/day = 71 kg/hr

Aerator Sizing

After determining N$_a$, the equivalent oxygen transfer to tapwater (N) at standard conditions in kg/hr can be calculated using the following equation:

$$N = \frac{N_a}{\alpha \left[\dfrac{C_{SW} - C_L}{C_S} \right] (1.025)^{T-20}}$$

Where α = Ratio of $\dfrac{\text{oxygen transfer to waste}}{\text{oxygen transfer to tap water}}$

C_{SW} = oxygen saturation value of the waste in mg/l calculated from the expression:

$$C_{SW} = \beta (C_{SS}) P$$

Where β = ratio of $\dfrac{\text{oxygen saturation value of the waste}}{\text{oxygen saturation value of tap water}}$

C_{SS} = oxygen saturation value of tapwater at the specified waste temperature

P = ratio of $\dfrac{\text{barometric pressure at the plant site}}{\text{barometric pressure at sea level}}$

C_L = dissolved oxygen concentration to be maintained in the waste (mg/l)

C_S = oxygen saturation value of tapwater at 20°C and 1 atmosphere pressure
     = 9.17 mg/l

T = pond water temperature (°C)

The design conditions are:

N_a = 71 kg O$_2$/hr

α = 0.9

β = 0.9

(continued)

Table 2 - 14 (continued)

P = 1.0 atmosphere
C_L = 2.0 mg/l
T = 15°C
C_{SW} = 0.9 (10.15) 1.0 = 9.14 mg/l

$$N = \cfrac{71}{0.9 \left[\cfrac{9.14 - 2.0}{9.17}\right] 0.884} = \cfrac{71}{0.9 \times 0.78 \times 0.884} = 114 \text{ kg/hr}$$

Assume 1.4 kg O_2/hp-hr for the aerator for estimating purposes (taken from catalogs).
Total hp required is $114/1.4$ = 81.4 break horsepower (bhp).
Aerator drive units should produce at least 81.4 bhp at the shaft.
  Assume 90% efficiency for gear reducer.
  Therefore, total motor horse power = 91 hp.

Ammonia and total Kjeldahl nitrogen reduction in aerated lagoons can occur through several processes:

1. Gaseous ammonia stripping to the atmosphere.
2. Ammonia assimilation in biomass.
3. Biological nitrification.
4. Biological denitrification.
5. Sedimentation of insoluble organic nitrogen.

The rate of gaseous ammonia losses to the atmosphere depends mainly on the pH value, temperature, hydraulic loding rate, and the mixing conditions in the pond. An alkaline pH value shifts the equilibrium equation $NH_3 + H_2O \leftrightarrow NH_4^+ + OH^-$ toward gaseous ammonia production, while the mixing conditions affect the magnitude of the mass transfer coefficient. Temperature affects both the equilibrium constant and mass transfer coefficient.

Ammonia nitrogen assimilation into biomass depends on the biological activity in the system and is affected by several factors such as temperature, organic load, detention time, and wastewater characteristics.

Biological nitrification depends on adequate environmental conditions for nitrifiers to grow and is affected by several factors such as temperature, dissolved oxygen concentration, pH value, detention time, and wastewater characteristics.

Within bottom sediments under anoxic conditions, denitrification can take place, and the rate of denitrification is affected by temperature, redox potential, and sediment characteristics. In well-designed aerated lagoons with good mixing conditions and distribution of dissolved oxygen, denitrification will be negligible.

Model Selection

In the past, design criteria for ammonia nitrogen and total Kjeldahl nitrogen (TKN) removals in aerated lagoons did not exist. The purpose of this section is to develop empirical equations that can be used to estimate ammonia nitrogen and TKN removals in aerated lagoon systems.

Two of the proposed models are based on plug flow hydraulics through multiple ponds in series and first-order kinetics. Other models evaluated consist of relationships between the fraction of ammonia nitrogen and TKN removals and the hydraulic detention time and areal removal rates versus areal loading rates. All models were developed to be used to predict the ammonia nitrogen and TKN concentrations in the final pond effluent.

It was assumed that the overall apparent reaction rate for all of the processes accounting for ammonia nitrogen and TKN removals can be described by a first-order reaction. To incorporate the effect of the surface area on ammonia stripping, as well as the effect of the detention time on the other processes, the plug flow model (Equation 2–6) was modified as shown in Equation 2–36.

$$C_e = C_o e^{-k_1 t} \tag{2–6}$$

$$C_e = C_o e^{-k(A/V)t} \tag{2–36}$$

C_o, C_e = influent and effluent nitrogen concentrations, M/L^3
A = total surface area, L^2
V = total volume, L^3
k = reaction rate, L/time
k_1 = reaction rate, time^{-1}
t = detention time, time
M = mass
L = length

The reaction rates k and k_1 incorporate several reaction rates and are assumed to be dependent on temperature and other factors such as mixing conditions, and pH value. In the following sections, the reaction rates k and k_1 are determined for five full-scale aerated lagoons, and the effects of temperature on the relationships shown in Equation 2–6 and 2–36 are evaluated.

In addition to the above model, the relationship between the fraction of ammonia nitrogen and TKN removed and detention time is examined. The influence of temperature and pH value is also incorporated into this relationship.

Areal ammonia nitrogen and TKN removals were compared with the areal ammonia nitrogen, TKN, and BOD$_5$ loading rates applied to the aerated lagoon systems. Other simple and complex relationships were evaluated [i.e., complete mix model (Equation 2–7), Wehner–Wilhelm equation (Equation 2–15), Monod growth kinetics, models containing multiple limiting relations, polynomial equations, and multiple regression analyses], but were discarded because of a poor fit of the field data to these models.

The Aerated Lagoon Systems

Five aerated lagoons have been monitored for ammonia nitrogen and TKN removal, temperature, hydraulic loading rates and detention time.[19-23] The aerated pond systems were described earlier and are located in Bixby, Oklahoma, Lake Koshkonong, Wisconsin, North Gulfport, Mississippi, Pawnee, Illinois, and Windber, Pennsylvania. All the systems experience significant weather changes, the exception being the North Gulfport system which is exposed to a semi-tropical climate. The mean air temperatures at the four systems with significant weather changes range from − 10 to 3°C (14–37°F) in January and from 22 to 27°C (71–81°F) in July. At North Gulfport the air temperature ranges from a mean of 11°C (52°F) in January to 28°C (82°F) in August.

The system in North Gulfport contains surface aerators, and the average depth of the aerated cells is 1.92 m (6.3 ft). Diffused-air aeration systems are used in the other ponds, and the average depth is 3.05 m (10 ft). The Bixby and North Gulfport systems consist of two aerated cells in series and the others, three cells in series. The aerated cells of the North Gulfport system are followed by settling ponds and a chlorine contact pond. The operating conditions for these systems and the influent wastewater characteristics entering these systems were significantly different, as shown in Table 2–15.

The Koshkonong wastewater is a typical weak domestic wastewater; the Bixby, North Gulfport, and Windber wastewaters are medium strength; and the Pawnee wastewater is a typical strong domestic wastewater. Windber wastewater has a low pH value when compared with typical domestic wastewater, and it has a low concentration of organic nitrogen. When compared with the other diffused air systems, the Windber system removed less ammonia nitrogen and TKN, and this is probably attributable to the low pH values of the wastewater and the high hydraulic loading rates applied to the system.

Ammonia Nitrogen and TKN Removal

The annual percentage ammonia nitrogen and TKN removals in the Bixby and Pawnee aerated lagoons ranged from 82 to 94%, and in the Koshkonong and North Gulfport systems ranged from 49 to 68%. There was no ammonia nitrogen or TKN removal in the Windber system. A sum-

Table 2-15
Wastewater Characteristics and Operating
Conditions for the Five Aerated Lagoons

PARAMETER	SYSTEM				
	Pawnee	Bixby	Koshkonong	Windber	North Gulfport
Wastewater characteristics					
BOD_5 mg/l	473	368	85	173	178
COD mg/l	1026	635	196	424	338
TKN mg/l	51.41	45.04	15.30	24.33	26.5
NH_4-N mg/l	26.32	29.58	10.04	22.85	15.7
Alkalinity mg/l	242	154	397	67	144
pH	6.8-7.4	6.1-7.1	7.2-7.4	5.6-6.9	6.7-7.5
Operating conditions					
Hydraulic loading rate, m/day	0.018	0.0221	0.0335	0.0563	0.109
Organic loading rate, kg BOD_5/ha·day	151	161	87	285	486
Detention time, days	144	108	73	48	18

mary of the mean annual ammonia nitrogen and TKN removals for all five systems are presented in Table 2–16. In the Koshkonong and North Gulfport systems, significant ammonia nitrogen conversion occurred through nitrification. Nitrate and nitrite nitrogen measurements were not made at the Bixby system; however, the high mixed liquor dissolved oxygen concentrations throughout the entire year of the study indicate that nitrification probably occurred.

Figures 2–39 and 2–40 show the annual percentage removals of am-

Table 2-16
Mean Annual Ammonia Nitrogen
and TKN Removals in Five Aerated Lagoons

PARAMETER	SYSTEM				
	Pawnee	Bixby	Koshkonong	Windber	North Gulfport
Temperature range (°C)	2-26	6-29	1-24	2-26	11-29
Mean range (°C)	12	18	11	16	22
pH range	7-9.3	6.3-9.2	7.4-7.9	6.6-8.5	6.8-7.5
Dissolved oxygen range (mg/l)	1.4-16.0	3.5-13.5	6.7-15.3	2.4-15	0.8-9.3
TKN removal (%)	87	82	50	0	59
NH_4-N removal (%)	94	87	49	0	68
Nitrification (%)[a]	4	—	62	0	30

[a]The ratio of nitrite nitrogen + nitrate nitrogen produced to the amount of ammonia nitrogen removed.

FIGURE 2-39

NH$_3$-N and TKN removals versus the hydraulic loading rate [aerated wastewater lagoons (diffused air) (mean annual data)].

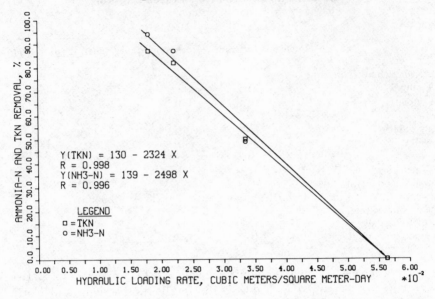

FIGURE 2-40

NH$_3$-N and TKN removals versus the hydraulic detention time [aerated wastewater lagoons (diffused air) (mean annual data)].

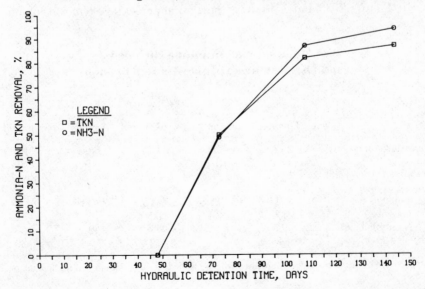

Table 2-17
Ammonia Nitrogen and TKN Removal in Aerated
Lagoons during the Months of December-March

PARAMETER	LOCATION				
	Pawnee	Bixby	Koshkonong	Windber	North Gulfport
Temperature range (°C)	2-8	5-17	1-4	2-14	11-19
Mean temperature (°C)	4.6	9.7	2.1	8.5	14.9
pH range	7.4-8.7	6.3-9.2	7.4-7.6	6.8-8.5	6.8-7.3
DO range (mg/l)	12.2-16.0	6.7-13.5	12.7-15.1	6.5-15.0	3.0-9.3
Hydraulic loading (m/day)	0.0166	0.0205	0.0345	0.0582	0.1063
Detention time (days)	154	116	73	47	18
TKN removal (%)	80	76	31	0	61
NH_4-N removal (%)	83	74	24	8	73

monia nitrogen and TKN for the four diffused-air aeration systems as a function of hydraulic loading rate and detention time, respectively. The percentage removals increased when the hydraulic loading rates decreased or the detention times increased. The ammonia nitrogen and TKN percentage removals at the North Gulfport site do not conform to the relationships shown in Figures 2–39 and 2–40. This lack of agreement is probably because of the different type of aeration system and the higher liquid temperatures in Mississippi.

The seasonal effects on the performance of the aerated lagoons are shown in Tables 2–17 and 2–18 and in Figures 2–41 through 2–44. With the exception of the Windber system, in which little ammonia nitrogen and

Table 2-18
Ammonia Nitrogen and TKN Removal in Aerated
Lagoons during the months of June-September

PARAMETER	LOCATION				
	Pawnee	Bixby	Koshkonong	Windber	North Gulfport
Temperature range (°C)	18-26	24-29	16-25	20-26	26-29
Mean temperature (°C)	20.8	26.9	21.5	22.9	28.1
pH range	7.0-8.2	6.5-8.1	7.5-7.8	6.9-7.8	7.1-7.4
DO range (mg/l)	1.8-4.7	3.5-9.5	6.7-9.3	2.4-8.4	0.9-3.1
Hydraulic loading (m/day)	0.0176	0.0240	0.0366	0.0570	0.1222
Detention time (day)	145	99	69	48	16
TKN removal (%)	89	89	60	0	59
NH_4-N removal (%)	98	98	63	0	69

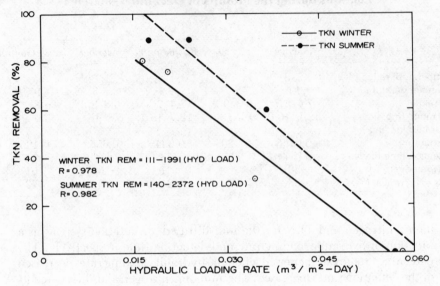

FIGURE 2-41
Relationship between TKN removal and the hydraulic loading rate for the summer and winter performance.

WINTER TKN REM = 111−1991 (HYD LOAD)
R = 0.978
SUMMER TKN REM = 140− 2372 (HYD LOAD)
R= 0.982

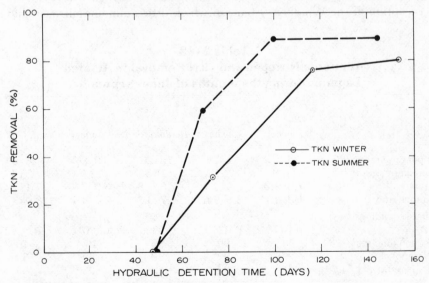

FIGURE 2-42
Relationship between TKN removal and the hydraulic detention time for the summer and winter performance.

76

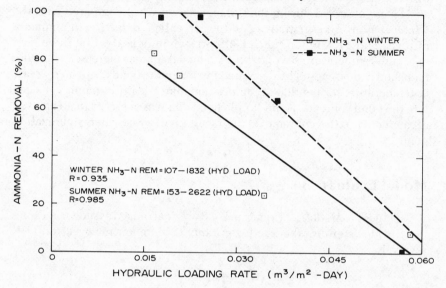

FIGURE 2-43
Relationship between ammonia nitrogen removal and the hydraulic loading rate for the summer and winter performance.

WINTER NH$_3$-N REM = 107 - 1832 (HYD LOAD)
R = 0.935
SUMMER NH$_3$-N REM = 153 - 2622 (HYD LOAD)
R = 0.985

NH$_3$-N WINTER
NH$_3$-N SUMMER

AMMONIA - N REMOVAL (%)

HYDRAULIC LOADING RATE (m^3/m^2 -DAY)

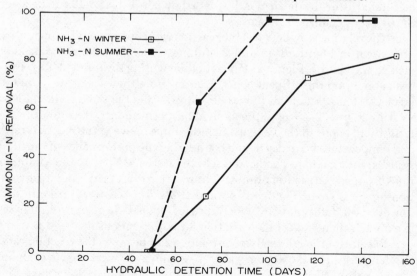

FIGURE 2-44
Relationship between ammonia nitrogen removal and the hydraulic detention time for the summer and winter performance.

NH$_3$-N WINTER
NH$_3$-N SUMMER

AMMONIA - N REMOVAL (%)

HYDRAULIC DETENTION TIME (DAYS)

77

TKN removal occurred throughout the year, and the North Gulfport system, in which the percentage removals did not change significantly throughout the year, greater percentage removals of ammonia nitrogen and TKN were observed during the summer season when compared with the winter season. An essentially constant percentage reduction in ammonia nitrogen and TKN in the North Gulfport system indicates that there is an optimum temperature beyond which removals do not increase. This phenomenon can be caused by decreased mixing in the ponds with an increase in temperature, or benthal decomposition may release ammonia nitrogen into the pond contents. A similar phenomenon was observed in an analysis of ammonia nitrogen removal in facultative wastewater stabilization ponds.[14]

Model Evaluations

Plug Flow Models. Equation 2–36 was rearranged as shown in Equation 2–37 to determine the kinetic constants for ammonia nitrogen and TKN removal:

$$k = \frac{-\left(\ln \frac{C_e}{C_o}\right)}{\frac{A}{V} t} \qquad (2\text{--}37)$$

The reaction rate constants for ammonia nitrogen and TKN were calculated using the mean monthly data shown in Tables 2–19 and 2–20 for the five aerated lagoon systems.

The relationships between the reaction rate constants and temperature are shown in Figures 2–45 and 2–46 for two combinations of data that yielded the best fit. Figure 2–45 was obtained using the influent TKN concentrations and the effluent concentrations from the various stages of the lagoon systems. Figure 2–46 was developed using the same combinations used in Figure 2–45 plus the individual performance data for the intermediate ponds in the systems. Relationships between the reaction rates and temperature were developed considering the performance of the total systems, intermediate ponds, and selected systems. None of the combinations produced satisfactory results. The reaction rate and temperature relationships shown in Figures 2–45 and 2–46 are statistically significant; however, the data points are widely distributed, and the regression analyses explain less than 50% of the variation for both relationships.

Because of the relatively poor relationship between the reaction rates and temperatures, the data in Tables 2–19 and 2–20 were used to evaluate the plug flow (Equation 2–6) and modified plug flow (Equation 2–36)

TABLE 2-19
Total Kjeldahl Nitrogen Data for Five Aerated Lagoons

TOTAL KJELDAHL NITROGEN, MG/L				TEMPERATURE, DEG C			FLOW	DETENTION TIME, DAY			
INF	POND #1	POND #2	POND #3	POND #1	POND #2	POND #3	RATE MGD	POND #1	POND #2	POND #3	LOCATION
42.90	-9.90	7.73	-9.90	6.0	5.0	-9.9	0.1111	66.6	66.6	-9.9	BIXBY, OKLAHOMA
44.36	-9.90	19.20	-9.90	-9.9	-9.9	-9.9	0.1084	68.3	68.3	-9.9	BIXBY, OKLAHOMA
64.80	-9.90	22.20	-9.90	17.0	16.0	-9.9	0.1228	60.3	60.3	-9.9	BIXBY, OKLAHOMA
40.63	-9.90	9.48	-9.90	19.0	19.0	-9.9	0.1613	45.9	45.9	-9.9	BIXBY, OKLAHOMA
40.20	-9.90	5.61	-9.90	21.0	20.0	-9.9	0.1410	52.5	52.5	-9.9	BIXBY, OKLAHOMA
40.80	-9.90	6.48	-9.90	26.0	26.0	-9.9	0.1106	66.9	66.9	-9.9	BIXBY, OKLAHOMA
47.48	-9.90	4.90	-9.90	29.0	29.0	-9.9	0.1235	59.9	59.9	-9.9	BIXBY, OKLAHOMA
36.33	-9.90	3.04	-9.90	28.0	28.0	-9.9	0.2221	33.3	33.3	-9.9	BIXBY, OKLAHOMA
44.20	-9.90	3.88	-9.90	25.0	24.0	-9.9	0.1399	52.9	52.9	-9.9	BIXBY, OKLAHOMA
47.12	-9.90	4.42	-9.90	14.0	13.0	-9.9	0.1242	59.6	59.6	-9.9	BIXBY, OKLAHOMA
47.38	-9.90	6.04	-9.90	10.0	9.0	-9.9	0.1339	55.3	55.3	-9.9	BIXBY, OKLAHOMA
44.31	-9.90	8.31	-9.90	7.0	7.0	-9.9	0.1484	49.9	49.9	-9.9	BIXBY, OKLAHOMA
17.89	17.17	13.97	10.21	4.0	3.0	2.0	0.2314	26.1	26.1	26.1	LAKE KOSHKONONG, WISC.
21.34	17.99	15.38	13.83	1.0	1.0	1.0	0.2093	28.9	28.9	28.9	LAKE KOSHKONONG, WISC.
18.03	18.19	13.41	13.10	1.0	1.0	1.0	0.2350	25.7	25.7	25.7	LAKE KOSHKONONG, WISC.
6.37	7.29	6.76	5.20	4.0	3.0	3.0	0.3214	18.8	18.8	18.8	LAKE KOSHKONONG, WISC.
11.29	10.27	7.06	3.38	11.0	11.0	12.0	0.2980	20.3	20.3	20.3	LAKE KOSHKONONG, WISC.
12.34	11.89	7.71	5.10	12.0	13.0	14.0	0.2700	22.4	22.4	22.4	LAKE KOSHKONONG, WISC.
17.10	18.87	12.97	7.54	21.0	23.0	23.0	0.2671	22.7	22.7	22.7	LAKE KOSHKONONG, WISC.
18.08	18.75	14.06	5.95	23.0	25.0	25.0	0.2647	22.9	22.9	22.9	LAKE KOSHKONONG, WISC.
15.03	21.17	14.61	4.66	22.0	23.0	23.0	0.2657	22.8	22.8	22.8	LAKE KOSHKONONG, WISC.
15.66	19.13	16.13	8.21	16.0	17.0	17.0	0.2586	23.4	23.4	23.4	LAKE KOSHKONONG, WISC.
9.76	16.10	14.03	5.51	12.0	12.0	12.0	0.2497	24.2	24.2	24.2	LAKE KOSHKONONG, WISC.
20.76	18.33	13.16	8.56	7.0	6.0	5.0	0.1236	48.9	48.9	48.9	LAKE KOSHKONONG, WISC.
29.70	19.40	13.10	-9.90	12.0	11.0	-9.9	0.7367	7.1	10.7	-9.9	NORTH GULFPORT, MISS.
29.10	19.10	13.30	-9.90	14.0	12.0	-9.9	0.8610	6.0	9.2	-9.9	NORTH GULFPORT, MISS.
21.40	19.50	7.20	-9.90	19.0	19.0	-9.9	0.6324	8.2	12.5	-9.9	NORTH GULFPORT, MISS.
30.90	20.40	9.50	-9.90	16.0	16.0	-9.9	0.6331	8.2	12.5	-9.9	NORTH GULFPORT, MISS.
30.40	19.20	9.60	-9.90	22.0	22.0	-9.9	0.6806	7.6	11.6	-9.9	NORTH GULFPORT, MISS.
29.40	23.50	12.10	-9.90	23.0	23.0	-9.9	0.7396	7.0	10.7	-9.9	NORTH GULFPORT, MISS.
26.60	9.30	9.50	-9.90	28.0	28.0	-9.9	0.7729	6.7	10.2	-9.9	NORTH GULFPORT, MISS.
20.60	9.00	9.90	-9.90	29.0	29.0	-9.9	0.9049	5.7	8.7	-9.9	NORTH GULFPORT, MISS.
25.60	8.10	9.50	-9.90	29.0	29.0	-9.9	0.8897	5.8	8.9	-9.9	NORTH GULFPORT, MISS.
23.60	12.60	11.00	-9.90	27.0	26.0	-9.9	0.7228	7.2	10.9	-9.9	NORTH GULFPORT, MISS.
24.40	8.40	12.50	-9.90	24.0	24.0	-9.9	0.5589	9.3	14.1	-9.9	NORTH GULFPORT, MISS.
26.30	10.30	12.20	-9.90	18.0	19.0	-9.9	0.6630	7.8	11.9	-9.9	NORTH GULFPORT, MISS.
24.93	-9.90	-9.90	0.00	5.0	8.0	-9.9	0.2644	66.6	31.0	15.9	PAWNEE, ILLINOIS
28.55	-9.90	-9.90	5.00	5.0	10.0	12.0	0.2067	85.1	39.7	20.3	PAWNEE, ILLINOIS
46.86	-9.90	-9.90	4.69	12.0	13.0	12.0	0.2270	77.5	36.1	18.5	PAWNEE, ILLINOIS
49.29	-9.90	-9.90	5.02	19.0	18.0	19.0	0.2137	82.4	38.4	19.7	PAWNEE, ILLINOIS
53.58	-9.90	-9.90	5.13	23.0	23.0	22.0	0.2037	86.4	40.3	20.6	PAWNEE, ILLINOIS
38.43	-9.90	-9.90	7.26	26.0	23.0	19.0	0.2103	83.7	39.0	20.0	PAWNEE, ILLINOIS
63.57	-9.90	-9.90	4.43	19.0	18.0	21.0	0.1990	88.4	41.2	21.1	PAWNEE, ILLINOIS
80.20	-9.90	-9.90	4.55	11.0	9.0	13.0	0.2032	86.6	40.4	20.7	PAWNEE, ILLINOIS
73.67	-9.90	-9.90	3.84	5.0	5.0	5.0	0.1819	96.8	45.1	23.1	PAWNEE, ILLINOIS
53.43	-9.90	-9.90	5.26	3.0	3.0	2.0	0.2021	87.1	40.6	20.8	PAWNEE, ILLINOIS
56.29	-9.90	-9.90	10.36	3.0	-9.9	3.0	0.1529	115.1	53.6	27.5	PAWNEE, ILLINOIS
48.13	-9.90	-9.90	14.71	8.0	8.0	8.0	0.2305	76.4	35.6	18.2	PAWNEE, ILLINOIS
21.91	-9.90	-9.90	14.43	19.0	15.0	14.0	0.7000	31.4	40.0	14.3	WINDBER, PENNSYLVANIA
29.50	-9.90	-9.90	19.50	14.0	8.0	8.0	0.8000	27.5	35.0	12.5	WINDBER, PENNSYLVANIA
16.70	-9.90	-9.90	22.33	12.0	4.0	2.0	1.5000	14.7	18.7	6.7	WINDBER, PENNSYLVANIA
13.21	-9.90	-9.90	18.58	12.0	6.0	5.0	1.5000	14.7	18.7	6.7	WINDBER, PENNSYLVANIA
14.88	-9.90	-9.90	14.76	13.0	10.0	8.0	1.4000	15.7	20.0	7.1	WINDBER, PENNSYLVANIA
18.09	-9.90	-9.90	16.61	14.0	12.0	11.0	1.5000	14.7	18.7	6.7	WINDBER, PENNSYLVANIA
27.73	-9.90	-9.90	24.76	20.0	17.0	17.0	1.1000	20.0	25.5	9.1	WINDBER, PENNSYLVANIA
34.14	-9.90	-9.90	32.77	26.0	24.0	24.0	1.0000	22.0	28.0	10.0	WINDBER, PENNSYLVANIA
25.07	-9.90	-9.90	33.62	25.0	23.0	23.0	1.4000	15.7	20.0	7.1	WINDBER, PENNSYLVANIA
26.41	-9.90	-9.90	28.16	24.0	22.0	22.0	1.5000	14.7	18.7	6.7	WINDBER, PENNSYLVANIA
46.00	-9.90	-9.90	34.11	22.0	20.0	20.0	1.3000	16.9	21.5	7.7	WINDBER, PENNSYLVANIA
18.36	-9.90	-9.90	23.25	17.0	14.0	13.0	1.4000	15.7	20.0	7.1	WINDBER, PENNSYLVANIA

VALUE OF —9.9 EQUALS NO DATA

models, excluding and considering the influence of temperature on the reactions. The results of the evaluation of the plug flow and modified plug flow models are shown in Figures 2–47 to 2–52. The results obtained using the total Kjeldahl nitrogen concentrations in the influent and the effluent

TABLE 2-20
Ammonia Nitrogen Data for Five Aerated Lagoons

INF	AMMONIA NITROGEN, MG/L POND #1	POND #2	POND #3	TEMPERATURE, DEG C POND #1	POND #2	POND #3	FLOW RATE MGD	DETENT TIME, DAYS POND #1	POND #2	POND #3	LOCATION
31.04	-9.90	4.59	-9.90	6.0	5.0	-9.9	0.1111	66.6	66.6	-9.9	BIXBY,OKLAHOMA
27.94	-9.90	13.87	-9.90	-9.9	-9.9	-9.9	0.1084	68.3	68.3	-9.9	BIXBY,OKLAHOMA
25.07	-9.90	14.76	-9.90	17.0	16.0	-9.9	0.1228	60.3	60.3	-9.9	BIXBY,OKLAHOMA
23.71	-9.90	3.01	-9.90	19.0	19.0	-9.9	0.1613	45.9	45.9	-9.9	BIXBY,OKLAHOMA
24.41	-9.90	1.23	-9.90	21.0	20.0	-9.9	0.1410	52.5	52.5	-9.9	BIXBY,OKLAHOMA
26.50	-9.90	1.27	-9.90	26.0	26.0	-9.9	0.1106	66.9	66.9	-9.9	BIXBY,OKLAHOMA
36.29	-9.90	0.83	-9.90	29.0	29.0	-9.9	0.1235	59.9	59.9	-9.9	BIXBY,OKLAHOMA
25.07	-9.90	0.51	-9.90	28.0	28.0	-9.9	0.2221	33.3	33.3	-9.9	BIXBY,OKLAHOMA
33.45	-9.90	0.17	-9.90	25.0	24.0	-9.9	0.1399	52.9	52.9	-9.9	BIXBY,OKLAHOMA
40.35	-9.90	0.22	-9.90	14.0	13.0	-9.9	0.1242	59.6	59.6	-9.9	BIXBY,OKLAHOMA
32.58	-9.90	0.11	-9.90	10.0	9.0	-9.9	0.1339	55.3	55.3	-9.9	BIXBY,OKLAHOMA
28.51	-9.90	1.00	-9.90	7.0	7.0	-9.9	0.1484	49.9	49.9	-9.9	BIXBY,OKLAHOMA
13.66	16.04	12.80	9.19	4.0	3.0	2.0	0.2314	26.1	26.1	26.1	LAKE KOSHKONONG,WISC.
16.12	16.56	13.77	12.51	1.0	1.0	1.0	0.2093	28.9	28.9	28.9	LAKE KOSHKONONG,WISC.
12.86	16.43	10.59	10.88	1.0	1.0	1.0	0.2350	25.7	25.7	25.7	LAKE KOSHKONONG,WISC.
4.71	6.40	5.36	3.60	4.0	3.0	3.0	0.3214	18.8	18.8	18.8	LAKE KOSHKONONG,WISC.
7.74	8.27	5.09	1.56	11.0	11.0	12.0	0.2980	20.3	20.3	20.3	LAKE KOSHKONONG,WISC.
7.63	7.84	2.89	0.66	12.0	13.0	14.0	0.2700	22.4	22.4	22.4	LAKE KOSHKONONG,WISC.
12.03	15.74	10.46	5.06	21.0	23.0	23.0	0.2671	22.7	22.7	22.7	LAKE KOSHKONONG,WISC.
11.79	14.43	11.00	2.51	23.0	25.0	25.0	0.2647	22.9	22.9	22.9	LAKE KOSHKONONG,WISC.
7.03	17.10	10.83	1.37	22.0	23.0	23.0	0.2657	22.8	22.8	22.8	LAKE KOSHKONONG,WISC.
7.34	15.37	12.83	4.94	16.0	17.0	17.0	0.2586	23.4	23.4	23.4	LAKE KOSHKONONG,WISC.
4.40	13.12	11.12	3.38	12.0	12.0	12.0	0.2497	24.2	24.2	24.2	LAKE KOSHKONONG,WISC.
15.11	16.39	11.66	7.46	7.0	6.0	5.0	0.1236	48.9	48.9	48.9	LAKE KOSHKONONG,WISC.
19.00	11.60	7.60	-9.90	12.0	11.0	-9.9	0.7367	7.1	10.7	-9.9	NORTH GULFPORT,MISS.
16.50	11.50	7.00	-9.90	14.0	12.0	-9.9	0.8610	6.0	9.2	-9.9	NORTH GULFPORT,MISS.
13.80	12.20	0.90	-9.90	19.0	19.0	-9.9	0.6324	8.2	12.5	-9.9	NORTH GULFPORT,MISS.
15.20	12.70	3.40	-9.90	16.0	16.0	-9.9	0.6331	8.2	12.5	-9.9	NORTH GULFPORT,MISS.
15.10	11.30	2.80	-9.90	22.0	22.0	-9.9	0.6806	7.6	11.6	-9.9	NORTH GULFPORT,MISS.
19.30	15.00	6.30	-9.90	23.0	23.0	-9.9	0.7396	7.0	10.7	-9.9	NORTH GULFPORT,MISS.
13.80	1.80	3.10	-9.90	28.0	28.0	-9.9	0.7729	6.7	10.2	-9.9	NORTH GULFPORT,MISS.
11.60	3.20	4.50	-9.90	29.0	29.0	-9.9	0.9049	5.7	8.7	-9.9	NORTH GULFPORT,MISS.
16.80	1.40	3.40	-9.90	29.0	29.0	-9.9	0.8897	5.8	8.9	-9.9	NORTH GULFPORT,MISS.
14.20	6.20	6.40	-9.90	27.0	26.0	-9.9	0.7228	7.2	10.9	-9.9	NORTH GULFPORT,MISS.
13.50	0.70	6.50	-9.90	24.0	24.0	-9.9	0.5589	9.3	14.1	-9.9	NORTH GULFPORT,MISS.
20.00	7.50	9.70	-9.90	18.0	19.0	-9.9	0.6630	7.8	11.9	-9.9	NORTH GULFPORT,MISS.
12.00	-9.90	-9.90	0.00	5.0	8.0	9.0	0.2644	66.6	31.0	15.9	PAWNEE,ILLINOIS
13.32	-9.90	-9.90	0.86	5.0	10.0	12.0	0.2067	85.1	39.7	20.3	PAWNEE,ILLINOIS
20.29	-9.90	-9.90	0.39	12.0	13.0	12.0	0.2270	77.5	36.1	18.5	PAWNEE,ILLINOIS
31.43	-9.90	-9.90	0.60	19.0	18.0	19.0	0.2137	82.4	38.4	19.7	PAWNEE,ILLINOIS
23.62	-9.90	-9.90	0.71	23.0	23.0	22.0	0.2037	86.4	40.3	20.6	PAWNEE,ILLINOIS
26.43	-9.90	-9.90	0.57	26.0	23.0	19.0	0.2103	83.7	39.0	20.0	PAWNEE,ILLINOIS
36.14	-9.90	-9.90	0.48	19.0	18.0	21.0	0.1990	88.4	41.2	21.1	PAWNEE,ILLINOIS
34.56	-9.90	-9.90	0.39	11.0	9.0	13.0	0.2032	86.6	40.4	20.7	PAWNEE,ILLINOIS
37.00	-9.90	-9.90	0.38	5.0	5.0	5.0	0.1819	96.8	45.1	23.1	PAWNEE,ILLINOIS
34.00	-9.90	-9.90	0.49	3.0	3.0	2.0	0.2021	87.1	40.6	20.8	PAWNEE,ILLINOIS
29.29	-9.90	-9.90	4.41	3.0	-9.9	3.0	0.1529	115.1	53.6	27.5	PAWNEE,ILLINOIS
17.72	-9.90	-9.90	6.07	8.0	8.0	8.0	0.2305	76.4	35.6	18.2	PAWNEE,ILLINOIS
20.41	-9.90	-9.90	12.04	19.0	15.0	14.0	0.7000	31.4	40.0	14.3	WINDBER,PENNSYLVANIA
37.24	-9.90	-9.90	16.61	14.0	8.0	8.0	0.8000	27.5	35.0	12.5	WINDBER,PENNSYLVANIA
15.17	-9.90	-9.90	22.29	12.0	4.0	2.0	1.5000	14.7	18.7	6.7	WINDBER,PENNSYLVANIA
12.32	-9.90	-9.90	18.60	12.0	6.0	5.0	1.5000	14.7	18.7	6.7	WINDBER,PENNSYLVANIA
13.66	-9.90	-9.90	14.28	13.0	10.0	8.0	1.4000	15.7	20.0	7.1	WINDBER,PENNSYLVANIA
19.80	-9.90	-9.90	20.11	14.0	12.0	11.0	1.5000	14.7	18.7	6.7	WINDBER,PENNSYLVANIA
31.51	-9.90	-9.90	28.00	20.0	17.0	17.0	1.1000	20.0	25.5	9.1	WINDBER,PENNSYLVANIA
30.00	-9.90	-9.90	30.97	26.0	24.0	24.0	1.0000	22.0	28.0	10.0	WINDBER,PENNSYLVANIA
22.40	-9.90	-9.90	32.75	25.0	23.0	23.0	1.4000	15.7	20.0	7.1	WINDBER,PENNSYLVANIA
23.78	-9.90	-9.90	26.72	24.0	22.0	22.0	1.5000	14.7	18.7	6.7	WINDBER,PENNSYLVANIA
29.06	-9.90	-9.90	26.74	22.0	20.0	20.0	1.3000	16.9	21.5	7.7	WINDBER,PENNSYLVANIA
18.88	-9.90	-9.90	25.98	17.0	14.0	13.0	1.4000	15.7	20.0	7.1	WINDBER,PENNSYLVANIA

VALUE OF —9.9 EQUALS NO DATA

from each of the individual ponds for the modified plug flow model is shown in Figure 2–47. The results for the same set of data incorporating the best-fit temperature correction factor is shown in Figure 2–48. The relationship obtained with the plug flow model without the area/volume correction factor

FIGURE 2-45
Reaction rate versus temperature relationship [aerated wastewater lagoons (diffused air), total Kjeldahl nitrogen (ponds 1, 2, and 3)].

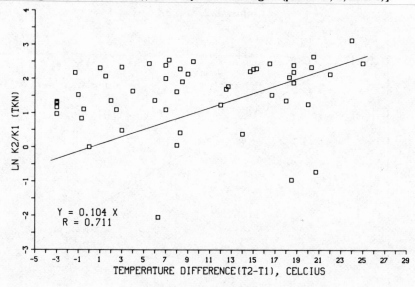

FIGURE 2-46
Reaction rate versus temperature relationship [aerated wastewater lagoons (surface + diffused air), ammonia nitrogen (all data)].

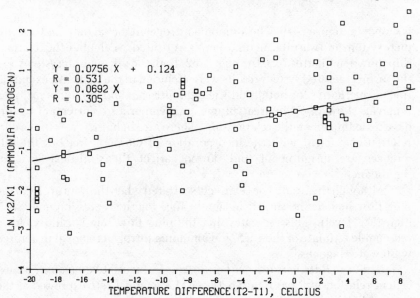

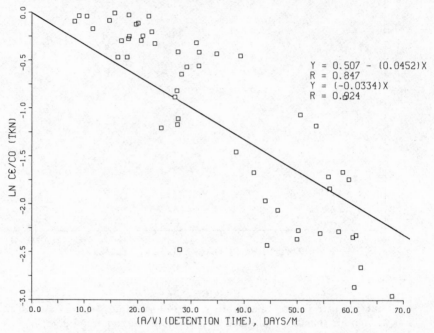

FIGURE 2-47

Effluent/influent versus (area/volume) (detention time) [aerated wastewater lagoons (diffused air), total Kjeldahl nitrogen (ponds 1, 2, and 3)].

is shown in Figure 2–49. The ammonia nitrogen removal data for the complete system and the intermediate ponds were used to calculate the relationship shown in Figures 2–50 through 2–52. Both the diffused-air systems and the surface-aerated system results were included in the plots. An examination of the figures for both total Kjeldahl nitrogen and ammonia nitrogen removals show that the area/volume and temperature correction factors have little impact on the relationship. There are slight increases in the correlation coefficient with the incorporation of temperature, but the differences are insignificant and do not affect the statistical level of significance.

Although the results are relatively scattered when the data are fit to the plug flow model, the statistical significance for the regression analysis is high (5% level). This indicates that the plug flow relationship yields a reasonable estimate of the TKN and ammonia nitrogen removal in aerated wastewater lagoons.

Removal Rate versus Loading Rate. In facultative lagoons, an excellent relationship was found between the removal rate expressed as the mass of BOD_5 removed per unit of surface area and the loading rates ex-

FIGURE 2-48
Effluent/influent versus (area/volume) (detention time) (1.02) [T-20]
[aerated wastewater lagoons (diffused air), total Kjeldahl nitrogen
(ponds 1, 2, and 3)].

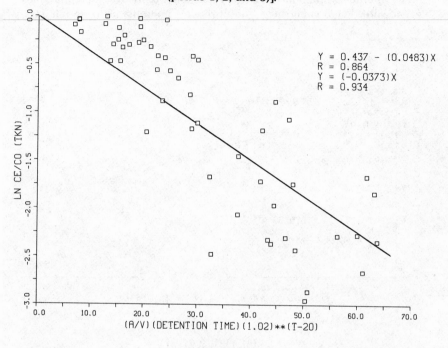

Y = 0.437 - (0.0483)X
R = 0.864
Y = (-0.0373)X
R = 0.934

FIGURE 2-49
The ln effluent/influent versus detention time [aerated wastewater
lagoons (diffused air), total Kjeldahl nitrogen (ponds 1, 2, and 3)].

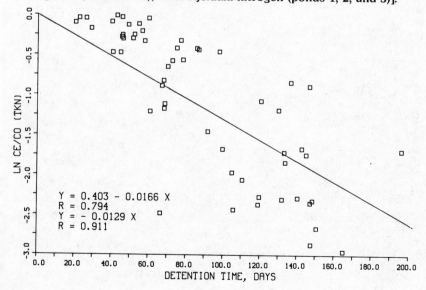

Y = 0.403 - 0.0166 X
R = 0.794
Y = - 0.0129 X
R = 0.911

83

FIGURE 2-50
The ln effluent/influent versus (area/volume) (detention time) [aerated wastewater lagoons (surface + diffused air), ammonia nitrogen (all data)].

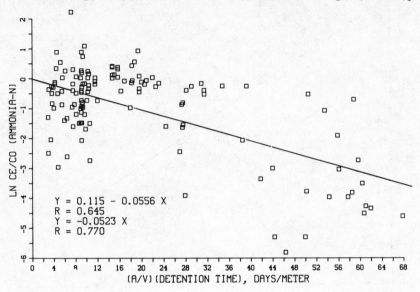

FIGURE 2-51
The ln effluent/influent versus (area/volume) (detention time) (1.03) [superscript T-20] [aerated wastewater lagoons (surface + diffused air), ammonia nitrogen (all data)].

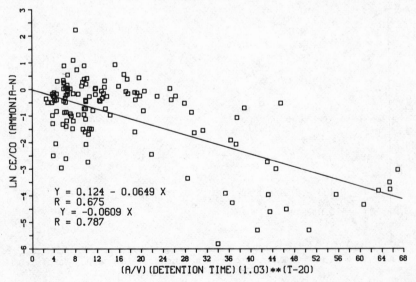

FIGURE 2-52

The ln effluent/influent versus detention time [aerated wastewater lagoons (surface + diffused air), ammonia nitrogen (all data)].

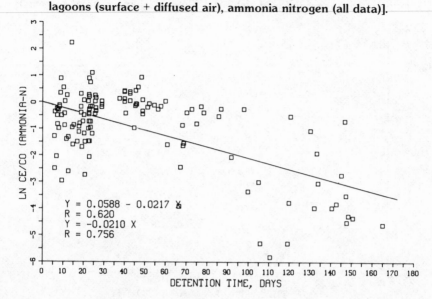

pressed in terms of mass of BOD_5 applied per unit of surface area. Similar relationship were developed using the mass of ammonia nitrogen and TKN removed and applied to the aerated lagoons. The relationships for the aerated lagoons were based on the volume of the lagoon cells rather than the surface area of the cells. The removal rates and loading rates were calculated for various stages of the system, and the removals were achieved by the total system, considering only the influent and effluent concentrations. The differences in the types of aeration equipment were also considered. All analyses of the TKN and ammonia nitrogen removal data yielded statistically significant results, but the best results were obtained when the influent and the final effluent concentrations were considered. Separation of the data by types of aeration systems yielded statistically significant relationships, but the best statistical fit of the data was obtained when the surface and diffused-air systems were combined. The relationships between the removal rates and loading rates for TKN and ammonia nitrogen are shown in Figure 2–53 to 2–60.

The combined data for the surface and diffused air aeration systems considering only the influent and final effluent from the systems for both TKN and ammonia nitrogen (Figures 2–53 and 2–57) resulted in a highly significant (1% level) relationship. The regression analyses accounted for greater than 88% of the variation for the TKN and ammonia nitrogen removals. In an attempt to verify the results presented in Figures 2–53 and 2–57, the removal rates in the intermediate ponds were calculated for the

FIGURE 2-53

Relationship of organic loading and removal rates [aerated wastewater lagoons (surface + diffused air), total Kjeldahl nitrogen (total systems)].

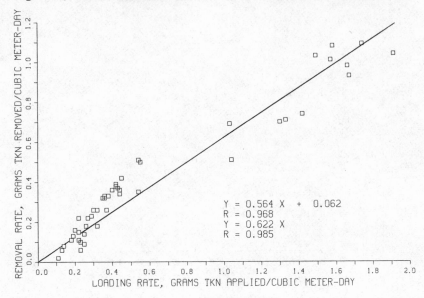

FIGURE 2-54

Relationship of organic loading and removal rates [aerated wastewater lagoons (surface + diffused air), total Kjeldahl nitrogen (intermediate ponds)].

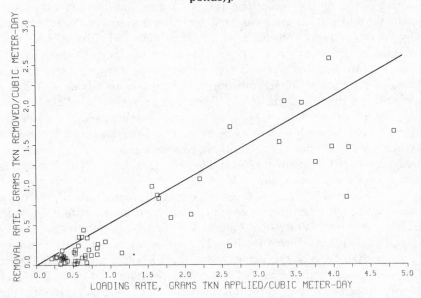

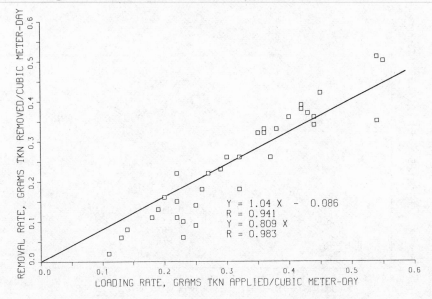

FIGURE 2-55
Relationship of organic loading and removal rates [aerated wastewater lagoons (diffused air), total Kjeldahl nitrogen (total systems)].

$$Y = 1.04 \ X \ - \ 0.086$$
$$R = 0.941$$
$$Y = 0.809 \ X$$
$$R = 0.983$$

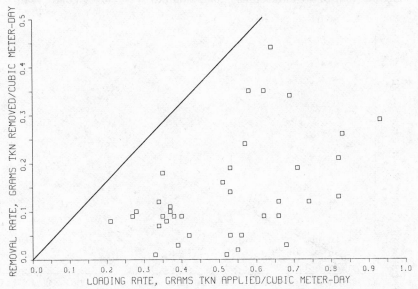

FIGURE 2-56
Relationship of organic loading and removal rates [aerated wastewater lagoons (diffused air), total Kjeldahl nitrogen (intermediate ponds)].

FIGURE 2-57
Relationship of organic loading and removal rates [aerated wastewater lagoons (surface + diffused air), ammonia nitrogen (total systems)].

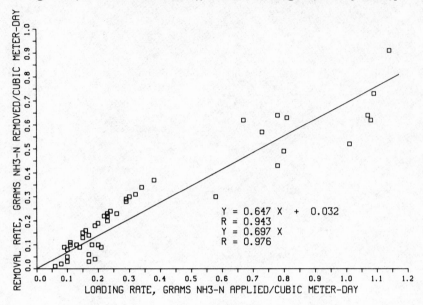

$$Y = 0.647\ X\ +\ 0.032$$
$$R = 0.943$$
$$Y = 0.697\ X$$
$$R = 0.976$$

FIGURE 2-58
Relationship of organic loading and removal rates [aerated wastewater lagoons (surface + diffused air), ammonia nitrogen (intermediate ponds)].

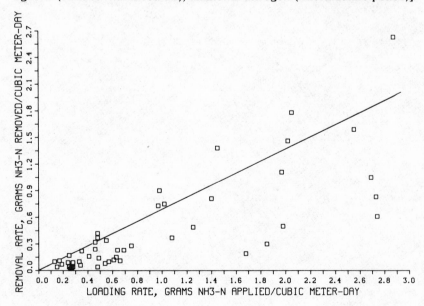

FIGURE 2-59
Relationship of organic loading and removal rates [aerated
wastewater lagoons (diffused air), ammonia nitrogen (total systems)].

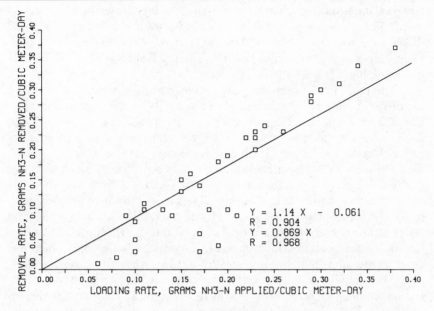

FIGURE 2-60
Relationship of organic loading and removal rates [aerated wastewater
lagoons (diffused air), ammonia nitrogen (intermediate ponds)].

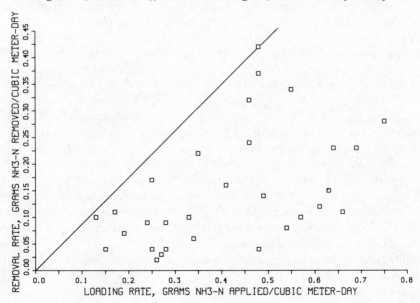

surface aerated and diffused-air aerated ponds, and plotted in Figures 2–54 and 2–58. Although the data are rather scattered at the higher loading rates, the performance in the intermediate ponds tends to follow the same trend as that exhibited in the total system. The lines drawn on Figures 2–54 and 2–58 are the line of best fit for the data presented in Figures 2–53 and 2–57. A least-squares fit of the data calculated for the intermediate ponds yields slopes that are not significantly different from the slopes of the lines drawn on Figures 2–54 and 2–58.

The relationship between the organic loading and removal rates for the diffused-air aerated systems considering only the influent and final effluent from the aerated lagoon systems is presented in Figures 2–55 and 2–59. Again, a highly significant (1 % level) relationship was obtained between the removal rate and the loading rate applied to the aerated lagoons. The performance of the intermediate ponds of the diffused-air systems was evaluated, and the results are plotted in Figures 2–56 and 2–60. The removal rates obtained in the intermediate ponds of the diffused air systems differed significantly when compared with the results obtained in the total systems. The lines shown in Figures 2–56 and 2–60 are the lines of best fit for the diffused-air relationships shown in Figures 2–55 and 2–59. Removal rates in the intermediate ponds of the diffused-air systems were significantly lower than those observed for the total systems. This is contrary to the results obtained when considering the diffused-air and surface-air systems combined. It is quite possible that this variation is simply attributable to the wide fluctuations in performance observed at the lower loading rates. Unfortunately, other detailed data were unavailable for comparison.

Removal Rate versus BOD Loading Rates.   Nitrification in activated sludge systems is related to the BOD_5 loading rates applied to the systems. To determine if a relationship existed between the TKN and ammonia nitrogen removal rates and the BOD_5 loading rates applied to the aerated wastewater lagoons, the data were plotted as shown in Figures 2–61 to 2–68. Various combinations of data were tried, and the best fit of the data was obtained when the surface aeration and diffused-air aeration systems were evaluated together and the performance of the total system was considered. The relationships for TKN and ammonia nitrogen removals versus the BOD_5 loading rates are shown in Figures 2–61 and 2–65. The regression analyses account for greater than 83 % of the variation in the relationship. To further evaluate the relationship between removal rates and BOD_5 loading rates, the performance of the intermediate ponds of the systems was also evaluated. The results of these analyses are shown in Figures 2–62 and 2–66. The lines drawn on the two figures represent the equation of the line for the least-squares fit of the results shown in Figures 2–61 and 2–65. An examination of the two figures shows that the intermediate ponds perform essentially the same as the total systems; however, at the higher loading rates there is a significant decline in removal rates for both TKN and ammonia nitrogen.

FIGURE 2-61
TKN removal versus BOD_5 loading rates [aerated wastewater lagoons (surface + diffused air), (total systems)].

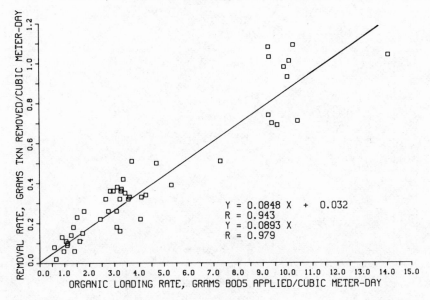

FIGURE 2-62
TKN removal versus BOD_5 loading rates [aerated wastewater lagoons (surface + diffused air), (intermediate ponds)].

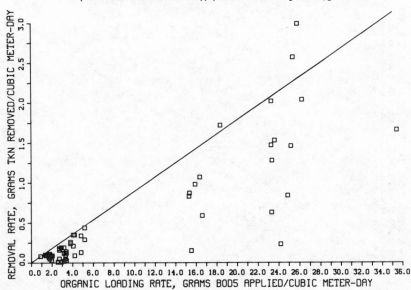

FIGURE 2-63
TKN removal versus BOD$_5$ loading rates [aerated wastewater lagoons (diffused air), (total systems)].

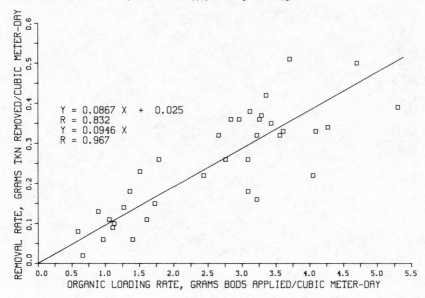

FIGURE 2-64
TKN removal versus BOD$_5$ loading rates [aerated wastewater lagoons (diffused air), (intermediate ponds)].

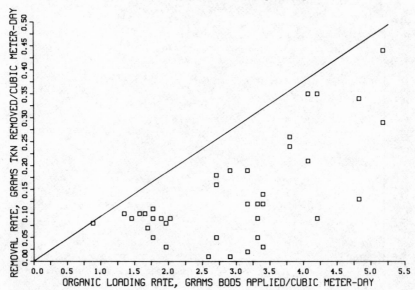

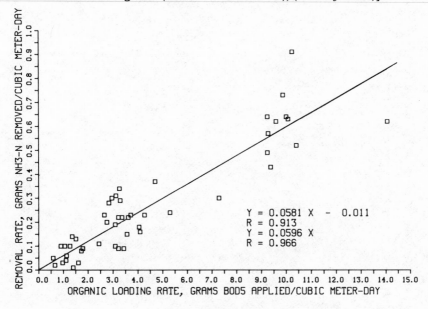

FIGURE 2-65
Ammonia nitrogen removal versus BOD_5 loading rates [aerated wastewater lagoons (surface + diffused air), (total systems)].

$$Y = 0.0581 \ X \ - \ 0.011$$
$$R = 0.913$$
$$Y = 0.0596 \ X$$
$$R = 0.966$$

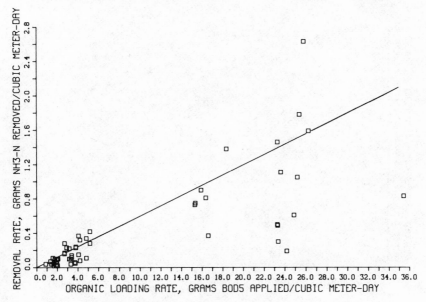

FIGURE 2-66
Ammonia nitrogen removal versus BOD_5 loading rates [aerated wastewater lagoons (surface + diffused air), (intermediate ponds)].

93

FIGURE 2-67

Ammonia nitrogen removal versus BOD_5 loading rates [aerated wastewater lagoons (diffused air) (total systems)].

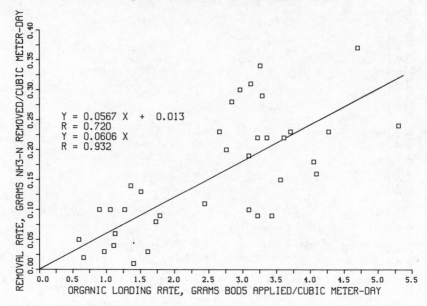

FIGURE 2-68

Ammonia nitrogen removal versus BOD_5 loading rates [aerated wastewater lagoons (diffused air) (intermediate ponds)].

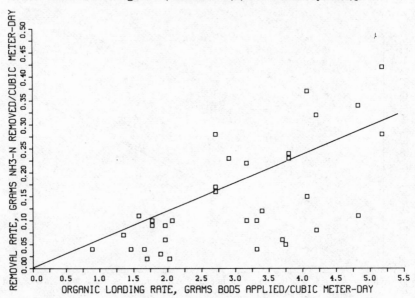

The results obtained with the aerated wastewater lagoons utilizing diffused-air aeration systems were evaluated independently, and the results are shown in Figures 2–63 and 2–67. A highly significant relationship was obtained for the diffused-air systems for the lower BOD_5 loading rates; however, the relationship was not as good as the results obtained when both the data for the diffused-air and surface aerated air systems were combined. An attempt was made to verify the results obtained with the diffused-air systems by calculating the performance in the intermediate ponds. For both the TKN and ammonia nitrogen removals in the intermediate ponds, the results were poorer than those experienced when considering the combined diffused-air and surface aerated systems.

Fraction Removed. The relationships between the TKN and ammonia nitrogen removal and the detention times in the aerated lagoons are shown in Figures 2–69 to 2–71. These relationships show the results when considering the influent TKN and ammonia nitrogen concentrations along with the effluent from each of the aerated lagoons with air supplied by diffused-air systems. As shown in Figure 2–69, there was a slight improvement in the correlation coefficient when a temperature factor was incor-

FIGURE 2-69
TKN removal versus detention time and temperature correction [aerated wastewater lagoons (diffused air) total Kjeldahl nitrogen (ponds 1, 2, and 3)].

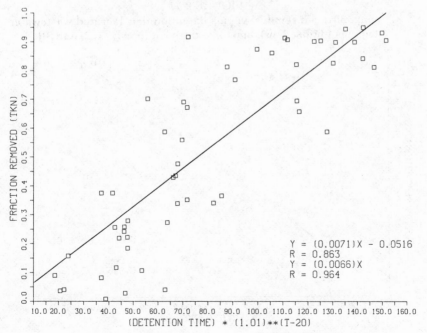

$$Y = (0.0071)X - 0.0516$$
$$R = 0.863$$
$$Y = (0.0066)X$$
$$R = 0.964$$

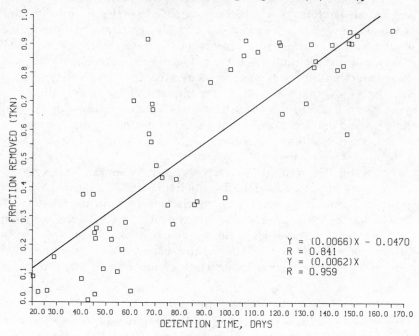

FIGURE 2-70
TKN removal versus detention time [aerated wastewater lagoons (diffused air) total Kjeldahl nitrogen (ponds 1, 2, and 3)].

$Y = (0.0066)X - 0.0470$
$R = 0.841$
$Y = (0.0062)X$
$R = 0.959$

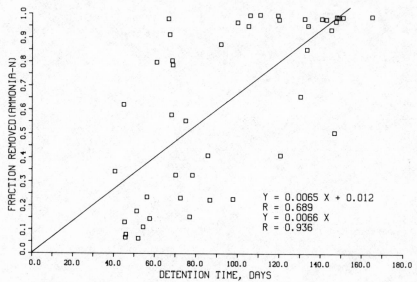

FIGURE 2-71
Ammonia nitrogen removal versus detention time [aerated wastewater lagoons (diffused air), ammonia nitrogen (ponds, 1, 2, and 3)].

$Y = 0.0065 X + 0.012$
$R = 0.689$
$Y = 0.0066 X$
$R = 0.936$

porated; however, the increase was statistically insignificant. When the fraction removed was plotted versus detention time (including data from both the diffused-air and surface-aerated systems) and the fraction removed obtained with the intermediate pond, the results were poor. The relationships obtained with the TKN data fit the fraction removed versus detention time model better than the ammonia nitrogen removal data. The differences were not statistically significant, but an examination of Figures 2–70 and 2–71 shows that the TKN relationship is less scattered than the ammonia nitrogen relationship.

The relationship between the fraction of TKN removed and the hydraulic detention time for the North Gulfport system (surface aeration) was not linear, as observed for the diffused-air aerated lagoons. This difference is probably attributable to the water-temperature differences between the North Gulfport system and the other four systems, the type of aeration system (surface versus diffused air), and the hydraulic detention times in the North Gulfport system. The maximum theoretical hydraulic detention time observed at the North Gulfport system was approximately equivalent to the lowest observed at the other four systems (Tables 2–19 and 2–20). Even with these short hydraulic detention times, the North Gulfport system achieved higher percentages of TKN reductions than were observed in the diffused-air systems at hydraulic detention times twice those at North Gulfport. It is impossible to distinguish between the influence of the higher liquid temperatures, the type of aeration equipment, and the possibility that the actual hydraulic detention times more closely approached the theoretical hydraulic detention times in the North Gulfport system. If the hydraulic characteristics of the North Gulfport system were superior to the other systems, it is possible that the actual hydraulic detention times in the North Gulfport lagoon system were greater than those in the diffused-air systems. Assuming that this was the difference in the systems, a combination of higher temperatures and adequate hydraulic detention times could have accounted for the better performance observed at the relatively short detention times in the North Gulfport system. As the detention times at the North Gulfport facility approached 25 days, the fraction of TKN removed stabilized at a value of approximately 0.65. A plausible explanation for this plateau of efficiency is that the system reaches the maximum efficiency possible within the temperature and hydraulic limits imposed on the system.

There was no indication that a minimum detention time was required for TKN removal to occur in the diffused-air systems; however, the minimum theoretical detention time in the diffused air systems exceeded 20 days, and all but four data points exceeded 40 days. It is possible that such a minimum value exists for diffused-air aerated lagoon systems, but it is impossible to determine such a value with the available data.

Design of Anaerobic Lagoons

Introduction

Anaerobic lagoons have been used for treatment of municipal, agricultural, and industrial wastewater. The primary function of anaerobic lagoons is to stabilize large concentrations of organic solids contained in wastewater and not necessarily to produce a high-quality effluent. Most often anaerobic lagoons are operated in series with aerated or facultative lagoons. A three-cell lagoon system produces a stable, high-quality effluent throughout its design life. Proper design and operation of an anaerobic lagoon should consider the biological reactions that stabilize organic waste material.

In the absence of oxygen, insoluble organics are hydrolyzed by extracellular enzymes to form soluble organics (i.e., carbohydrates such as glucose, cellobiose, xylose). The soluble carbohydrates are biologically converted to volatile acids. These organic (volatile) acids are predominantly acetic, proprionic, and butyric. The group of facultative organisms that transforms soluble organic molecules to short-chain organic acids are known as acid formers or acid producers. The next sequential biochemical reaction that occurs is the conversion of the organic acids to methane and carbon dioxide by a group of strict, anaerobic bacteria known as methane formers or methane producers. The sequence of biochemical reactions is depicted in Figure 2–72.

Anaerobic decomposition of carbohydrate to bacterial cells with formation of organic acids can be illustrated as:

$$5(CH_2O)_x \rightarrow (CH_2O)_x + 2CH_3COOH + \text{energy} \qquad (2\text{-}38)$$

The acid formed by the first reaction is neutralized by the bicarbonate buffer present in solution:

$$2CH_3COOH + 2NH_4HCO_3 \rightarrow 2CH_3COONH_4 + 2H_2O + 2CO_2 \quad (2\text{-}39)$$

During the growth of methane bacteria, ammonia acetate (CH_3COONH_4) is decomposed to methane and regeneration of the bicarbonate buffer, NH_4HCO_3:

$$2CH_3COONH_4 + 2H_2O \rightarrow 2CH_4 + 2NH_4HCO_3 \qquad (2\text{-}40)$$

If sufficient buffer is not available, the pH will decrease, which will inhibit the third reaction.

The facultative acid formers are not as sensitive to ambient environmental factors such as pH value, heavy metals, and sulfides. Acid formers are normally very plentiful in the system and are not the rate-limiting step. The rate-limiting step in anaerobic digestion is the methane fermentation process. Methane-producing bacteria are quite sensitive to such factors as

FIGURE 2-72
Sequential representation of the anaerobic digestion of organic waste. Courtesy of the American Chemical Society, Washington, D.C. (Ref. 30).

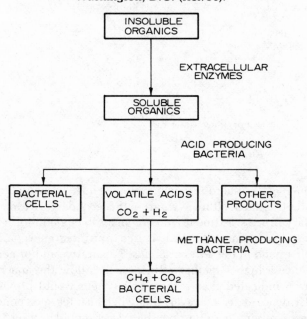

pH changes, heavy metals, detergents, alterations in alkalinity, ammonia-nitrogen concentration, temperature, and sulfides. Furthermore, methane-fermenting bacteria have a slow growth rate.

Environmental factors that affect methane fermentation are presented in Table 2–21.[31] In contrast, work by Kotze et al.,[32] Chan and Pearson,[33] Hobson et al.,[34] Ghosh et al.,[35] and Ghosh and Klass[36] provide some evidence that the hydrolysis step may become rate limiting in the digestion of particulates and cellulosic feeds. Design and operation of anaerobic lagoons should be founded on the fundamental biochemical and kinetic principles that govern the process. Most anaerobic lagoons, however, have been empirically designed.

The purpose of anaerobic lagoons is the decomposition and stabilization of organic matter. Water purification is not the primary function of anaerobic lagoons. Anaerobic lagoons are used as sedimentation basins to reduce organic loads on subsequent treatment units. The following material is a compilation of the state-of-the-art in anaerobic lagoon design.

Table 2-21
Environmental Factors for Methane Fermentation[30]

Variable	Optima	Extreme
Temperature (°C)	30-35	25-40
pH	6.8-7.4	6.2-7.8
Oxidation-reduction potential (MV)	−520-−530	−490-−550
Volatile acids (mg/l as acetic)	50-500	2000
Alkalinity (mg/l as $CaCO_3$)	2000-3000	1000-5000

Source: Compiled from statements in Ref. 30.

Design

Municipal. Anaerobic lagoons (ponds) have been used in municipal wastewater treatment either as a part of the secondary treatment system or as a tertiary process. Effluent from an anaerobic pond is of poor quality; therefore, when an anaerobic pond is used in an integrated ponding system, it is the first pond in the series of ponds. Facultative and/or aerobic ponds that produce a higher quality effluent will follow the anaerobic pond. McKinney[37] indicated that a three-cell system would provide the best system. Evidence exists, however, that when an integrated ponding system is designed, a minimum of four ponds in series should be used.[29,38]

A minimum hydraulic retention time of 5 days per pond is suggested.[38] The first pond is a deep anaerobic pond (10–15 ft) in which partial BOD removal and denitrification is accomplished, followed by an algal growth or a facultative pond, which is followed by at least two effluent clarification ponds.[38]

As previously stated, design of anaerobic lagoons is based on empirical information. Table 2–22 presents the design and operational conditions cited in the literature for anaerobic lagoons treating municipal wastewater. Municipal anaerobic lagoons are primarily designed on areal organic mass loadings (mass of BOD_5/unit of surface area·day) and hydraulic detention times. As shown in Table 2–22, there exists a wide range of values for parameters used to design anaerobic ponds. Anaerobic ponds should be designed on the basis of volumetric loading (mass of BOD_5/volume·day), because anaerobic biochemical reactions are a function of detention time not surface area.[31]

However, most design BOD_5 mass loading information concerning municipal anaerobic lagoons is based on volumetric loadings. An empirical equation for design of anaerobic lagoons derived from experiments with septic tanks in the United States and Zambia was presented by Vincent et

Table 2-22
Design and Operational Parameters for Anaerobic Lagoons Treating Municipal Wastewater

Ref.	Areal BOD₅ Mass Loading (lb/acre·day)[a]		BOD₅ Removal Efficiency (%)		Temperature	Depth (ft)	Hydraulic Detention Time (days)	VSS Loading (lb/acre·day)[a]	pH
	Summer	Winter	Summer	Winter					
41	360		75			3-4		360	
41	280		65			3-4		280	7.1–8.2
41	100		86			3-4			
41	170		52			3-4			
41	560	400	89		>15°C desirable	3-4			
40	400	100	70	60	≅32°C				
42	900–1200	675	60–70			3-5	2–5		
43						8-10	30–50		
44	220–600						15–160		
45	500		70			8-12	5		
31						8-12	2 (summer) 5 (winter)		

[a]lb/acre·day = 0.8903 (kg/hectare·day).

al.[39] Equation 2–41 is valid for tropical and subtropical regions. The reduction of BOD_5 can be predicted by the following relationship:

$$L_p = \frac{L_o}{k_n \left(\frac{L_p}{L_o}\right)^n R + 1} \tag{2-41}$$

where

L_o = influent BOD_5 (mg/l)
L_p = effluent BOD_5 (mg/l)
R = detention times (days)
n, k_n = empirical exponent and design coefficient

Oswald[40] stated that organic mass loading criteria is dependent on geographical location. Geographical location will influence the physical, chemical, and biological makeup of the water and its temperature. Temperature affects the kinetic rates of decomposing organic material. Temperatures less than 15°C have drastically inhibited reduction of organic matter.[40] Data obtained by Oswald showed that an anaerobic pond loaded at 47 kg BOD_5/ha·day (42 lb BOD_5/acre·day) produced 52.5 m³ of gas/ha·day (750 ft³ of gas/acre·day) at 15°C; approximately 280 m³ of gas/ha·day (4000 ft³ of gas/acre·day) at 20°C; and about 455 m³ of gas/ha·day (6500 ft³ of gas/acre·day) at 23°C. Pond temperatures less than 15°C resulted in a relatively constant gas production of 0.62–0.75 m³/kg of BOD_5 applied (10–12 ft³/lb BOD_5 applied). Anaerobic pond temperature, therefore, should be maintained at levels greater than 15°C and preferably around 32°C.[40]

Biochemical reaction rates are temperature dependent. To ensure adequate time for necessary biochemical reactions to occur, hydraulic detention times must also be a function of temperature. In general, minimum design detention times in the summer months are 2 days and increase to 5 days during the cooler winter months (Table 2–22).

Depths of anaerobic ponds are a minimum of 1–1.5 m (3–5 ft). Most anaerobic lagoons evaluated in Australia have been 1–1.5 m (3–5 ft) deep.[41,42,46] Parker and Skerry[46] advocate the use of shallow ponds to enhance mixing of the active sludge with the influent, thereby increasing the destruction of BOD. Lagoon evaluations in the United States, however, have favored deeper ponds, 2.4–3.7 m (8–12 ft) deep (Table 2–22). Deeper ponds allow for greater solids storage, less energy loss from the lagoon, more uniform temperatures with depth, and less oxygen transfer.[40]

A major problem associated with anaerobic ponds is the production of odors. Odor can be controlled in ponds by providing an aerobic zone at the surface to oxidize the volatile organic compounds that cause the odors. Recirculation from an aerobic pond to the primary anaerobic pond can alleviate odors by providing dissolved oxygen from the aerobic pond effluent which overlays the anaerobic pond and oxidizes sulfide odors.[40] To avoid contact of anaerobic fermentation processes with oxygen, influent wastewater may enter at the center of the pond into a design chamber in which the sludge accumulates to some depth, as shown in Figure 2–73. This type of inlet design has proved advantageous.[47] Mixing of influent with active sludge will enhance BOD removal efficiency and reduce odor.[46]

Fisher and Gloyna[48] reported on the ability of anaerobic lagoons to stabilize excess activated sludge. Data obtained from a field pond and three laboratory units indicate that anaerobic lagoons can effectively treat excess activated sludge at mass loadings of 63 kg BOD_5/ha·day (56 lb BOD_5/day·acre) with a depth of 1.52 m (5 ft). Twice this organic loading can be handled by an anaerobic pond if the pond depth is maintained at 1.8–2.7 m (6–9 ft). This is recommended for areas with climates similar to Austin, Texas.

In summary, information provided in the literature indicates that areal BOD_5 mass loading design criteria for anaerobic lagoons should be determined for a specified geographic or climatological region. Design areal BOD_5 loadings, however, are approximately 450 kg BOD_5/ha·day (400 lb BOD_5/acre·day) in the warmer summer months and 112 kg BOD_5/ha·day (100 lb BOD_5/acre·day) during cooler winter months. In general, anaerobic lagoons yield a 70 % reduction in influent BOD_5 concentrations. Hydraulic detention times for an anaerobic pond should be a minimum of 5 days. In the United States anaerobic ponds are designed with

FIGURE 2-73
Some methods of creating a digestion chamber in the bottom of a waste pond. Courtesy of Center for Research for Water Resources, The University of Texas at Austin (Ref. 40).

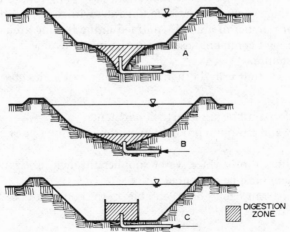

DIGESTION ZONE

a depth of 2.4–3.7 m (8–12 ft). Ponds have been shown to work effectively, however, at depths of 1–1.5 m (3–5 ft). Temperatures greater than 15°C and desirably about 32°C should be maintained within the pond.

Agricultural Wastes

Anaerobic lagoons have been used to treat waste discharges from all phases of the vast agricultural industry. Design of anaerobic lagoons for treating agricultural waste are discussed according to the type of waste discharged: (1) treatment of livestock manure, (2) treatment of meat-processing waste, and (3) cannery waste.

Treatment of Livestock Manure. Intensification of animal production in large confined areas has become a major agricultural industry. The feeding of high areal densities of dairy and beef cattle, chickens, and hogs has become convenient and economical. As an example, the population of beef cattle contained in feedlots was approximately 14 million in 1976.[49] Feedlots represent the largest single source of solid waste generated in the United States (over 2,000,000,000 tons annually).[50] The livestock farmer has a mammoth task in managing waste generated from his or her operation. Anaerobic lagoons have been considered a suitable treatment system for manure waste. Manure lagoons treat large amounts of organic solid matter. The objective of the lagoon is stabilization of the organic matter rather than water purification. Slurries introduced to the lagoon often contain just enough water to transport the solids to the lagoon. In contrast, municipal wastewater lagoons handle large quantities of water containing relatively light solids concentrations (approximately 0.1% solids). The areal BOD_5 mass loading rate to municipal anaerobic ponds may be as high as 450 kg/ha·day (400 lb/acre·day); whereas mass loading rates to lagoons treating manure may be 1120 to 1350 kg/ha·day (1000–1200 lb/acre·day).[51] Most authors normalize the mass loading rates with respect to the volume of the pond or number of animals confined in the feeding area. Areal mass loadings are not commonly used for designing anaerobic lagoons treating livestock manure.

There are four criteria that must be met by the designer of manure ponds.[52]

1. Odors must be minimal to nonexistent.
2. Fly and mosquito production must be controllable or must not occur.
3. Pollution of surface water supplies through infiltration of pond liquor must be prevented.
4. Appearance must be acceptable, or the pond must be screened and hidden.

The first and second criteria are a function of mass loading to the lagoon. The third criteria is related to the land on which the pond is placed, and the fourth is related to the location and loading of the lagoon.

Table 2–23 presents the mass of wet manure produced per gram of animal per day and the solids concentration of wet manures for poultry, swine, and dairy and beef cattle. The production of manure per unit weight of animal is highly correlated to animal weight.[49]

The BOD_5, COD, and nutrient contents of fowl manures is outlined in Table 2–24. Reported values vary widely. Data do not indicate operating conditions of the livestock confinement, however. Each manure is different in its physical and chemical characteristics, and normal design parameters do not adequately delineate the treatment process required for livestock manures.

State agencies have set design criteria for anaerobic lagoons (Table 2–25). The major parameter used for designing anaerobic lagoons is the BOD_5 loading. The average state design loading is 0.193–0.240 kg/m³·day (12 to 15 lb BOD_5/1000 ft³·day). The lagoons should be 3.05–4.57 m (10–15 ft) deep.[53] Experience in treating livestock waste manures has indicated that suggested values of design parameters are highly variable (Table 2–26). Furthermore, the critical recommended design parameter varies between authors (i.e., lb BOD_5/acre·day; lb BOD_5/1000 ft³·day; lb VS/1000·ft³·day; ft³ of volume/animal; surface area, ft²/animal). Anaerobic lagoons have functioned successfully at volatile solid loadings from 0.064 to 1.12 kg VS/m³·day (4–70 lb VS/1000 ft³·day). Suggested design VS loading for poultry manure lagoons are from 0.064 to 0.161 kg VS/m³·day (4 to 10 lb VS/1000 ft³·day). Swine waste lagoons have operated satisfactorily up to 0.321 kg VS/m³·day (20 lb VS/1000 ft³·day). Volatile solid loadings from 0.161 to 1.12 kg VS/m³·day (10–70 lb VS/1000 ft³·day) have been adequately treated by anaerobic lagoons stabilizing dairy manures. Desirable mass loadings to the lagoon depend on the proximity of the operation to populous areas. If the operation is isolated and odors pose no problem, higher loadings can be used in designing the lagoon. However, as mass loading rates increase, a corresponding increase in the rate of accumulation of solids will develop. Therefore, the lagoon will fill at a faster rate, which will mandate more frequent sludge removal. A population of purple sulfur bacteria once established in sufficient number in an anaerobic lagoon can reduce odors stimulated by heavy loadings.[54]

The temperature of the lagoon system is vital in regulating the operating efficiency. Anaerobic lagoons operate efficiently at temperatures greater than 15°C. When the temperature drops below 15°C, biological activity within the lagoon is minimal, and the lagoon merely acts as a settling basin and stores the organic mass.[53,56,59,60] Anaerobic lagoons treating manure from dairy cattle are covered with a crust of grease and inert

Table 2-23
Physical Characteristics of Livestock Defecation

Animal	Wet Manure (g/g of Animal-day)	Total Solids (g/g of Animal-day)	Volatile Solids (g/g of Animal-day)
Poultry	0.0234	0.011	—
	0.027-0.087	0.011-0.022	0.0084-0.017
	0.074	0.021	—
	—	0.014	0.0098
	—	0.013	0.0101
	—	0.013-0.019	—
	0.072	0.0086	0.0054
	0.083	—	—
Average	0.062	0.014	0.0096
Swine	0.084	0.011	—
	0.028-0.095	0.008-0.016	0.0068-0.0136
	0.087	0.016	—
	—	0.0080	0.0063
	—	0.0097	0.0080
	—	0.0050	0.0035
	—	0.0071	—
	—	0.0048	0.0033
	0.074	0.0059	0.0047
	—	0.0099	0.0070
Recommended Value	0.074	0.0089	0.0054
Cattle (Dairy)	0.071	0.0114	—
(Dairy)	0.058	0.0087	—
(Dairy)	—	0.0104	0.0083
(Dairy)	—	0.0068	0.0057
(Dairy)	—	0.0075	—
(Dairy)	0.124	0.0025	0.0018
(Beef)	0.082	0.0197	—
(Beef)	0.039-0.074	0.0095-0.0114	—
(Beef)	0.063	0.0095	—
(Beef)	0.067	0.0090	0.0069
(Beef)	—	0.0036	0.0032
(Beef)	0.063	0.0050	0.0040
(Beef)	—	0.0091	—
Average Dairy	0.084	0.0079	0.0053
Average Beef	0.066	0.0095	0.0047
Sheep	0.072	0.016	—
Ducks	—	0.016	—

Source: From Ref. 49. Courtesy of American Society of Civil Engineers, New York, New York.

106

Table 2-24
Nutrient and Sanitary Characteristics of Animal Wastes

ANIMAL	CHARACTERISTICS OF FOWL MANURES (G/G OF ANIMAL-DAY × 10⁻³)					
	BOD	COD	Ammonia Nitrogen	Total Nitrogen	Phosphorus (P_2O_5)	Potassium (K_2O)
Poultry	—	—	—	1.12	0.72	0.36
	—	—	—	0.27-1.27	0.22-1.00	0.11-0.42
	—	—	—	1.1	—	—
	3.33	11.1	0.13	0.67	0.58	—
	2.91	11.2	0.52	0.70	0.60	0.27
	3.33-7.11	—	—	—	—	—
	1.33-2.22	—	—	—	—	—
	—	—	—	0.23	0.37	—
	3.74	7.1	—	0.58	0.72	—
	4.27	—	0.12	—	—	—
Average	3.46	9.8	0.26	0.74	0.60	0.30
Ducks	2.0-4.0	—	—	8.0	0.6-1.6	—
Swine	—	—	—	0.51	0.32	0.62
	—	—	—	0.42-0.60	0.29-0.32	0.34-0.62
	—	—	—	0.53	—	—
	2.0	7.6	0.24	0.32	0.25	0.11
	4.3	5.4	—	0.64	—	—
	—	4.7	—	—	—	—
	2.5	—	—	—	—	—
	2.2	—	—	0.70	—	—
	5.6	—	—	—	—	—
	—	—	—	0.41	0.55	—
	3.1	6.4	—	—	—	—
	2.0	5.2	—	—	—	—
	3.2	9.3	—	0.44	0.67	—
Recommended Value	3.1	6.4	0.24	0.51	0.42	0.40
Beef cattle	—	—	—	0.36	0.115	0.274
	—	—	—	0.35-0.44	0.11-0.12	0.27-0.34
	—	—	—	0.29	—	—
	1.11-2.22	10.0	—	0.26	—	—
	1.02	3.26	0.11	0.26	—	—
	—	—	—	0.41	0.25	—
	1.87	1.50	—	0.16	0.31	—
	1.84	—	—	—	—	—
Average	1.61	9.42	0.11	0.32	0.18	0.29
Sheep				0.86		
				0.34	0.25	
Average				0.60	0.25	

(continued)

Table 2-24 (*continued*)

ANIMAL	CHARACTERISTICS OF FOWL MANURES $(G/G\ OF\ ANIMAL\text{-}DAY \times 10^{-3})$					
	BOD	COD	Ammonia Nitrogen	Total Nitrogen	Phosphorus (P_2O_5)	Potassium (K_2O)
Dairy cattle	—	—	—	—	0.30	—
	—	1.53	19.1	—	—	—
	0.31	1.53	8.4	—	0.38	0.12
	—	1.32	5.8	0.23	0.37	—
	—	0.44	—	—	0.49	—
	—	0.95	5.7	—	0.16	0.11
Average	0.31	1.15	9.8	0.23	0.34	0.12

SOURCE: From Ref. 49. Courtesy of American Society of Civil Engineers, New York, New York.

solids.[51,53,56] This cover, in conjunction with the depth of the lagoon (3.05–4.57 m or 10–15 ft), provides insulation for the lagoon and relatively higher temperatures during winter operation.

The depth of the lagoon stabilizes the anaerobic biological activity by minimizing oxygen transfer through the lagoon, more stable temperatures, and more solids storage capacity. It is usually more economical to construct a deep lagoon because of less land area required. Normal depths of anaerobic lagoons treating livestock waste are 2.44–4.57 m (8–15 ft).

Infiltration rates from livestock waste anaerobic lagoons of 0.5–2.1 in./day have been reported in the literature.[51,52,60] Seepage of lagoon water to the groundwater can govern whether or not a lagoon can be constructed. Salts and organic material passing through the soil could pollute subsurface

Table 2-25
State Agencies Anaerobic Lagoon Criteria

STATE	BOD (LB/1000 FT³)[a]	SUSPENDED SOLIDS (LB/1000 FT³)[a]	DETENTION (DAYS)	EXPECTED REDUCTION (%)
Georgia	3,[b]15[c]	1.25,[b]8[c]	—	60–80
Illinois	15–20	—	5	60
Iowa	12–15	—	5–10	60–80
Montana	2–10	—	10 minimum	70
Nebraska	12–15	—	3–5	75
South Dakota	15	—	—	60
Texas	25–100	100–400	5–30	50–100

SOURCE: From Ref. 53.

[a] lbs/1000 ft³ = 62.305 (kg/m³).
[b] No recirculation provided.
[c] 1:1 recirculation provided.

Table 2-26
Design and Operational Criteria for Anaerobic Lagoons Treating Livestock Manure Waste

Ref.	Type of Waste	Areal BOD$_5$ Mass Loading (LB BOD$_5$/ACRE·DAY)[a]	BOD Removal Efficiency (%)	Volumetric Loading LB BOD$_5$/1000 FT^3·DAY[b]	Temp. (°C)	Depth (FT)	Detention Time (DAYS)	VSS Loading (LBS VSS/1000 FT^3·DAY)[b]	pH	Loading Rate
51	Poultry	350	90	1.04		8-10-12		3.94	7.0	13.6 ft^3/5 lb chicken
51	Dairy cattle	511		1.87		8-10-12		10.4	7.0	795 ft^3/1000 lb cow
51	Swine	388		1.53		8-10-12		4.95	6.9	124 ft^3/100 lb hog
55	Poultry	935	86		24	8		70	6.8	
56	Dairy cattle			20	≥17	4	7.35	5-10	7.0	
57	Poultry					5-8				
58	Swine					≥3				15 ft^2/hog
59	Swine		80	10	>15	4	60	20	7.2-7.5	
60	Dairy (milking parlor waste)		85	8.6	29			12.3	6.9-6.3	
61	Swine	230 lb COD/acre-day				3-10		0.4 to 4.0		4 to 20 ft^2/hog

[a] lb BOD$_5$/acre·day = 0.8903 (kg BOD$_5$/ha·day).
[b] lb/1000 ft^3·day = 62.305 (kg/m^3·day).

109

water supplies. Therefore, seepage rates must be determined prior to lagoon construction and preventive measures implemented.

Treatment of Meat-Processing Wastewaters

Meat-processing operations are divided into two distinct categories, (1) slaughterhouse and (2) packinghouse. Slaughterhouses are killing and dressing plants. Little or no processing of by-products occurs within a slaughterhouse. Packinghouse operations involve not only the slaughtering of the animals but also normally involves cooking, curing, smoking, and pickling of meat, manufacture of sausage, and rendering of edible and inedible fats.[62] Sources of wastewater from meat-processing operations include: killing floor, carcass dressing, casing cleaning, rendering, hog-hair removal, hide processing, meat processing, cooling room, general cleanup, and animal holding pens. The quantity of wastewater produced is a function of the methods employed to recover and process the by-products. Anaerobic lagoons have been used successfully to stabilize slaughterhouse and meat-packing wastewater discharges.

Slaughterhouse waste. Slaughterhouse waste primarily comes from the animal-holding pens, killing floor, rendering, and sanitary sewage. Lagoon systems treating slaughterhouse waste normally consist of three ponds in series. The first pond is anaerobic, followed by a facultative and/or aerobic pond(s). The anaerobic pond preceding an aerobic pond ensures that the subsequent oxidation ponds will function without the production of odors.[63] Design criteria for anaerobic lagoons treating slaughterhouse waste has been generally based on volumetric BOD_5 mass loadings (Table 2-27).

Criteria used for lagoon design have been fairly consistent. Anaerobic lagoons are designed on an organic mass loading of 0.241 kg/m³·day (15 lb BOD_5/1000 ft³·day). The lagoons are normally deep (4.57 m or 15 ft) to minimize oxygen transfer and energy loss. Hydraulic detention times are normally 5–7 days.

Two anaerobic lagoons operated in series have not improved operating performance.[64] If high recirculation rates are provided, however, a second pond may be warranted to provide a more clarified effluent.[53]

Lagoons obtain optimum performance once a thick crust consisting primarily of grease and paunch manure is formed on the surface of the lagoon.[64-67] This grease scum layer not only prevents oxygen transfer at the surface but also insulates the lagoon from atmospheric temperature. Rollag and Dornbush[64] observed a drop of 6°C(10°F) in the temperature of the lagoon when the grease scum layer was disrupted. An anaerobic lagoon treating waste generated at a hog abattoir at Logansport, Indiana had to create a cover on the surface by wasting grease from the abattoir and supplementing this material with straw. Once the cover was established, the

Table 2-27
Design and Operational Criteria for Anaerobic Ponds Treating Slaughterhouse Waste

Ref.	Type of Animal	BOD Removal Efficiency (%)	Volumetric Mass Loading (LB BOD$_5$/1000 FT^3·DAY)	Temperature (°C)	Detention Time (DAYS)	VSS Loading (LB VSS/1000 FT^3·DAY)[a]	pH	Depth[b] (FT)
64	Beef	58.2	20	24	7			10-15
65	Hog	65	15		7		7-7.25	15
66	Beef	87	15	22-27	5	37	6.4-7.8	15
67	Hog	78	14.7 (Actual 31.4)	27	7			14

[a] lbs/1000 ft^3·day = 62.305 (kg/m^3·day).
[b] ft = 0.3049 (m).

111

lagoon temperature was maintained above 27°C (80°F).[67] In addition, the thick surface crust aids in preventing odors from escaping the lagoon.

Startup of an anaerobic lagoon may be enhanced by initially seeding the lagoon with anaerobic microorganisms obtained from municipal anaerobic digestors or nearby anaerobic systems. Rollag and Dornbush[64] have estimated the amount of seed required on the basis of 0.1 lb dry solids/gpd average flow (0.012 kg/day·l).

Meat-Packing Operations. The volume and strength of wastes generated from the meat-packing industry can vary, depending on the amount of water used and the process. Design criteria for anaerobic lagoons treating these wastes also vary as noted in Table 2–28. Design criteria are based on:

1. Areal mass loadings (kg BOD_5/ha·day or lb BOD_5/acre·day)
2. Volumetric mass loadings (kg BOD_5/m³·day or lb BOD_5/1000 ft³· day)
3. Animal units/volume·week

Coerver[68] determined design loads to a lagoon in terms of the total waste from slaughtering one hog (termed hog unit). Estimations of the hog unit equivalents for the waste from slaughtering animals other than hogs were as follows:

Animals Slaughtered	Waste Equivalent
One hog	1 hog unit
One sheep	1 hog unit
One calf (less than 600 lb live weight)	1 hog unit
One medium beef (600–900 lb live weight)	3 hog units
One heavy beef (over 900 lb live weight)	5 hog units

Areal mass BOD_5 loadings vary from 450 to 1528 kg BOD_5/ha·day (400–1360 lb BOD_5/acre·day). Satisfactory lagoon performance has been achieved at volumetric BOD_5 loadings of 0.177–0.321 kg BOD_5/m³·day (11–20 lb BOD_5/1000 ft³·day). As previously stated, anaerobic fermentation processes are a function of detention time; therefore, volumetric loadings provide a more appropriate design criteria. Hydraulic detention times are from 3 to 6 days.

There is a significant divergence in opinion concerning design depth of the lagoon. Anaerobic lagoons treating packinghouse waste in Louisiana are shallow (0.61–1 m or 2–3 ft depth);[68] whereas anaerobic lagoons in northern climates have been constructed at depths of 5.5–6.1 m (18–20 ft).[69] In general, design depths should be 2.4–6.1 m (8–20 ft). Deeper ponds provide less surface area for reaeration, greater energy conservation, and less light penetration.

Anaerobic lagoon investigations concluded that maximum performance efficiency was not observed from the system until a thick crust had

Table 2-28

Design and Operational Criteria for Anaerobic Lagoons Treating Meat Packinghouse Waste

Ref.	Areal Loading (lb BOD$_5$/acre·day)[a]	BOD$_5$ Removal Efficiency (%)	Volumetric Loading (lb BOD$_5$/1000 ft^3 day)[b]	Temperature (°C)	Depth (ft)[c]	Detention Time (days)	Surface Area (acres)[d]	pH	Design Loadings
68		92			2-3			7.8-8.5	500 hog units/ acre-ft· week
69	410	80-90	11 to 15		18-20	3		7.0	
70					8-17				
Wilson & Co., Inc.					3-4				
Albert Lea, Minn.									
Moultri, Ga.			14		14	6	1.4		
New Zealand			12			1.5			
Virginia	1360	80			8		0.25		
Idaho	520	66			8				
71		85	11	>24	14	5	1.4		
62		75	20	>24	12-15	4			

[a] lb BOD$_5$/acre·day = 0.8903 (kg BOD$_5$/ha·day).
[b] lb BOD$_5$/1000 ft^3· day = 62.305 (kg BOD$_5$/m^3·day).
[c] ft = 0.3049 (m).
[d] acres = 0.4047 (ha).

113

formed on the surface.[62,68] This crust formed from primary greases and paunch manure maintains anoxic conditions, higher temperature, and minimal release of obnoxious odors to the atmosphere. Grease removal prior to lagoon treatment should be achieved to retard massive accumulation of grease in the lagoon; however, *it must not be so efficient that an adequate grease cover will not form on the liquid surface.*

Lagoon operating temperatures should be maintained greater than 24°C to achieve maximum degradation of organic matter. Lagoons must be designed to maximize anaerobic biochemical reactions and minimize energy losses.

Cannery Waste. A single lagoon is normally inadequate to treat waste discharges from canneries. Anaerobic lagoons are used as the first cell of a series lagoon system. Fruit and vegetable cannery wastes contain large concentrations of sugars or starches and, therefore, are acidic or rapidly develop high acidity and low pH. Complete mixing of the waste with the contents of the anaerobic pond is required to prevent development of localized areas of high acidity, which can rapidly extend and stop the anaerobic fermentation process.[72] A square lagoon with multiple inlet and outlet structures should facilitate mixing of the lagoon contents with the influent waste stream.

Research has shown that appropriate mixture of cannery waste and sewage effluent to establish a proper BOD_5/nitrogen ratio and regular daily seeding with digested sludge and supernatant liquor cannery wastes can be effectively treated by anaerobic lagoons.[39] A ratio of BOD_5/nitrogen of 50:1 has been recommended for proper performance of the lagoon.[41] Cases in which the BOD_5-to-nitrogen ratio is 100:1.25 have shown a reduction in BOD_5 removal.[41,72]

In Australia areal BOD_5 loading is used for the design of lagoons treating cannery waste. Maximum organic loadings for anaerobic lagoons treating fruit cannery wastes are from 674 to 900 kg BOD_5/ha·day (600–800 lb BOD_5/acre·day).[41] The BOD_5 contents of the lagoon are 400–600 mg/l. Vegetable cannery wastes are treated in lagoons designed with a maximum summer loading of 393 kg BOD_5/ha·day (350 lb BOD_5/acre·day). Winter loads are 168–225 kg BOD_5/hectare·day (150 to 200 lb BOD_5/acre·day). The BOD_5 concentration in the lagoon is 80–100 mg/l. The lagoon depth is normally 1.37 m (4.5 ft). Lagoons in Australia are generally shallower than lagoons designed in the United States [maximum depths of 1.83–2.44 m (6–8 ft) compared to 3.66–4.57 m (12–15 ft) or greater in the United States].

Industrial Wastes

Anaerobic lagoons have been used in upgrading waste discharges from chemical oil refinery and petrochemical industries. Oil refinery and petrochemical industries employ anaerobic lagoons as a pretreatment process prior to aerobic treatment. The essential feature is to construct small ponds in series to allow successional changes to occur.[73] Initially, oil contains no aquatic organisms. Through organism succession, the environment is altered to sustain microbial growth and produce a high-quality effluent. A large amount of nutrients are released in the anaerobic pond.[73] The release of these nutrients decreases with detention time. Utilization of these nutrients by facultative and/or aerobic ponds is prohibited by toxicants contained in the effluent. Toxicity of these substances decreases with increases in hydraulic detention time.[73] Depths of anaerobic lagoons treating oil refinery wastes are approximately 4.3 m (14 ft), with surface areas of 0.1–0.8 ha (0.25 to 2 acres). Hydraulic detention times vary from 2 to 15 days.[73]

Problems associated with petrochemical industrial waste are high BOD, odors, elevated temperatures, and surfactants.[74,75] Anaerobic lagoons prior to aerobic treatment can absorb major waste quality changes without altering the quality of the system contents. Furthermore, the large surface area allows cooling of the wastewater for subsequent aerobic processes. Surfactants and some influent biologically oxygen demanding substances can be anaerobically stabilized, thereby reducing the oxygen demand and operating problems of the aerobic process.[74] Hovious et al.[74] showed that the required depth of the lagoon was a function of influent sulfate concentrations. A low organically loaded pond (0.144–0.161 kg $COD/m^3 \cdot day$ or 9–10 lb $COD/day \cdot 1000$ ft^3) treating a petrochemical waste high in acetic acid has shown that as much as 4000 mg/l sulfate ion could be tolerated in a 0.75-m (2.5-ft) deep lagoon without inhibitory levels of sulfide produced. A petrochemical waste of 3000 mg COD/l can be adequately treated with an anaerobic lagoon having a hydraulic retention time of 10 days.[74] This system should achieve a 40% COD removal and a 50% BOD_5 removal under winter conditions in a Gulf Coast location.

Anaerobic lagoons have successfully stabilized low sulfate chemical wastes.[75] The chemical industry report by Woodley and Brown[75] produced the following wastes which were treated by an anaerobic lagoon:

1. Methylamines still bottoms.
2. Alkaterge condensate.
3. Amino bufanol still bottoms.
4. Caustic washes of fermentation equipment.
5. Sewage.
6. Floor washings.

Areal BOD_5 loadings varied from 211 to 4530 kg/ha·day (188–4031 lb/acre·day) (0.0161–0.369 kg $BOD_5/m^3 \cdot day$ or 1–23 lb $BOD_5/1000$ ft$^3 \cdot day$). No odors were observed during any of the BOD loading rates. Lack of odors is attributed to low influent sulfate concentrations. A 90% reduction of BOD_5 was achieved during warm summer months. The lagoon depth was approximately 1.22 m (4 ft), and it was recommended that anaerobic lagoons be designed with a minimum depth of 3.66 m (12 ft).[76] No significant effect on lagoon operation was observed at pH levels from 6.7 to 9.4. To maintain proper performance of the lagoon system the following BOD_5 to nutrient ratios were presented as operational guidelines:

Influent BOD/nitrogen,	20:1
Influent BOD/potassium,	100:1
Influent BOD/phosphorus,	100:1

Layout and Construction

Dimensions.   A minimum of two anaerobic lagoons will allow operation with one pond temporarily out of service. Recommended liquid depth of the lagoon should be 2.44–4.57 m (8–15 ft). Surface area provided for anaerobic lagoons is normally not greater than 0.8 ha (2 acres). Lagoon side slopes have been constructed as steep as 1:1 and as flat as 3:1.[62] Side slopes of 1.5:1 and 2:1 (Figure 2–74) provide stable interior slopes.[60,62,66,69] Outside dike slopes should not be greater than 3:1 to facilitate grass mowing.[62,75] Slope depends on soil stability. Generally anaerobic lagoons are rectangular with L/W ratios of 2:1.

Inlets and Outlets.   Lagoon inlet inverts are placed at an elevation of 0.6–1 m (2–3 ft) from the bottom.[62,66] Positioning of the inlet near the bottom permits influent wastewater to mix with active sludge solids. As

FIGURE 2-74
Cross section of an anaerobic lagoon (Ref. 62).

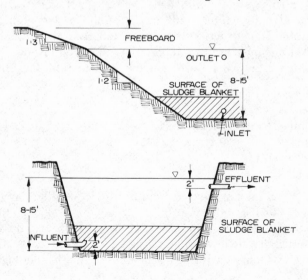

Table 2-29
Typical Design Parameters for Aerobic Stabilization Ponds

Parameter	Aerobic (High-rate) Pond	Aerobic Pond[a]	Aerobic (Maturation) Pond	Aerated Lagoons
Flow regime	Intermittently mixed	Intermittently mixed	Intermittently mixed	Completely mixed
Pond size, (ha)	0.25-1	<4 multiples	1-4	1-4 multiples
Operation[b]	Series	Series of parallel	Series or parallel	Series or parallel
Detention time, d[b]	4-6	10-40	5-20	3-10
Depth, m	0.30-0.45	1-1.5	1-1.5	2-6
pH	6.5-10.5	6.5-10.5	6.5-10.5	6.5-8.0
Temperature range (°C)	5-30	0-30	0-30	0-30
Optimum temperature (°C)	20	20	20	20
BOD_5 loading, (kg/ha·day)[c]	80-160	40-120	≤ 15	
BOD_5 conversion	80-95	80-95	60-80	80-95
Principal conversion products	Algae, CO_2, bacterial cell tissue	Algae, CO_2, bacterial cell tissue	Algae, CO_2, bacterial cell tissue, NO_3	CO_2, bacterial cell tissue
Algal concentration, (mg/l)	100-260	40-100	5-10	
Effluent suspended solids, (mg/l)[d]	150-300	80-140	10-30	80-250

Source: Ref. 28. From *Wastewater Engineering: Treatment, Disposal, Reuse* by Matcalf & Eddy, Inc. Copyright © 1979, 1972 by McGraw-Hill, Inc. Used with the permission of McGraw-Hill Book Company.

[a] Conventional aerobic ponds designed to maximize the amount of oxygen produced rather than the amount of algae produced.
[b] Depends on climatic conditions.
[c] Typical values (much higher values have been applied at various locations). Loading values are often specified by state control agencies.
[d] Includes algae, microorganisms, and residual influent suspended solids. Values are based on an influent soluble BOD_5 of 200mg/l and, with the exception of the aerobic ponds, an influent suspended-solids concentration of 200 mg/l.

Note:
$ha \times 2.4711 = acre$
$m \times 3.2808 = ft$
$kg/ha \cdot d \times 0.8922 = lb/acre \cdot d$

Table 2-30
Typical Design Parameters for Anaerobic and Facultative Stabilization Ponds

PARAMETER	AEROBIC-ANAEROBIC (FACULTATIVE) POND	AEROBIC-ANAEROBIC (FACULTATIVE) POND	ANAEROBIC POND	AERATED LAGOONS
Flow regime	—	Mixed surface layer	—	Completely mixed
Pond size, (ha)	1-4 multiples	1-4 multiples	0.2-1 multiples	1-4 multiples
Operation[a]	Series or parallel	Series or parallel	Series	Series or parallel
Detention time, (days)[a]	7-30	7-20	20-50	3-10
Depth (m)	1-2	1-2.5	2.5-5	2-6
pH	6.5-9.0	6.5-8.5	6.8-7.2	6.5-8.0
Temperature range (°C)	0-50	0-50	6-50	0-40
Optimum temperature (°C)	20	20	30	20
BOD_5 loading (kg/ha·day)[b]	15-80	50-200	200-500	
BOD_5 conversion	80-95	80-95	50-85	80-95
Principal conversion products	Algae, CO_2, CH_4, bacterial cell tissue	Algae, CO_2, CH_4, bacterial cell tissue	CO_2, CH_4, bacterial cell tissue	CO_2, bacterial cell tissue
Algal concentration (mg/l)	20-80	5-20	0-5	
Effluent suspended solids (mg/l)[c]	40-100	40-60	80-160	80-250

SOURCE: Ref. 28. "From Wastewater Engineering: Treatment, Disposal, Reuse by Metcalf & Eddy, Inc. Copyright © 1979, 1972 by McGraw-Hill, Inc. Used with the permission of McGraw-Hill Book Company."

[a] Depends on climatic conditions.

[b] Typical values (much higher values have been applied at various locations). Loading values are often specified by state control agencies.

[c] Includes algae, microorganisms, and residual influent suspended solids. Values are based on an influent soluble BOD_5 of 200 mg/L and, with the exception of the aerobic ponds, an influent suspended-solids concentration of 200 mg/L.

Note:
$ha \times 2.4711 = acre$
$m \times 3.2808 = ft$
$kg/ha \cdot d \times 0.8922 = lb/acre \cdot d$
$mg/L = g/m^3$

previously mentioned, Oswald[40] has proposed the construction of a special digestion chamber surrounding the inlet. The digestion chamber design was either a conical shape enlargement with increasing depth or a submerged baffle surrounding the digestion zone (Figure 2-73). A single inlet for a lagoon bottom width of 6.1–9.1 m (20–30 ft) is normally accepted.[62] Submerged outlets should be provided. The outlet invert is normally 0.61 m (2 ft) below the surface of the liquid.[62,66] Multiple outlets are recommended to reduce discharge velocities, which may scour the sludge blanket.

Wastewater containing high concentrations of grease may clog inlet pipes; therefore, a 0.305-m (12-in.) diameter inlet pipe with provisions for easy cleanout should be installed.[62] Because of the corrosive nature of anaerobic lagoon effluent, splitter and outlet boxes are covered with corrosion-resistant open grating and fiberglass.

Recirculation. Recirculation of sludge solids is provided in a two-stage anaerobic lagoon system. Sludge solids are recirculated from the bottom of the second pond to the inlet of the first-stage pond. Recirculation of solids stimulates an anaerobic contact process. No substantial advantage has been observed by recirculating sludge solids in anaerobic lagoon systems.[53,62]

Other Design Parameters

Summaries of the more commonly used design parameters for various types of wastewater stabilization ponds are presented in Tables 2-29 and 2-30. In general, the design of the pond systems discussed above were based on the parameters shown in Tables 2-29 and 2-30. Many modifications were employed, and these vary from state to state, usually related to climatic conditions.

References

1. Environmental Protection Agency. 1977. *Performance Evaluation of Existing Lagoons-Peterborough, New Hampshire.* EPA–600/2-77–085. EPA Technology Series. Municipal Environmental Research Laboratory, Cincinnati, Ohio.

2. Environmental Protection Agency. 1977. *Performance Evaluation of Kilmichael Lagoon.* EPA–600/2-77–109. EPA Technology Series. Municipal Environmental Research Laboratory, Cincinnati, Ohio.

3. Environmental Protection Agency. 1977. *Performance Evaluation of an Existing Lagoon System at Eudora, Kansas.* EPA–600/2-77–167. EPA Technology Series. Municipal Environmental Research Laboratory, Cincinnati, Ohio.

4. Environmental Protection Agency. 1977. *Performance Evaluation of an Existing Seven Cell Lagoon System.* EPA–600/2-77–086. EPA Technology Series. Municipal Environmental Research Laboratory, Cincinnati, Ohio.

5. Canter, L. W., and A. J. Englande. 1970. States' Design Criteria for Waste Stabilization Ponds. *JWPCF*, **42**, 10, 1840–1847.

6. McGarry, M. G., and M. B. Pescod. 1970. *Stabilization Pond Design Criteria for Tropical Asia.* 2nd International Symposium for Waste Treatment Lagoons, Missouri Basin Eng. Health Council, Kansas City, Missouri. Pp. 114–132.

7. Larsen,T. B. 1974. *A Dimensionless Design Equation for Sewage Lagoons.* Ph.D. Dissertation, University of New Mexico, Albuquerque, New Mexico.

8. Gloyna, Ernest F. 1976. Facultative Waste Stabilization Pond Design. In: *Ponds as a Wastewater Treatment Alternative.* Water Resources Symposium No. 9, edited by E. F. Gloyna, J. F. Malina, Jr., and E. M. Davis. Center for Research in Water Resources, University of Texas, Austin, Texas. Pp. 143–157.

9. Thirumurthi, Dhandapani. 1969. Design Principles of Waste Stabilization Ponds. *J. Sanit. Eng. Div. Am Soc. Civ. Eng.* 95, SA2, 311–330.

10. Wehner, J. F., and R. H. Wilhelm. 1956. Boundary Conditions of Flow Reactor. *Chem. Eng. Sci.* 6, 89–93.

11. Stratton, F. E. 1968. Ammonia Nitrogen Losses from Streams. *J. Sanit. Eng. Div., ASCE*, SA6.

12. Stratton, F. E. 1969. Nitrogen Losses from Alkaline Water Impoundments. *J. Sanit. Eng. Div., ASCE*, SA2.

13. King, D. L. 1978. The Role of Ponds in Land Treatment of Wastewater. *Proceedings International Symposium on Land Treatment of Wastewater*, held in Hanover, New Hampshire. Pp. 191–198.

14. Pano, A., and E. J. Middlebrooks. 1981. Ammonia Nitrogen Removal in Facultative Wastewater Stabilization Ponds. Report submitted to Center for Environmental Research Information, U.S. Environmental Protection Agency, Cincinnati, Ohio.

15. Walter, C. M., and S. L. Bugbee. 1974. *Progress Report-Blue Springs Lagoon Study Blue Springs, Missouri, from Upgrading Wastewater Stabilization Ponds to Meet New Discharge Standards.* Edited by E. J. Middlebrooks, Utah State University, UWRL PRWG 159-1, November.

16. Russel, J. S., E. J. Middlebrooks, and J. H. Reynolds. 1980. Wastewater Stabilization Lagoon-Intermittent Sand Filter Systems. EPA 600/2–80–032. Municipal Environmental Research Laboratory, U.S. Environmental Protection Agency, Cincinnati, Ohio.

17. Bhagat, S. K., and D. E. Proctor. 1969. Treatment of Dairy Manure by Lagooning. *J. Water Poll. Contr. Fed.*, **41**, 5, 785–795.

18. Ramani, R. 1976. Design Criteria for Polishing Ponds. In *Proceeding Water Resources Symposium No. 9.* Edited by E. F. Gloyna, T. F. Malina, and E. M. Davis, University of Texas, Austin.

19. Reid, G. W., and L. Straebin. 1979. *Performance Evaluation of Existing Aerated Lagoon System at Bixby, Oklahoma.* EPA–600/2–79–014. Municipal Environmental Research Laboratory, U.S. Environmental Protection Agency, Cincinnati, Ohio.

20. Gurnham, C. F., B. A. Rose and W. T. Fetherston. 1979. *Performance Evaluation of the Existing Three-Lagoon Wastewater Treatment Plant of Pawnee Il-*

linois. EPA–600/2–79–043. Municipal Environmental Research Laboratory, U.S. Environmental Protection Agency, Cincinnati, Ohio.

21. Englande, A. J., Jr. 1980. *Performance Evaluation of the Aerated Lagoon System at North Gulfport, Mississippi*. EPA–600/2–80–006. Municipal Environmental Research Laboratory, U.S. Environmental Protection Agency, Cincinnati, Ohio.

22. Polkowski, L. B. 1979. *Performance Evaluation of Existing Aerated Lagoon System at Consolidated Koshkonong Sanitary District, Edgerton, Wisconsin*. EPA–600/2–79–182. Municipal Environmental Research Laboratory, U.S. Environmental Protection Agency, Cincinnati, Ohio.

23. Earnest, C. M., E. A. Vizzini, D. L. Brown, and J. L. Harris. 1978. *Performance Evaluation of the Aerated Lagoon System at Windber, Pennsylvania*. EPA–600/2–78–023. Municipal Environmental Research Laboratory, U.S. Environmental Protection Agency, Cincinnati, Ohio.

24. Boulier, G. A., and T. J. Atchison. 1974. *Aerated Facultative Lagoon Process*. Hinde Engineering Company. Highland Park, Illinois.

25. Bartsch, E. H., and C. W. Randall. 1971. Aerated Lagoons—A Report on the State of the Art. *J. Water Poll. Contr. Fed.*, **43**, 4, 699–708.

26. Lewis, R. F. 1977. Personal communication, Environmental Protection Agency, Cincinnati, Ohio.

27. McKinney, R. E. 1970. *Second International Symposium for Waste Treatment Lagoons*. June 23–25, 1970, Kansas City, Missouri.

28. Metcalf & Eddy, Inc. 1979. *Wastewater Engineering*. McGraw-Hill, New York.

29. Oswald, W. J. 1976. *Syllabus on Waste Pond Design, Algae Project Report*. Sanitary Engineering Research Laboratory, University of California, Berkeley, California.

30. Andrews, J. F., and S. P. Graef. 1970. Dynamic Modeling and Simulation of the Anaerobic Digestion Process. *Advances in Chemistry*, Vol 105, 126–162, American Chemical Society, 1155 Sixteenth St., N. W., Washington, D.C. 20036.

31. Malina, J. F., Jr., and R. A. Rios. 1976. Anaerobic Ponds. In *Ponds as a Wastewater Treatment Alternative*. Edited by E. F. Gloyna, J. F. Malina, Jr., and E. M. Davis. Water Resources Symposium Number Nine. University of Texas Press, Austin, Texas. p. 131.

32 Kotzé, J. P., P. G. Thiel, D. F. Toerien, W. H. J. Attingh, and M. L. Siebert. 1968. A Biological and Chemical Study of Several Anaerobic Digesters. *Water Res.* **2**, 3, 195–213.

33. Chan, D. B., and E. A. Pearson. 1970. *Comprehensive Studies of Solid Waste Management: Hydrolysis Rate of Cellulose in Anaerobic Fermentation*. SERL Report No. 70–3. University of California, School of Public Health, Berkeley, California.

34. Hobson, P. N., S. Bousfield, and R. Summers. 1974. Anaerobic Digestion of Organic Matter. *Crit. Rev. Environ. Control*, **4**, 131–191.

35. Ghosh, S., J. R. Conrad, and D. L. Klass. 1974. Development of an Anaerobic

Digestion-Based Refuse Disposal Reclamation System. Paper presented at the 47th Annual Conference of the Water Pollution Control Federation, Denver, Colorado, October, 6–11.

36. Ghosh, S., and D. L. Klass. 1974. Conversion of Urban Refuse to Substitute Natural Gas by the BIOGAS ® Process. Paper presented at the Proceedings 4th Mineral Waste Utilization Symposium, Chicago, Illinois. May 7–8.

37. McKinney, R. E. 1974. State of the Art of Lagoon Wastewater Treatment. Symposium Proceedings Upgrading Wastewater Stabilization Ponds to Meet New Discharge Standards. Utah State University, Logan, Utah 84322.

38. Ramani, R. 1976. Design Criteria for Polishing Ponds. In *Ponds as a Wastewater Treatment Alternative*. Edited by E. F. Gloyna, J. F. Malina, Jr., and E. M. Davis. Water Resources Symposium Number Nine, University of Texas, Austin, Texas. p. 159.

39. Vincent, J. L., W. E. Algie, and G.v.R. Marais. 1961. A System of Sanitation for Low Cost High Density Housing. *Proceedings of a Symposium on Hygiene and Sanitation in Relation to Housing*. Niamei, London.

40. Oswald, W. J. 1968. Advances in Anaerobic Pond Systems Design. In *Advances in Water Quality Improvement*. Edited by E. F. Gloyna and W. W. Eckenfelder, Jr. University of Texas Press, Austin, Texas. p. 409.

41. Parker, C. D. 1970. Experiences with Anaerobic Lagoons in Australia. *Second International Symposium for Waste Treatment Lagoons*. Kansas City, Missouri. June 23–25. p. 334.

42. Parker, C. D., H. L. Jones, and N. C. Greene. 1959. Performance of Large Sewage Lagoons at Melbourne, Australia. *Sewage Industrial Wastes*, 31, 2, 133.

43. Eckenfelder, W. W. 1961. *Biological Waste Treatment*. Pergamon Press, London.

44. Cooper, R. C. 1968. Industrial Waste Oxidation Ponds. *Southwest Water Works J*. 5, 21.

45. Oswald, W. J., C. G. Golueke, and R. W. Tyler. 1967. Integrated Pond Systems for Subdivisions. *JWPCF*, 39, 8, 1289.

46. Parker, C. D., and G. P. Skerry, 1968. Function of Solids in Anaerobic Lagoon Treatment of Wastewater. *JWPCF*, 40, 2, 192.

47. Dornbush, J. N. 1970. State of the Art—Anaerobic Lagoons. *Second International Symposium for Wastewater Treatment Lagoons*. Kansas City, Missouri, June 23–25, p. 382.

48. Fisher, C. P., and E. F. Gloyna. 1965. Treatment of Activated Sludge in Stabilization Ponds. *JWPCF*, 37, 11, 1511.

49. Agricultural Waste Management Task Committee. 1976. *State of the Art Report—Animal Waste Management*. American Society of Civil Engineers, Environmental Engineering Division, Solid Waste Management Committee.

50. Kreis, D. R., et al. 1972. *Characteristics of Rainfall Runoff From a Beef Cattle Feedlot*. Robert S. Kerr, Water Research Center. Ada, Oklahoma. EPA–R2–72–061.

51. Hart, S. A., and M. E. Turner. 1965. Lagoons for Livestock Manure. *JWPCF* 37, 11, 1578.

52. Hart, S. A., and M. E. Turner. 1968. Waste Stabilization Ponds for Agricultural Wastes. In *Advances in Water Quality Improvement*. Edited by E. F. Gloyna and W. W. Eckenfelder, Jr. University of Texas Press, Austin, Texas. p. 457.

53. White, J. E. 1970. Current Design Criteria for Anaerobic Lagoons. *Second International Symposium for Waste Treatment Lagoons*. Kansas City, Missouri. June 23–25. p. 360.

54. van Lotringen, T. J. M., and J. B. Gerrish. 1978. H_2S Removal by Purple Sulfur Bacteria in Swine Waste Lagoons. *Proceedings of the 33rd Industrial Waste Conference*. Purdue University, Lafayette, Indiana. May 9–11. p. 440.

55. Nemerow, N. L. 1978. *Industrial Water Pollution: Origins, Characteristics, and Treatment*. Addison-Wesley, Reading, Mass.

56. Bhagat, S. K., and D. E. Proctor. 1969. Treatment of Dairy Manure by Lagooning. *JWPCF*, **41**, 5, 785.

57. Dornbush, J. N. and Anderson, J. R. 1964. Lagooning of Livestock in South Dakota. *Proceedings of the 19th Industrial Waste Conference*. Purdue University, Lafayette, Indiana, p. 317.

58. Ricketts, R. L. 1960. Lagoons for Hog Feeding Floors. *Proceedings of Symposium on Waste Stabilization Lagoons*. Kansas City, Missouri, Aug. 1–5.

59. Oldham, W. K., and L. Nemeth. 1973. Anaerobic Lagoons for Treatment of High-Strength Organic Wastes. *JWPCF*, **45**, 11, 2397.

60. Loehr, R. C., and J. A. Ruf. 1968. Anaerobic Lagoon Treatment of Milking-Parlor Wastes. *JWPCF*, **40**, 1, 83.

61. Clark, C. E. 1965. Hog Waste Disposal by Lagooning. *ASCE J. Sanit. Engr. Div.* 4467, SA6, 27.

62. Hammer, M. J., and C. D. Jacobson. 1970. *Anaerobic Lagoon Treatment of Packinghouse Wastewater*. Presented at the Second International Symposium for Waste Treatment Lagoons. Kansas City, Missouri. June 23–25.

63. Ludwig, H. F. 1964. Industry's Idea Clinic. *JWPCF*, **36**, 8, 937.

64. Rollag, D. A., and J. N. Dornbush. 1966. Design and Performance Evaluation of an Anaerobic Stabilization Pond System for Meat-Processing Wastes. *JWPCF*, **38**, 11, 1805.

65. Wymore, A. H., and J. E. White. 1968. Treatment of a Slaughterhouse Waste Using Anaerobic and Aerobic Lagoons. *Proceedings of the 23rd Industrial Waste Conference*. Purdue University, Lafayette, Indiana. May 7–9. p. 601.

66. Enders, K. E., M. J. Hammer, and C. L. Weber, 1968. Anaerobic Lagoon Treatment of Slaughterhouse Waste. *Water Sewage Works*, **6**, 283.

67. Niles, C. F., and H. P. Gordon. 1970. Operation of an Anaerobic Pond on Hog Abattoir Wastewater. *Proceedings of the 25th Industrial Waste Conference*. Purdue University, Lafayette, Indiana. May 5–7. p. 612.

68. Coerver, J. F. 1964. Anaerobic and Aerobic Ponds for Packinghouse Waste Treatment in Louisiana. *Proceedings of the Nineteenth Industrial Waste Conference*, Purdue University, Lafayette, Indiana. May 5–7.

69. Stanley, D. R. 1966. Anaerobic and Aerobic Lagoon Treatment of Packing Plant Wastes. *Proceedings of the 21st Industrial Waste Conference*. Purdue University, Lafayette, Indiana. May 3–5. p. 275.

70. Steffen, A. J. 1963. Stabilization Ponds for Meat Packing Wastes. *JWPCF*, **35**, 4, 440.

71. Sollo, F. W. 1960. Pond Treatment of Meat Packing Plant Wastes. *Proceedings of Symposium on Waste Stabilization Lagoons*. Kansas City, Missouri. August 1–5.

72. Parker, C. D. 1976. Pond Design for Industrial Use in Australia with Reference to Food Wastes. In *Ponds as a Wastewater Treatment Alternative*, Water Resources Symposium Number Nine. University of Texas, Austin, Texas. p. 285.

73. Dorris, T. C., D. Patterson, and B. J. Copeland. 1963. Oil Refinery Effluent Treatment in Ponds. *JWPCF*, **35**, 7, 932.

74. Hovious, J. C., R. A. Conway, and C. W. Ganze. 1973. Anaerobic Lagoon Pretreatment of Petrochemical Wastes. *JWPCF*, **45**, 1, 71.

75. Woodley, R. A., and T. F. Brown. 1971. Anaerobic Lagoon Treatment of Low Sulfate Chemical Wastes. *Proceedings of the 26th Industrial Waste Conference*. Purdue University, Lafayette, Indiana. May 4–6. p. 844.

76. Voege, F. A., and D. R. Stanley. 1963. Industrial Waste Stabilization Ponds in Canada. *JWPCF*, **35**, 8, 1019.

CHAPTER
3

Alternatives for Upgrading Lagoon Effluents

WASTEWATER STABILIZATION PONDS are effective in reducing BOD_5 and have only one basic disadvantage, that is, high concentrations of solids in the effluents. These solids leave the lagoon along with the other constituents and can create problems in receiving streams. Concentrations of suspended solids can exceed 100 mg/l, but, as shown in the detailed performance data presented in Chapter 2, such high concentrations are usually limited to 2 to 4 months during the year. However, during some months of the year, the suspended solids (SS) concentration exceeds the standard specified by most regulatory agencies. With the new standards proposed in the September 2, 1976, issue of the Federal Register (see Chapter 1), small flow systems were excluded from the SS effluent requirements, provided that these solids are in the form of algae. In areas in which water-quality-limited streams occur, it is presumed that algae removal will be required.

A discussion is presented of the various alternatives available for upgrading existing lagoons and designing original systems to meet water quality standards. Many good sources of information exist on upgrading lagoons.[1-3]

Land Application Systems

Land application of wastewater has been the subject of numerous EPA publications.[4-8] These reports describe the type of systems available, provide engineering criteria, information on costing and performance expectations, and assess potential health issues.

There are a variety of land application systems using wastewater effluents for different purposes. If treatment or further renovation is the major goal, the systems can be classified in three major categories, according to the rate of application and the pathway for treated water:

Slow Rate (SR). Similar to conventional irrigation practice, application of the wastewater is via sprinklers, surface flooding, and so on. Application rates vary from 0.61 to 6.1 m (2–20 ft) per year, and SR is the most commonly used land treatment concept. The type of wastewater applied to the land varies from raw sewage to tertiary effluent. Surface vegetation is a critical component in the renovative process, and all types of crops have been used: forests, greenbelts, golf courses, and the like. The applied water percolates downward to underdrains and/or the groundwater table. Soil types used most often are medium to fine textured, with moderate permeability. Climate constraints on crops and the like may require winter storage of the wastewater.

Rapid Infiltration (RI). RI usually consists of intermittent flooding with wastewater of shallow basins in relatively coarse textured soils of rapid permeability, and the application rates vary from 6.1 to 183 m (20–600 ft) per year. Vegetation is not a critical factor in the renovative process, and water movement is via percolation to underdrains, recovery wells, or groundwater. The type of wastewater applied varies from raw sewage to tertiary effluent, and climate is not usually a constraint on application.

Overland Flow (OF). Application techniques similar to those employed in SR are used, but the wastewater is applied to gently sloping fields on essentially impermeable soils. The water moves via sheet flow down the slope to collection ditches at the toe of the slope, and vegetation is an essential component in systems using water tolerant grasses. Application rates vary from 3 to 21 m (10 to 70 ft) of water per year, and the wastewater applied varies from raw sewage to secondary effluent. Winter wastewater storage may be needed in cold climates. Overland flow is a relatively new concept compared to SR and RI, and the first successful performance was with industrial effluents (i.e., food processing). The process has been successfully demonstrated with municipal wastewaters.

These basic concepts are illustrated in Figure 3–1. Tables 3–1 and 3–2 summarize comparative site characteristics and design features. All three methods are potentially usable alternatives for polishing municipal wastewater stabilization pond effluents. The selection of a system for a particular situation is dependent on the availability of land, with the characteristics summarized in Table 3–1 as well as cost considerations and public acceptance. The expected effluent quality from these land treatment systems is listed in Table 7–2 (Chapter 7), along with the estimated annual energy needs for a typical 3785 m³/day (1 mgd) system.

In humid climates slow rate systems are generally designed for the

Table 3-1
Comparison of Site Characteristics for Land Treatment Processes

	Slow Rate	Rapid Infiltration	Overland Flow
Grade	Less than 20% on cultivated land; less than 40% on noncultivated land	Not critical; excessive grades require much earthwork	Finish slopes 2-8%[a]
Soil permeability	Moderately slow to moderately rapid	Rapid (sands, sandy loams)	Slow (clays, silts, and soils with impermeable barriers)
Depth to groundwater	0.6-1 m (minimum)[b]	0.3 m during flood cycle; 1.5-3 m during drying cycle	Not critical[c]
Climatic restrictions	Storage often needed for cold weather and precipitation	None (possibly modify operation in cold weather)	Storage often needed for cold weather

Source: From Ref. 4.

[a] Steeper grades might be feasible at reduced hydraulic loadings.
[b] Underdrains can be used to maintain this level at sites with high groundwater table.
[c] Impact on groundwater should be considered for more permeable soils.

Table 3-2
Comparison of Typical Design Features for Land Treatment Processes

Feature	Slow Rate	Rapid Infiltration	Overland Flow
Application techniques	Sprinkler or surface[a]	Usually surface	Sprinkler or surface
Annual application rate (m)	0.5-6	6-125	3-20
Field area required (ha)[b]	23-280	3-23	6.5-44[c]
Typical weekly application rate (cm)	1.3-10	10-240	6-40[c]
Minimum preapplication treatment provided in the United States	Primary sedimentation[d]	Primary sedimentation	Grit removal and comminution[e]
Disposition of applied wastewater	Evapotranspiration and percolation	Mainly percolation	Surface runoff and evapotranspiration with some percolation
Need for vegetation	Required	Optional	Required

Source: From Ref. 4.

[a] Includes ridge-and-furrow and border strip.
[b] Field area in hectares not including buffer area, road, or ditches for 3,785 m^3/d (1 Mgal/d) flow.
[c] Range includes raw wastewater to secondary effluent, higher rate for higher level of preapplication treatment.
[d] With restricted public access; crops not for direct human consumption.
[e] With restricted public access.

FIGURE 3-1
Methods of land application (Ref. 4).

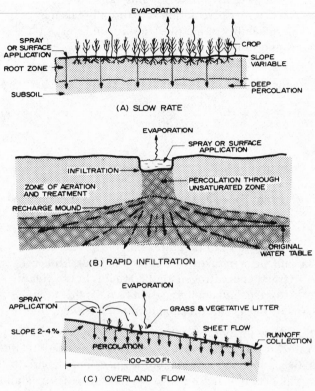

maximum possible hydraulic loading to minimize land requirements. Systems of this type have been successfully designed for forests, pastures, forage grasses, corn, and other crop production. In arid climates in which water conservation is more critical, wastewater is often applied at rates that just equal the irrigation needs of the crop. Water rights must also be given careful consideration in arid climates prior to diverting pond effluent to another location for land treatment.

Design of SR systems for typical municipal effluents is usually based on either the limiting permeability of the *in situ* soils or on meeting nitrate requirements in the groundwater, if the site overlies a potable aquifer. The wastewater hydraulic loading rate is calculated for both conditions, and the limiting value then controls design.

Hydraulic loading based on permeability can be calculated as:

$$Lw = Pw + Et - Pr \tag{3-1}$$

where

Lw = wastewater hydraulic loading rate
Pw = wastewater percolation rate (usually taken as 4–10% of SCS permeability of most restricting soil layer)
Et = evapotranspiration rate
Pr = precipitation rate

Hydraulic loading based on nitrogen limits is calculated as:

$$Lw_N = \frac{Cp(Pr - Et) + (A)U}{(1 - f)(Cn) - Cp} \qquad (3\text{--}2)$$

where

Lw_N = hydraulic loading based on nitrogen limitations (cm or in.)
A = constant (10 metric; 4.43 English units)
Cp = design concentration of total nitrogen in percolate, usually taken as 10 mg/l unless specified otherwise
Pr = precipitation rate (cm or in.)
Et = evapotranspiration rate (cm or in.)
U = nitrogen uptake of cover crop (kg/ha), (lb/acre)
f = fraction of nitrogen applied that is lost to volatilization, soil storage, and denitrification (usually 0.1–0.25)
Cn = nitrogen concentration in applied effluent (mg/l)

The time base for both equations can be days, weeks, months, or a year. The rates are then expressed as a depth of water over that time period. It is typical to conduct the first calculation on an annual basis to determine the limiting value, and then to repeat on a monthly basis for detailed design. Reference 4 presents additional detail on these design procedures.

Experience with rapid infiltration systems for polishing pond effluents is limited, but indications are that it is a suitable alternative. The lower portion of the hydraulic loading range in Table 3–2 is suggested to avoid algae clogging problems. The high hydraulic loadings preclude meeting stringent groundwater nitrate nitrogen requirements without special operational management. It is typical to locate the infiltration basins over nonpotable aquifers, or in such a way that percolate emerges as subflow into a nearby surface water, or to recover the percolate with wells or drains for other use.

An estimate of the annual hydraulic loading can be calculated with Equation 3–1. Reference 4 provides complete detail on process design. Successful operation is based on brief periods of basin flooding followed by drying to renew the infiltration capacity of the soil surface. A typical design for a moderate climate might be based on 2 days of flooding followed by 12 days of drying time using seven (or some multiple) basins. Colder climates will require lower hydraulic loading or longer drying times in the winter months. For most sites a thorough geohydrological study is recommended to ensure satisfactory soil conditions and to confirm design assumptions regarding rate and direction of subsurface water movement.

Infiltration testing is also recommended prior to final design. Based on actual test results the Pw value in Equation 3–1 can be taken as 10–15% of the steady-state infiltration test results for hydraulic loading design. In the western United States the legal aspects of water rights can be an issue for the location and use of rapid infiltration systems. Reference 4 provides further details on design, construction, and operation of this process.

Unlike the two other concepts, the overland flow process collects and usually discharges the treated effluent to surface waters. Treatment is dependent on the physical and biochemical activity in the near surface environment on the slope. The process is analogous to conventional attached growth biological systems (trickling filters, RBC, etc.), and reactions seem to follow first-order kinetics.

The overland flow process has been used successfully to treat industrial wastewaters from food-processing operations in a variety of locations in the United States. Both the U.S. EPA and the Army Corps of Engineers have conducted research and operated full-scale pilot units with municipal wastewaters ranging from raw sewage to lagoon and other secondary process effluents. A similar concept, termed grass filtration, has been in successful use in England for many decades for final polishing of conventional treatment plant effluents. Table 3–3 presents a summary of performance at systems in the United States receiving pond effluents. Table 3–4 is a summary of the English experience.

Most existing systems in the United States have been designed within the range of criteria presented in Table 3–2, with grades from 2 to 8%, slope lengths from 30 to 61 m (100–200 ft), and hydraulic loadings ranging from 3 to 18 m (10–60 ft) per year. A system designed for lagoon effluents would typically be near the midrange of hydraulic loadings. Sprinklers require longer slope lengths than gravity distribution but offer an advantage of more uniform distribution for high solids content wastewater. Any grade within the range specified should function successfully. The choice is

Table 3-3
Performance of Overland Flow Systems
Treating Stabilization Pond Effluents

Location	Slope Length (M)	Slope Grade (%)	Hydraulic Loading (CM/WK)	Average BOD (MG/L) In	Out	Average SS (MG/L) In	Out	Average N (MG/L) In	Out	Average P (MG/L) In	Out
Davis, CA	30	2	14	68	25	53	19				
Utica, MS	46	2–8	6.5	22	10	35	15	20	2	10	5
Carbondale, IL	60	7–12	44	29	7	28	24	16	6	3	1
Easley, SC	53	6	15	28	15	60	40	4	2	7	2

SOURCE: From Ref. 7.

Table 3-4
Performance Data for Irrigation over Grassland at Some British Treatment Plants

Total Area (Acres)	Loading[a] (g/ft²·day)	BOD (mg/L) In	Out	SS In	Out	Ammonia Nitrogen In	Out	Oxidized Nitrogen In	Out	Dissolved Oxygen In	Out	Comments
2.6	8.4[b]	24	10	56	15	10.5	8.4	24.3	21.9			No special vegetation. 97% removal of Coliaerogenes
2.5	2.3[b]	16	8	34	14	4.5	2.3	39.5	39.8			90% reduction in Coliaerogenes
0.7	1.1[b]	23	9	40	17	8.6	4.9	42.4	41.4			Special grass sown
1.7	2.0	18	11	25	10	13.5	9.3					Eight plots; two in use for 1 week, rested for 3 weeks
—	1.7	19	7	12	4	4.2	3.8	25.3	20.6	68	65	Spray irrigation. Small works serving population of 5000
24.0	2.0	18	5	17	10	25.6	25.6	6	6	2	4	Grass sown every 6 months
13.0	0.8	20	6	27	9	11.1	7.1	14.6	7.3			Plots underdrained; 2 plots fed alternately for 2 months
2.0	1.6	12	7	24	11	7.8	6.5	28.1	27.2	4.7	7.3	30% reduction in E. coli
16.0	2.1	15	7	11	6	24.1	22.4					

(Dissolved Oxygen values of 68/65 and 2/4 are Percent saturation.)

SOURCE: From Refs. 12, 13.

[a] Loading based on total area

[b] Area actually in use (loading based on this).

Notes: 1 acre = 0.4047 ha; 1 g/ft²·day = 0.04074 m³/m²·day

132

dependent on existing site topography. The final constructed surface must be smooth and maintained in that condition. Ruts and erosion channels will divert the sheet flow and degrade overall performance.

Rational procedures, based on process kinetics, are becoming available for design of overland flow systems.[7,11] Two similar approaches have been developed at the University of California, Davis, and at the U.S. Army Cold Regions Research and Engineering Laboratory. The basic design equation is:

$$\frac{M_L}{M_o} = Ae^{(-KL/S^{\frac{1}{3}}qm)} \tag{3-3}$$

where

M_o = mass of BOD applied to slope (kg)
M_L = mass of BOD in overland flow effluent (kg)
A = constant = 0.52
K = rate constant = 0.00234
S = grade of overland flow slope (m/m)
L = slope length, in flow direction (m)
qm = average flow rate on slope = q(applied) + q (runoff) $\div$ 2 (m³/hr per m of slope width)

The approach is based on providing adequate detention time on the slope for removal to occur. The equation provides a method to vary the hydraulic loading to suit existing site conditions and thereby minimize earthwork. Details on this design procedure, climatic constraints, and expected performance for other wastewater constituents are available in Reference 11.

Water-tolerant grasses are an essential component on the overland flow slope. Mixtures of Reed canary grass, Alta fescue, Redtop, Dallis grass, and Rye grass have been successfully used. The Rye grass will emerge quickly and stabilize the slope for erosion protection. Eventually, the water-tolerant types such as Reed canary grass will dominate. Regular harvest of this grass cover is essential for successful performance. If nitrogen removal is one of the design goals for the system, the harvested grass must be removed; otherwise the cuttings can remain on the slope.

Both overland flow and SR systems depend on vegetation and near-surface microbial activity. As a result, performance is inhibited in sustained cold weather, and storage of wastewater may be required during the extreme winter months. Estimates of the storage needs for overland flow and SR systems, based on climatic constraints, can be obtained from Reference 4. If the land treatment system is used only on a seasonal basis for polishing of pond effluents in the summer months, then these storage requirements do not apply.

Intermittent Sand Filtration to Upgrade Lagoon Effluents

Intermittent sand filtration is capable of polishing lagoon effluents at relatively low cost. It is not a new technique; rather, it is the application of an old technique to the problem of upgrading lagoon effluents. It is similar to the practice of slow sand filtration in potable water treatment or the slow sand filtration of raw sewage which was practiced during the early 1900s. Intermittent sand filtration of lagoon effluents is the application of lagoon effluent on a periodic or intermittent basis to a sand filter bed. As the wastewater passes through the sand filter bed, SS and organic matter are removed through a combination of physical straining and biological degradation processes. The particulate matter collects in the top 5–7.5 cm (2–3 in.) of the sand filter bed. This buildup of organic matter eventually clogs the top 5–7.5 cm (2–3 in.) of the sand filter bed and prevents the passage of the effluent through the sand filter. The sand filter is then taken out of service, and the top layer of clogged sand is removed. The sand filter is then put back into service, and the spent sand is either discarded or washed and used as replacement sand for the sand filter.

A historical perspective and review of intermittent sand filtration has been compiled by Hill et al.[14] and Harris et al.[15]

Single-Stage Filtration

The development of intermittent sand filtration to upgrade lagoon effluents has been conducted primarily at Utah State University.[14,16–22] The work at Utah State University has been conducted on laboratory-scale, pilot-scale, and field-scale filters; however, the principal work was conducted on six prototype scale filters shown in Figure 3–2. Each of these filters was operated at a different hydraulic loading rate for 12 months. The filter sand employed in the study was ungraded pit run sand with an effective size of 0.17 mm (0.007 in.) and a uniformity coefficient of 9.74 (see Table 3–5).

Table 3-5
Sieve Analysis of Filter Sand[a]

U.S. Sieve Designation Number	Size of Opening (mm)	Percent Passing (%)
3/8	9.5	100.0
4	4.76	92.1
10	—	61.7
40	0.42	27.0
100	0.149	6.2
200	0.074	1.7

[a] $e = 0.170$ mm; $u = 9.74$.

FIGURE 3-2
Cross section of a typical intermittent sand filter
(Ref. 16).

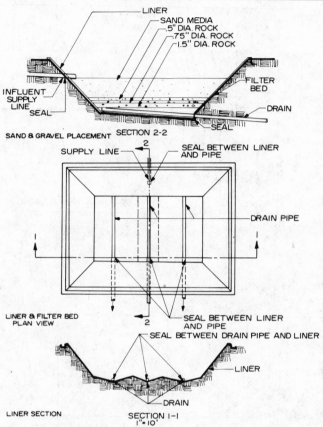

The BOD$_5$ removal performance for each of the six filters is shown in Figure 3–3. The overall average influent BOD$_5$ concentration was 19 mg/l. The influent BOD$_5$ concentration ranged from 4 mg/l to over 288 mg/l, exceeding 5 mg/l for 94 % of the time. Figure 3–3 shows the consistently high quality of filter effluent, which was unaffected by influent BOD$_5$ fluctuations. Effluent quality was below 5 mg/l 93 % of the time (except filter 2 during the winter). The effluent BOD$_5$ concentration never exceeded 12 mg/l and exceeded 7 mg/l only four times during the study.

The winter operation of filter 2 was different from the other filters. Filter 2 was constantly flooded during winter operation. This constant-flooded mode was an attempt to improve winter operation; however, the anaerobic condition that developed because of this constant flooding greatly reduced the filter efficiency. The effluent BOD$_5$ concentration for filter 2 exceeded 5 mg/l 92 % of the time. At the end of the winter operation,

FIGURE 3-3
Single-stage intermittent sand filtration BOD$_5$ removal performance
(Ref. 18).

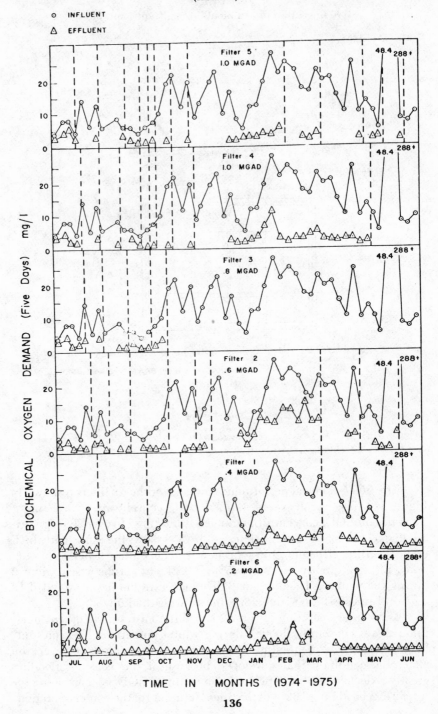

filter 2 was returned to normal operation, and as a result the filter effluent returned to normal when compared to the other filters in the study.

SS removal performance of the filters is shown in Figure 3–4. The influent exceeded 5 mg/l SS concentration 100% of the time and was greater than 30 mg/l 44% of the time. The average influent SS concentration was 31 mg/l, with a low of 6 mg/l and a high of 130 mg/l. The three filters in operation at the time of maximum influent SS (130 mg/l) had effluent SS concentrations of less than 3 mg/l.

The initial loading period (Figure 3–4) shows a relatively high concentration of SS in the filter effluent. This is primarily due to the "washing" of fine dirt (inorganic fines) from sand during the startup period. After this initial startup period, the filter effluent exceeded 5 mg/l suspended solids only 15% of the time and had averages of less than 6 mg/l. Seldom did the effluent exceed 10 mg/l. The anaerobic conditions of filter number 2 during winter operations caused its effluent to exceed 5 mg/l 83% of the time. The SS concentrations from each filter were consistently low.

The length of filter runs achieved during this study was a function of the hydraulic loading rate and the influent SS concentration. Filter run lengths ranged from 8 days at a hydraulic loading rate of 9360 m³/ha·day (1.0 mgad) during the summer months to 188 days at a hydraulic loading rate of 1872 m³/ha·day (0.2 mgad) during winter months.

Intermittent sand filters should be similar in design of those illustrated in Figure 3–2, with hydraulic loading rates of 2744–5616 m³/ha·day (0.4–0.6 mgad). Filter sands should have an effective size of 0.15–0.25 mm (0.006–0.010 in.) with a uniformity coefficient from 1.5 to 10.[23] The expected filter run length from intermittent sand filters designed on the above criteria will depend on the filter influent quality, but it should be a minimum of 30–60 days.

These results[18,19,24] clearly indicate that intermittent sand filtration of lagoon effluents can produce a final effluent with a BOD_5 concentration and a SS concentration of less than 15 mg/l consistently throughout the entire year, even during periods of subzero temperatures.

Effect of Sand Size on Single-Stage Performance

Tupyi et al.[22] studied the effect of various size filter sands on intermittent sand filter performance. The work employed the same filter system as Harris et al.[18] (see Figure 3–2), except the effective size of the filter sand employed for the various filters was 0.68, 0.40, 0.31, and 0.17 mm. Hydraulic loading rates ranging from 1871 m³/ha·day (0.2 mgad) to 28,061 m³/ha·day (3.0 mgad) were studied. In addition, application rates ranging from 0.008 m³/sec (0.29 cfs) to 0.048 m³/sec (1.68 cfs) were investigated.

The BOD_5 performance of all the intermittent sand filters with respect to various effective size sands, hydraulic loading rates, and application rates is recorded in Table 3–6 and illustrated in Figure 3–5. Yearly average

FIGURE 3-4
Single-stage intermittent sand filtration SS removal performance (Ref. 18).

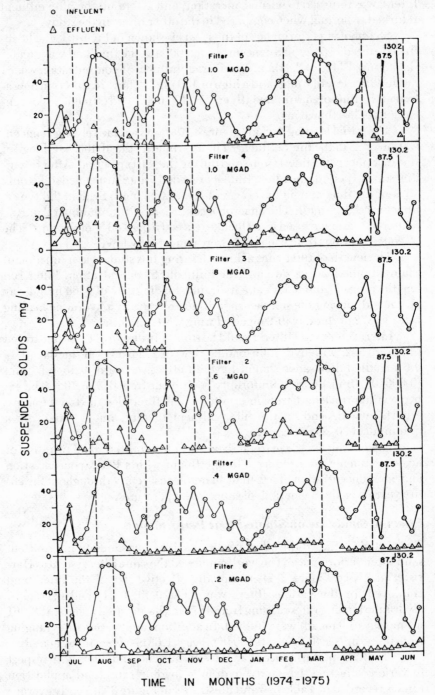

138

Table 3-6
A Summary of the BOD₅ Performance

Effective Size Filter Sand (mm)	Hydraulic Loading Rate (m³/H·D)	Application Rate (m³/sec)	Influent BOD₅ (mg/L)			Effluent BOD₅ (mc/L)			Average Percent Removal
			Minimum	Maximum	Average	Minimum	Maximum	Average	
0.17	1,870.8	0.048	2.6	22.3	10.9	0.3	3.7	1.1	90.1
0.17	3,741.6	0.048	2.6	22.3	11.5	0.1	7.4	2.6	77.2
0.31	9,353.9	0.048	4.5	20.8	13.8	5.0	11.2	7.8	43.5
0.31	9,353.9	0.008	9.7	9.7	9.7	6.1	6.1	6.1	33.7
0.40	9,353.9	0.048	2.6	22.3	10.5	3.7	17.8	8.2	21.9
0.40	9,353.9	0.008	4.5	19.8	12.2	3.7	11.0	5.4	56.0
0.40	14,030.9	0.048	9.8	11.5	10.7	3.9	6.0	5.0	53.3
0.40	18,707.9	0.048	2.6	22.3	11.3	2.6	23.3	8.6	23.9
0.40	28,061.9	0.048	9.8	9.8	9.8	4.5	4.5	4.5	54.6
0.68	9,353.9	0.048	2.6	22.3	11.6	4.1	17.3	8.2	28.8
0.68	9,353.9	0.008	3.7	20.8	13.2	2.9	15.4	7.9	39.8
0.68	14,030.9	0.048	4.4	12.9	7.8	4.0	6.8	5.7	27.5
0.68	18,707.9	0.048	2.6	22.3	11.5	3.2	15.6	9.0	21.0
0.68	28,061.9	0.048	6.3	12.9	9.4	2.9	11.9	6.8	27.1
						Loaded with primary lagoon effluent twice weekly			
0.40	9,353.9	0.008	9.1	75.6	26.8	3.7	28	10.5	60.8

Source: From Ref. 22.

139

FIGURE 3-5

The weekly BOD₅ performance by intermittent sand filters with various effective size sands.

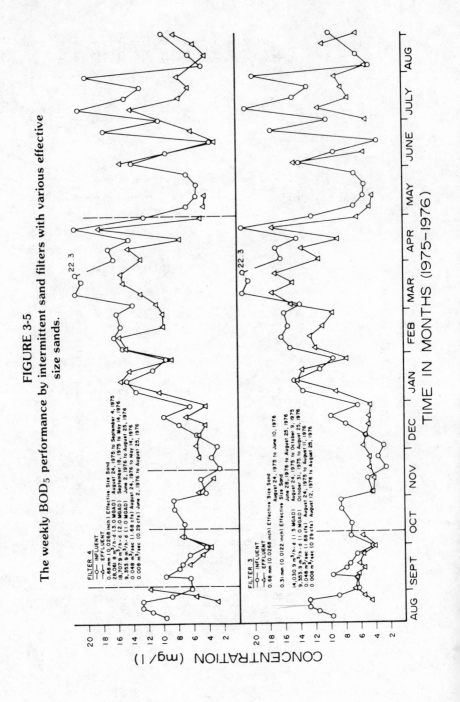

TIME IN MONTHS (1975-1976)

(continued)

FIGURE 3-5 Continued

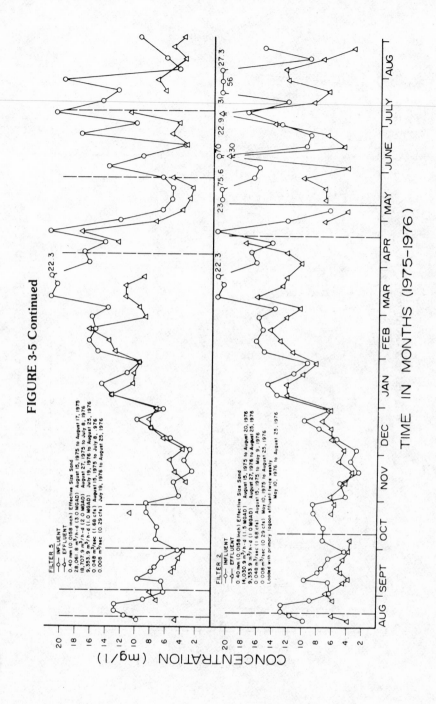

(continued)

141

FIGURE 3-5 Continued

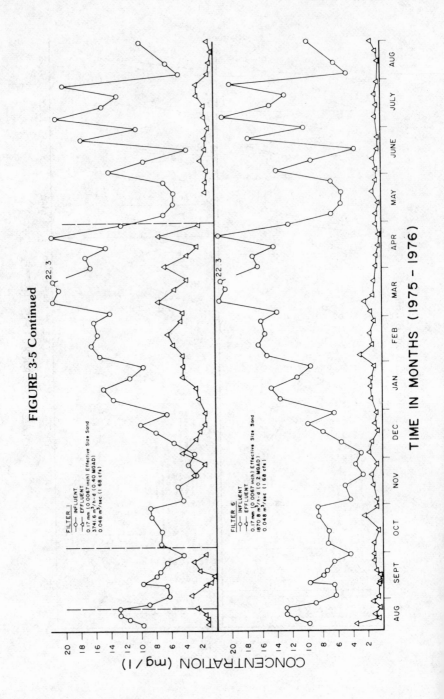

TIME IN MONTHS (1975 – 1976)

142

BOD_5 concentration in the effluent applied to the filters was 11 mg/l, with the daily BOD_5 concentration ranging from 3 mg/l to 22 mg/l throughout the study.

The BOD_5 removal performance of the 0.40-mm and 0.68-mm effective size sand (filters 2, 3, 4, and 5) with a high application rate of 0.048 m³/sec (1.68 cfs) was not adequate to produce an effluent that was consistently less than 10 mg/l. The 0.31 mm effective size sand (filter 3) produced a significant BOD_5 removal; however, the influent characteristics during this phase of study indicate that these results were inconclusive. Lowering the application rate on the 0.40-mm effective size sand (filter 5) and the 0.68-mm effective size sand (filter 4) appeared to increase BOD_5 removal; however, the zooplankton in the influent during that experiment make such a conclusion questionable. The 0.17-mm effective size sand (filters 1 and 6) was shown to be capable of high BOD_5 removal at low hydraulic loading rates of 3742 m³/ha·day (0.4 mgad) and 1871 m³/ha·day (0.2 mgad). No conclusion can be established with relation to the Federal Secondary Treatment Standards, which require an effluent BOD_5 of 30 mg/l or less, because the influent BOD_5 concentration did not exceed 23 mg/l during the entire study period.

SS removal by intermittent sand filters with various effective size filter sands, hydraulic loading rates, and application rates are shown in Table 3–7 and Figure 3–6. The mean yearly influent SS concentration (secondary lagoon effluent) was 23 mg/l, and the daily SS concentration ranged from 3 mg/l to 65 mg/l.

The 0.68, 0.40, and 0.31 mm effective size sand (filters 2, 3, 4, and 5) with a high application rate of 0.048 m³/sec (1.68 cfs) were unable to satisfy a 10 mg/l standard more than 50% of the time. Lowering the application rate to 0.008 m³/sec (0.29 cfs) on the 0.68 and 0.40 mm effective size sand filters (filters 4 and 5) increased SS removal performance, meeting a 10 mg/l standard a minimum of 67% of the time. The indication that influent SS significantly influence effluent SS concentrations preclude the use of these filter sands to satisfy stringent discharge standards. It appears that lower application rates increase SS removal, but a definite conclusion cannot be reached, because of the short period of study at the lower application rates and the heavy growth of *Daphnia* in the secondary lagoon effluent during the low application rate study.

The 0.40-mm effective size sand (filter 2) with a hydraulic loading rate of 9354 m³/ha·day (1.0 mgad) and a low application rate of 0.008 m³/sec (0.29 cfs) loaded with primary lagoon effluent twice weekly produced high SS removals. SS removal averaged 76% during the study and further indicates that application rate may have a definite effect on SS removal. However, operation of this filter does not represent normal single-stage intermittent sand filter operation, since lagoon effluent was applied to the filter only twice weekly, rather than daily.

Table 3-7
A Summary of the Suspended Solids Performance

Effective Size Filter Sand (mm)	Hydraulic Loading Rate (m³/H·D)	Application Rate (m³/sec)	Influent SS (mg/L)			Effluent SS (mg/L)			Average Percent Removal
			Minimum	Maximum	Average	Minimum	Maximum	Average	
0.17	1,870.8	0.048	3.2	74.3	23.0	0.6	23.8	2.7	88.2
0.17	3,741.6	0.048	3.2	74.3	20.8	0.3	17.6	3.5	83.0
0.31	9,353.9	0.048	8.2	64.8	27.8	7.7	29.2	15.4	44.6
0.31	9,353.9	0.008	20.0	20.1	20.1	10.2	21.0	15.6	22.4
0.40	9,353.9	0.048	3.2	51.5	18.7	1.2	30.8	13.1	30.1
0.40	9,353.9	0.008	12.2	35.9	21.8	1.8	16.0	7.5	65.5
0.40	14,030.9	0.048	34.3	44.9	39.6	10.7	12.5	11.6	70.7
0.40	18,707.9	0.048	3.2	64.8	18.1	1.2	39.8	11.6	35.9
0.40	28,061.9	0.048	44.9	44.9	44.9	83.0	83.0	83.0	0
0.68	9,353.9	0.048	3.2	51.5	15.8	1.6	25.2	11.2	29.3
0.68	9,353.9	0.008	8.9	74.3	34.1	2.9	39.7	15.5	54.7
0.68	14,030.9	0.048	17.7	51.6	38.2	7.5	30.0	19.6	48.6
0.68	18,707.9	0.048	3.2	51.5	16.7	3.2	24.4	13.1	21.6
0.68	28,061.9	0.048	33.2	51.6	44.9	19.3	57.8	35.4	21.1
			Loaded with primary lagoon effluent twice weekly						
0.40	9,353.9	0.008	10.5	70.7	34.0	3.2	18.0	7.9	76.7

SOURCE: From Ref. 22.

FIGURE 3-6

The weekly SS performance by intermittent sand filters with various effective size sands.

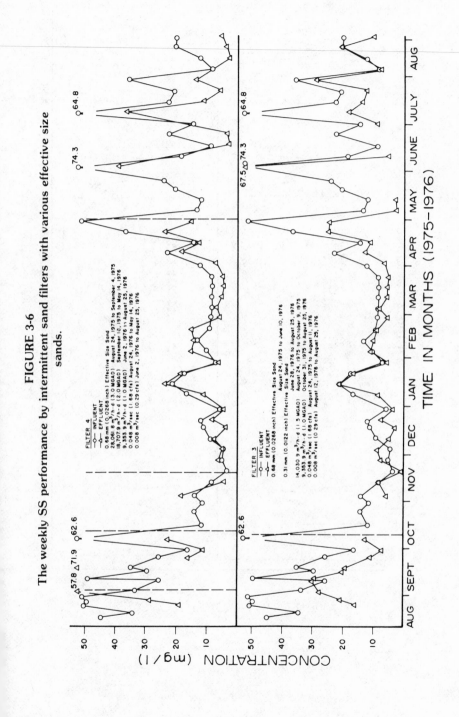

TIME IN MONTHS (1975-1976)

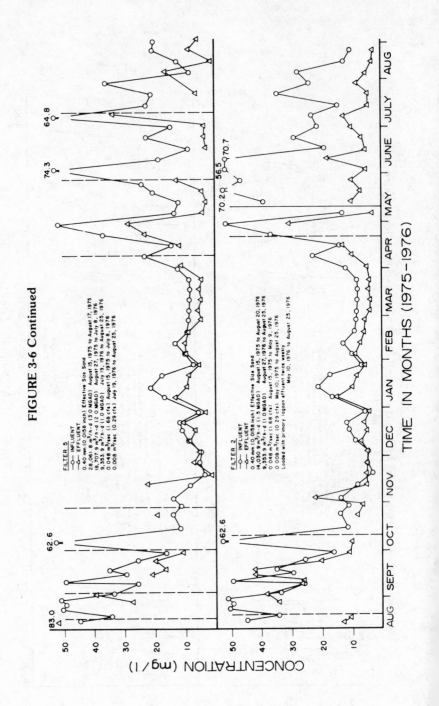

FIGURE 3-6 Continued

TIME IN MONTHS (1975-1976)

146

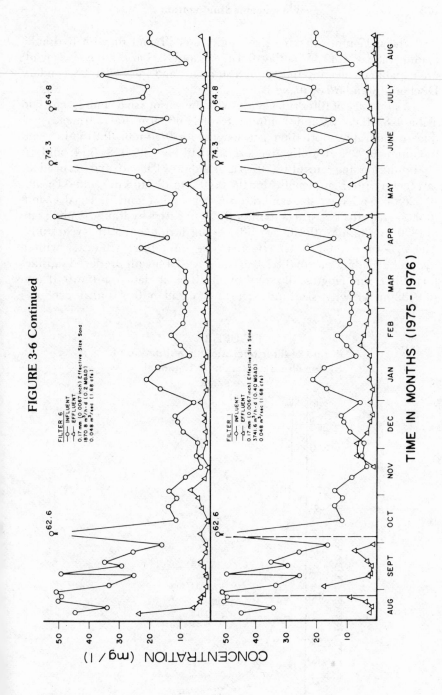

FIGURE 3-6 Continued

The 0.17-mm effective size sand (filters 1 and 6) with hydraulic loading rates of 3742 m³/ha·day (0.4 mgad) and 1871 m³/ha·day (0.2 mgad) are capable of meeting the 10 mg/l standard and the Federal Secondary Discharge Standard of 30 mg/l.

A summary of filter run lengths with the various size sands is shown in Table 3–8. High hydraulic loading rates of 28,061.9 m³/ha· (3.0 mgad) produce undesirable short filter run lengths for 0.40-mm (0.0158-in.) and 0.68-mm (0.0268-in.) effective size sand filters (filters 2, 3, 4, and 5). Hydraulic loading rates of 18,707.9 m³/ha·day (2.0 mgad) and less produce satisfactory filter run lengths for the 0.40-mm (0.0158-in.) and 0.68-mm (0.0268-in.) effective size sand filters (filters 2, 3, 4, and 5). The 0.17-mm (0.0067-in.) effective size sand filter (filter 6) with a hydraulic loading rate of 1870.8 m³/ha·day (0.2 mgad) did not plug during the entire 1-year study. The 0.17-mm (0.0067-in.) effective size sand filter (filter 1) with a hydraulic loading rate of 3741.6 m³/ha·day (0.4 mgad) produced satisfactory filter run lengths. Because of insufficient data for the 0.31-mm (0.0122-in.) effective size sand filter (filter 3) and the 0.68-mm (0.0268-in.)

FIGURE 3-7
Period of filter operation as a function of
SSL for filters containing 0.17 mm effective
size sand.

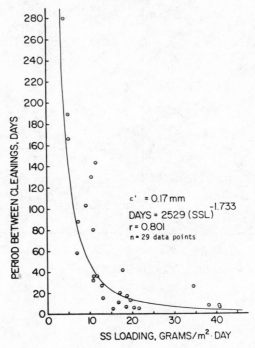

and 0.40-mm (0.0158 in.) effective size sand filters (filters 4 and 5) with hydraulic loading rates of 9353.9 m³/ha·day (1.0 mgad) and a low application rate of 0.008 m³/sec (0.29 cfs), no conclusion can be reached. However, data collected thus far suggest that filter run length may be increased by lowering the application rate.

Period of Filter Operation

The data discussed in the above sections were used to predict the length of filter operation for ISF systems containing 0.17-, 0.40-, and 0.68-mm effective size sands (Figures 3–7, 3–8, and 3–9). Length of filter operation (days) and daily mass surface loading rates (g/m²·day) are inversely related. The larger the daily mass surface loading rate (SSL), the shorter will be the period of filter operation before it will require maintenance.

It is not recommended that sands with effective sizes of greater than 0.35 mm be used to polish wastewater stabilization pond effluents if a 30

FIGURE 3-8

Period of filter operation as a function of SSL for filters containing 0.40 mm effective size sand.

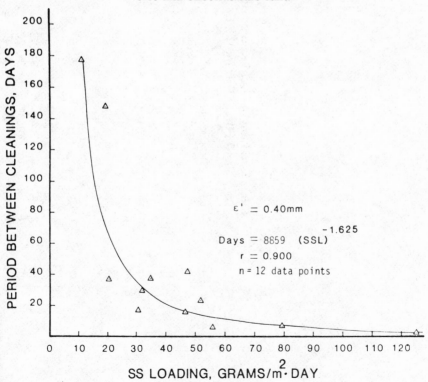

$\varepsilon' = 0.40$mm

Days $= 8859$ (SSL)$^{-1.625}$

$r = 0.900$

$n = 12$ data points

Table 3-8
Filter Run Lengths Achieved by the Various Effective Size Sand Filters During the Experimental Period

Effective Size Sand (mm)	Hydraulic Loading Rate (m³/ha day)	Application Rate (m³/sec)	SS Removal (kg)	Volatile SS Removal (kg)	Method of Rejuvenation	Consecutive Days of Operation
0.17	1,870.8	0.048	121.03	100.17	N.A.	280 and 94
0.17	3,741.6	0.048	14.19	10.26	Scraped	11
0.17	3,741.6	0.048	29.69	22.65	Scraped	36
0.17	3,741.6	0.048	55.95	53.47	Scraped and Rested 14 days	166
0.17	3,741.6	0.048	75.56	68.68	N.A.	103
0.31	9,353.9	0.048	44.43	48.29	N.A.	45
0.31	9,353.9	0.008	5.45	15.02	N.A.	14
0.40	9,353.9	0.048	40.92	63.31	Scraped	44
0.40	9,353.9	0.048	59.10	60.08	Scraped	177
0.40	9,353.9	0.048	20.47	19.73	N.A.	17

0.40	9,353.9	0.008	42.06	39.41	N.A.	37
0.40	14,030.9	0.048	20.03	17.31	Scraped	6
0.40	18,707.9	0.048	15.25	20.26	Scraped	7
0.40	18,707.9	0.048	28.33	37.35	Rested 22 days	18
0.40	18,707.9	0.048	0.00	2.86	Scraped	6
0.40	18,707.9	0.048	68.00	67.77	Scraped	148
0.40	18,707.9	0.048	87.01	65.71	Scraped	42
0.40	18,707.9	0.048	61.98	57.34	Scraped	23
0.40	28,061.9	0.048	0.00	17.89	Scraped	3
0.68	9,353.9	0.048	71.67	79.46	N.A.	196
0.68	9,353.9	0.008	124.20	98.16	N.A.	84
0.68	14,030.9	0.048	102.57	102.03	Scraped	46
0.68	18,707.9	0.048	51.31	42.43	Rested 19 days	23
0.68	18,707.9	0.048	0.00	11.93	Scraped	19
0.68	18,707.9	0.048	101.26	106.82	Scraped	152
0.68	18,707.9	0.048	14.95	12.85	N.A.	11
0.68	28,061.9	0.048	46.36	47.22	Scraped	11

Loaded with primary lagoon effluent twice weekly

0.40	9,353.9	0.008	62.25	72.74	N.A.	30

SOURCE: From Ref. 22.

FIGURE 3-9
Period of filter operation as a function of SSL for filters containing 0.68 mm effective size sand.

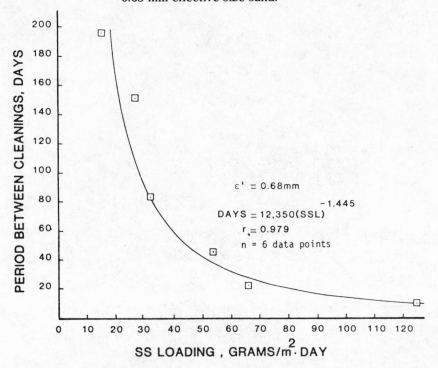

mg/l BOD_5 and SS effluent quality are required. The quality of effluent expected from the 0.40 and 0.68 mm effective size sands can be obtained from the report by Tupyi.[22]

Carbonate Precipitation Problems. There was a definite decrease in the period of operation for some of the filter units when compared with the run times predicted using Figures 3–7, 3–8, and 3–9. When high algal growth (and resultant high pH value) occurs in stabilization ponds treating hard waters, the bicarbonate-carbonate equilibrium system shifts toward increased carbonate concentrations and precipitation occurs: Ca^{2+} + $CO_3^{2-} \rightarrow CaCO_3 \downarrow$. The result is that the ISF develops a "plasterlike" surface crust, which shortens filter run time. Days to plug were plotted versus corresponding SSL values for the carbonate precipitation case, and the results are shown in Figure 3–10. Comparing the results shown in Figures 3–7 and 3–10 (both with 0.17-mm effective size sand) over the range of SSL levels studied in the field shows that filtration systems receiving hard-water stabilization pond effluents will operate for approximately one-half the time that a system treating effluents without calcium carbonate precipitation problems at SS loading rates of 10 g/m²·day. At SS loading rates greater

FIGURE 3-10

Period of filter operation as a function of SSL for lagoon effluents having calcium carbonate precipitation problems.

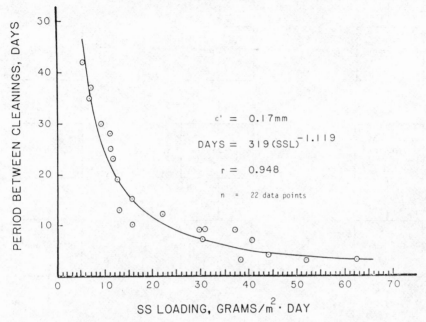

than 10 g/m²·day the period of operation between cleanings becomes essentially equal. At higher SS loadings, the formation of the crust requires more time than is available before the filter plugs from trapping solids.

Series Intermittent Sand Filtration

In an attempt to increase the length of filter run achievable with intermittent sand filtration of lagoon effluent, Hill et al [14] conducted laboratory-and pilot-scale studies on a series arrangement of intermittent sand filters. The experimental design employed in these studies is shown in Figure 3–11. Series intermittent sand filtration involves the passage of lagoon effluent through two or three intermittent sand filters arranged in series with progressively smaller effective size sands. Hill et al [14] investigated three different hydraulic loading rates on a pilot-scale basis.

The weekly BOD_5 removal performance of the pilot-scale series intermittent sand filters is illustrated in Figure 3–12. The influent BOD_5 concentration varied from 4.1 mg/l to 24.0 mg/l and averaged 10.7 mg/l during the study. The final effluent BOD_5 concentration varied from 0.6 mg/l at the 14,031 m³/ha·day (1.5 mgad) hydraulic loading rate to 4.2 mg/l at the 9353 m³/ha·day (1.0 mgad) hydraulic loading rate. At no time did the final ef-

FIGURE 3-11
Series intermittent sand filtration of lagoon effluents (Ref. 14).

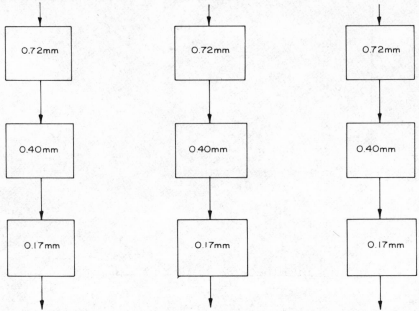

fluent BOD_5 concentration from the operation exceed 5.0 mg/l. Statistical analysis at the 1% level revealed that the effluent BOD_5 concentration was statistically identical at all three hydraulic loading rates for the 0.72-mm (0.0284-in.) and 0.40-mm (0.0158-in.) effective size sand filters. The effluent BOD_5 concentration from the 0.17-mm (0.0067-in.) effective size sand filter at the 14,031 m³/ha·d (1.5 mgad) hydraulic loading rate was significantly higher (1% level) than from the 9353.9 m³/ha·day (1.0 mgad) and 4677 m³/ha·day (0.5 mgad) hydraulic loading rates. However, the actual numerical differences were small.

The weekly SS removal performance of the series pilot-scale field filters is shown in Figure 3–13. The influent SS concentration ranged from 12 mg/l to 69 mg/l and averaged 32 mg/l for the study. The final average effluent SS concentration ranged from 9 mg/l for the 4677 m³/ha·day (0.5 mgad) hydraulic loading rate to 6 mg/l for the 14,031 m³/ha·day (1.5 mgad) hydraulic loading rate. There was no significant difference (1% level) between the final effluent SS concentrations among the different hydraulic loading rates.

An analysis of Figure 3–13 indicates that at the begining of the study the final effluent SS concentration was dependent on the influent concentration. This was due to the "filter washing" effect which takes place during

FIGURE 3-12
Weekly BOD₅ removal performance of pilot-scale series intermittent sand filtration (Ref. 14).

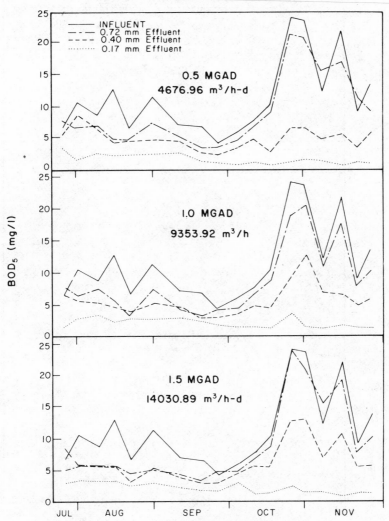

the initial startup of intermittent sand filters, when inert fines must be washed from the filter. When this washing was completed, excellent removals were obtained. At the end of the experimental phase, when removals were exceptional, the 0.17-mm (0.0067-in.) effective size sand filter effluent SS concentration was essentially independent of the influent concentration. The efficiency of removal of intermittent sand filters is in-

FIGURE 3-13
Weekly SS removal performance of pilot-scale series intermittent sand filtration (Ref. 14).

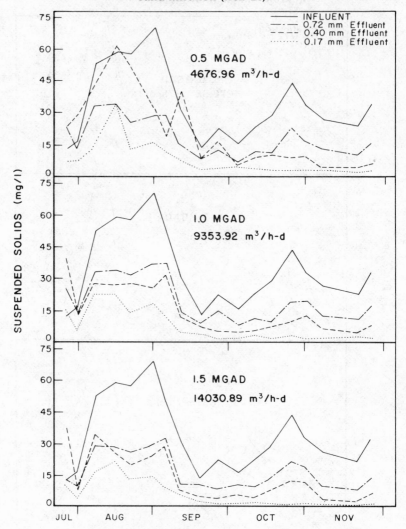

creased as the "schmutzdecke" (filtering skin) builds up on the surface of the filters.

One of the main advantages obtained with the use of a series intermittent sand filtration operation is the increased length of the filter runs. At the time operations were suspended because of freezing conditions (December 2, 1974), all three filter systems had operated for 131 consecutive days without plugging. Until the operations ceased, the applied influent loading passed completely through all three filters in the series within 4 hr. It is difficult to estimate the length of filter run that could have resulted if freezing had not occurred; however, based on the data available, filter runs of at least 131 days may be obtained with a hydraulic loading between 4677 m³/ha·day (0.5 mgad) and 14,031 m³/ha·day (1.5 mgad).

Hill et al.[14] conducted a series of both laboratory- and pilot-scale studies of series intermittent sand filtration to determine the effect of hydraulic loading rates on the length of filter run achievable with three-stage series intermittent sand filtration. The results are summarized in Figure 3–14. The results indicate that hydraulic loading rates greater than 28,080 m³/ha·day (3.0 mgad) significantly reduce the length of filtration run. However, Figure 3–14 neglects the effect of influent SS concentration on filter run length. Thus variations to Figure 3–14 will occur as the influent SS concentration varies.

The results of these studies indicate that a three-stage series intermittent sand filtration system should be designed with filter sands of effective sizes between 0.72 mm (0.0284 in.) and 0.17 mm (0.007 in.) arranged according to Figure 3–11. Hydraulic loading rates should not exceed 28,080 m³/ha·day (3.0 mgad) and preferably should be in the 14,031 m³/ha·day (1.5 mgad) range. Using this criteria, filter run lengths in excess of 131 days should be possible.

FIGURE 3-14

Effect of hydraulic loading rate on three-stage series intermittent sand filtration (Ref. 14).

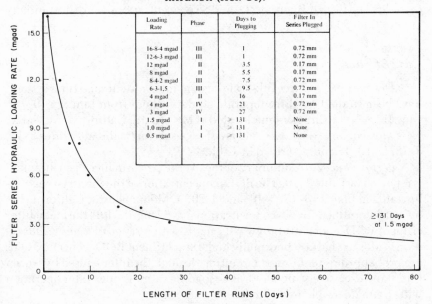

Filtration of Aerated Lagoon Effluents. Bishop et al.[21] conducted a pilot-scale single-stage intermittent sand filtration study to determine the feasibility of upgrading aerated lagoon effluent with intermittent sand filters. The results of the study clearly indicate that although BOD$_5$ removal was acceptable (i.e., effluent concentrations less than 30.0 mg/l), SS concentrations were not significantly reduced. Thus direct intermittent sand filtration of aerated lagoon effluent does not appear feasible. The aerated lagoon was treating wastewater from a milk-processing and cheese-manufacturing plant. Whether the passing of solids through the intermittent sand filter was attributable to the type of wastewater being treated is unknown. However, activated sludge processes treating milk wastewater produce fluffy, low-density solids which are easily dispersed when agitated. Field experience with filtering aerated lagoon effluents have been successful and are discussed later in this chapter.

The study[21] also investigated the single-stage intermittent sand filtration of a facultative lagoon that was preceded by an aerated lagoon. The results indicate that a high-quality effluent (less than 20 mg/l BOD$_5$ and SS) can be obtained. Thus milk waste aerated lagoon effluent may be upgraded by utilizing intermittent sand filtration, provided that the aerated lagoon is followed by a facultative lagoon before the effluent is applied to the intermittent sand filter.

Intermittent Sand Filtration of Anaerobic Lagoon Effluent. A laboratory-scale study was conducted to determine the feasibility of intermittent sand filtration of anaerobic lagoon effluent.[20,21] The results indicate that direct intermittent sand filtration of anaerobic lagoon effluent is not feasible. The filter effluent BOD$_5$ and SS concentrations were substantially greater than 30.0 mg/l in this study.

Case Studies

Description. Three full-scale lagoon–intermittent sand filter systems have been studied for three separate 30-day periods from January 1977 to June 1978.[25] The three systems located at Mt. Shasta, California, Moriarty, New Mexico, and Ailey, Georgia are shown schematically in Figures 3–15, 3–16, and 3–17 and described in Table 3–9.

Performance. A summary of the overall performance of each system is reported in Table 3–10. The BOD$_5$ concentration of the three systems is illustrated in Figures 3–18, 3–19, and 3–20. At Mount Shasta, California the mean filter influent BOD$_5$ concentration was 22 mg/l, while the mean filter effluent BOD$_5$ concentration was 11 mg/l, with a range of 2 mg/l. During tour 1 (Mount Shasta) abnormally high filter effluent BOD$_5$ concentrations are probably a result of short circuiting through the filter caused by frozen filters and excessively high hydraulic loading rates. Proper filter operation will eliminate problems due to freezing.[15]

FIGURE 3-15
Schematic flow diagram and aerial photograph of Mt. Shasta lagoon–intermittent sand filter system (Ref. 25).

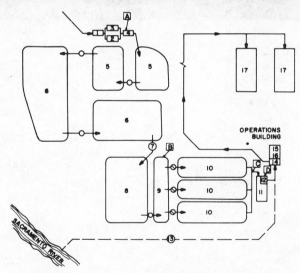

LEGEND		**9**	DOSING BASIN	
I	BAR RACK	10	INTERMITTENT SAND FILTERS	
2	COMMINUTER	II	CHLORINE CONTACT BASIN	
3	BYPASS BAR RACK	12	EFFLUENT PARSHALL FLUME	
4	INFLUENT PARSHALL FLUME	13	RIVER DISCHARGE LINE	
5	PRIMARY AERATED LAGOONS	14	OUTFALL PUMP STATION	
6	SECONDARY AERATED LAGOONS	15	CHLORINATOR / SULFONATOR	
7	LAGOON SYSTEM PARSHALL FLUME	16	AERATION BLOWERS	
8	BALLAST LAGOON	17	WATER RECLAMATION SYSTEM	
A	LAGOON INFLUENT SAMPLING POINT	C	FILTER EFFLUENT SAMPLING POINT	
B	LAGOON EFFLUENT SAMPLING POINT	D	CHLORINATED FILTER EFFLUENT POINT	

159

FIGURE 3-16

Schematic flow diagram and aerial photograph of Moriarty, New Mexico, lagoon–intermittent sand filter system (Ref. 25).

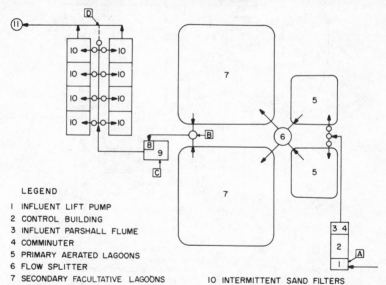

LEGEND

1 INFLUENT LIFT PUMP
2 CONTROL BUILDING
3 INFLUENT PARSHALL FLUME
4 COMMINUTER
5 PRIMARY AERATED LAGOONS
6 FLOW SPLITTER
7 SECONDARY FACULTATIVE LAGOONS
8 TABLET CHLORINATOR
9 DOSING BASIN WITH AUTOMATIC SYPHON
A LAGOON INFLUENT SAMPLING POINT
B LAGOON EFFLUENT SAMPLING POINT

10 INTERMITTENT SAND FILTERS
11 DISCHARGE OUTFALL LINE
O MANUAL VALVES
C CHLORINATED LAGOON EFFLUENT
SAMPLING POINT
D FILTER EFFLUENT SAMPLING POINT

FIGURE 3-17

Schematic flow diagram and aerial photograph of Ailey, Georgia, lagoon-intermittent sand filter system (Ref. 25).

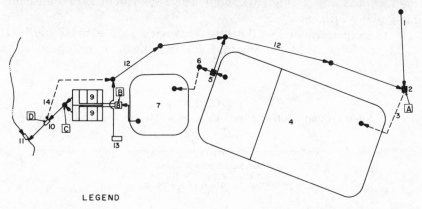

LEGEND

1 INFLUENT MAIN LINE	6 LIFT STATION #2	11 IN-STREAM PARSHALL FLUME
2 LIFT STATION #1	7 POLISHING POND	12 SEWER RETURN LINE
3 FORCED MAIN	8 DOSING BASIN	13 CONTROL BUILDING
4 OXIDATION POND	9 INTERMITTENT SAND FILTERS	14 CONTACT CHAMBER DRAIN
5 FLOW SPLITTER	10 CHLORINE CONTACT CHAMBER	LINE
A LAGOON INFLUENT SAMPLING POINT	C FILTER EFFLUENT SAMPLING POINT	
B LAGOON EFFLUENT SAMPLING POINT	D CHLORINATED FILTER EFFLUENT SAMPLING POINT	

161

Table 3-9

Design Criteria for Full-Scale Systems at Mt. Shasta, California, Moriarty, New Mexico, and Ailey, Georgia

PARAMETER	MT. SHASTA	MORIARTY	AILEY
Design Q (mgd)	1.2	0.4	0.08
Lagoon type	Aerated	Aerated/facultative	Facultative
Filter area (acre)	0.5	0.082	0.14
Number of filters	6	8	2
Hydraulic L. R. (mgad)	0.7	0.6	0.4
Effective size (mm)	0.37	0.20	0.50
Uniformity coefficient	5.1	4.1	4.0

SOURCE: From Ref. 25.

At Moriarty, New Mexico mean filter effluent BOD_5 concentrations were 20 mg/l and 21 mg/l during tours 1 and 3, respectively. However, during tour 2, the mean filter effluent BOD_5 concentration was reduced to 10 mg/l. This reduction was due to proper filter maintenance and lower influent BOD_5 concentrations.

At Ailey, Georgia the mean filter effluent BOD_5 concentration was 8 mg/l, with a range of 2 mg/l to 22 mg/l. This system produced a consistently high-quality effluent.

The SS performance for the three systems is illustrated in Figures 3–21, 3–22, and 3–23. At Mount Shasta, California the three tours produced a

FIGURE 3-18

Final effluent BOD_5 concentration at Mt. Shasta, California (Ref. 25).

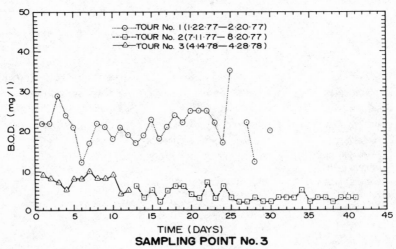

TIME (DAYS)
SAMPLING POINT No. 3

Table 3-10

Summary of Composite Mean Values for Each Sample Point at Each Sample Site During Three Tours

Parameter	Mt. Shasta Water Pollution Control Facility				Moriarty Wastewater Treatment Facility				Ailey Sewage Treatment Plant			
	Facility Inf.	Lagoon Eff.	Filter Eff.	Facility Eff.	Facility Inf.	Lagoon Eff.	Filter Eff.	Facility Eff.	Facility Inf.	Lagoon Eff.	Filter Eff.	Facility Eff.
BOD (mg/l)	114	22	11	8	148	30	17	17	67	22	8	6
S. BOD (mg/l)	41	7	4	5	74	17	16	16	17	10	6	5
SS (mg/l)	83	49	18	16	143	81	13	13	109	43	15	13
VSS (mg/l)	70	34	13	10	118	64	9	9	87	32	8	6
FC (col/100ml)	1.16×10^6	292	30	<2	4.24×10^6	290	18	34	2.17×10^6	55	8	<1
I-pH (pH)	6.9	8.7	6.8	6.6	8.0	8.9	8.0	8.0	7.3	8.9	7.1	6.8
I-DO (mg/l)	4.8	12.4	5.5	5.3	1.8	10.9	8.3	8.3	6.7	10.2	7.4	7.9
COD (mg/l)	244	100	87	68	305	84	43	43	160	57	32	25
S. COD (mg/l)	159	71	64	50	197	67	34	34	82	41	23	16
Akl (mg/l as CaCO₃)	95	75	51	42	436	293	260	260	93	84	76	69
TP (mg-P/l)	4.68	3.88	3.09	2.72	10.3	4.02	2.8	2.8	4.96	3.10	2.67	2.45
TKN (mg-N/l)	15.5	11.1	7.5	5.2	60	22	12.1	12.1	14.2	7.3	4.1	2.2
NH₃(mg-N/l)	10.8	5.56	1.83	1.76	38	16	9.16	9.16	5.5	0.658	0.402	0.31
Org-N (mg-N/l)	4.8	5.6	5.7	3.4	22	5.7	3.3	3.3	8.7	6.7	3.8	1.9
NO₂ (mg-N/l)	0.16	0.56	0.077	0.020	0.05	0.159	1.66	1.66	0.479	0.028	0.073	0.010
NO₃ (mg-N/l)	0.28	0.78	4.3	4.5	0.05	0.09	4.09	4.09	1.6	0.15	2.36	2.14
Total Algal Count (cells/ml)	N.A.	398022	144189	141305	N.A.	756681	32417	32417	N.A.	349175*	21583*	29360*
Flow (mgd)	0.637	N.A.	N.A.	0.488	0.096	N.A.	0.046	N.A.	N.A.	N.A.	N.A.	0.070

Source: From Ref. 25.

N.A. = Not available.

* For tours 1 and 2.

163

FIGURE 3-19
Final effluent BOD$_5$ concentration at Moriarty, New Mexico (Ref. 25).

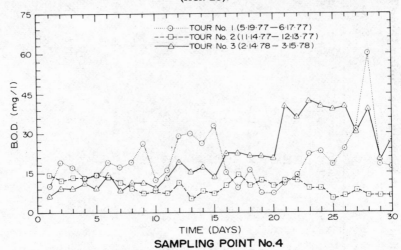

SAMPLING POINT No.4

mean filter effluent SS concentration of 17 mg/l, with a range of 1 mg/l to 49 mg/l. The high values are associated with the winter freezing problem.

At Moriarty, New Mexico the three-tour mean filter effluent SS concentration was 13 mg/l, with a range of 2 mg/l to 39 mg/l. The mean filter influent suspended solids concentration was 81 mg/l.

At Ailey, Georgia the three-tour mean filter effluent SS concentration

FIGURE 3-20
Final effluent BOD$_5$ concentration at Ailey, Georgia (Ref. 25).

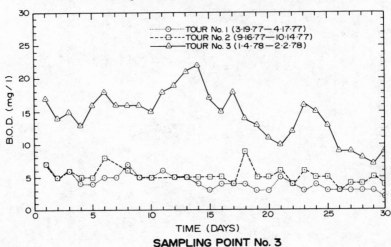

SAMPLING POINT No. 3

FIGURE 3-21
Final effluent SS concentration at Mt. Shasta, California (Ref. 25).

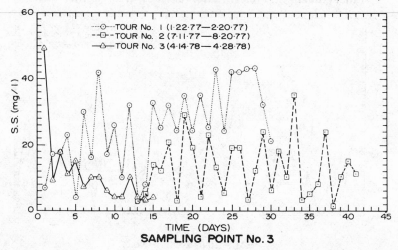

SAMPLING POINT No. 3

was 15 mg/l, with a range of 1 mg/l to 45 mg/l. The mean filter influent SS concentration was 55 mg/l.

Operation and Maintenance. Maintenance requirements for the three sites were basically the same, except for the increase in requirements for the Mt. Shasta facility because of its complex nature and larger size. Each of the sites used the same basic processes that required common routine maintenance to provide a normal design service life.

Table 3–11 presents a summary of the reported maintenance re-

FIGURE 3-22
Final effluent SS concentration at Moriarty, New Mexico (Ref. 25).

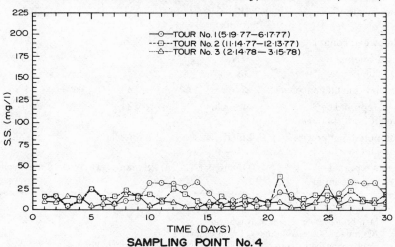

SAMPLING POINT No. 4

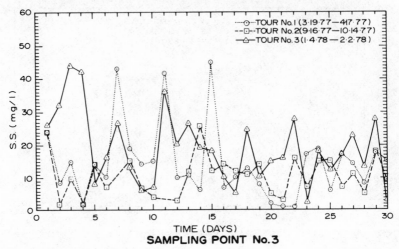

FIGURE 3-23
Final effluent SS concentration at Ailey, Georgia (Ref. 25).

SAMPLING POINT No. 3

Table 3-11
Summary of Reported Annual Maintenance

Job Description	Mt. Shasta WPCF	Moriarty WWTF	Ailey STP
Daily operation and maintenance (daily monitoring)	(1.0 hr) × 7 days × 52 wks = 364	(1.0 hr) × 7 days × 52 wks = 364	(0.5 hr) × 5 days × 52 wks = 130
Filter cleaning	54[a]	28[a]	None
Filter raking	12 raking 16 mixing	13	22
Filter weed control	N.A.	None	26
Miscellaneous maintenance	N.A.	11	None
Grounds maintenance	42	8	28
Total reported man-hours per year	488 plus	424	206
Computed manpower requirements	2.4 man-years[b]	1 man-year[b]	1 man-year[b]
Actual reported manpower input	2.0 man-years[c]	0.28 man-year[b]	0.14 man-year[b]

Source: From Ref. 25.

[a] Man-hours with mechanical assistance.
[b] Assuming 1500 man-hours = 1 man-year.
[c] Considering extra assistance for filter cleaning and weekend monitoring.

quirements for each of the three sites for a period of approximately 1 year. The Moriarty, New Mexico and Ailey, Georgia reports are the most complete and the most representative of requirements for maintenance of a lagoon–intermittent sand filter system.

The Mt. Shasta facility was seemingly plagued with problems either induced or accidentally caused by the operator. These problems many times resulted in extensive repairs and in some cases in complete overhaul of portions of the system. The aeration system at Mt. Shasta was also a source of many problems observed at the facility. Constant operational changes and problems in conjunction with the extensive maintenance activities have distorted the reported maintenance requirements for the Mt. Shasta facility.

In general, each of the three full-scale lagoon–intermittent sand filter systems consistently produced an effluent BOD_5 and SS concentration of less than 30 mg/l. Reported maintenance requirements varied with the skill and ability of the operator; however, operation and maintenance of lagoon–intermittent sand filter systems are less costly than conventional systems.

Dissolved Air Flotation

Several studies have shown the dissolved air flotation process to be an efficient and a cost effective means of algae removal from wastewater stabilization lagoon effluents. The performance obtained in several of these studies is summarized in Table 3–12.

Three basic types of dissolved air flotation are employed to treat wastewaters: total pressurization, partial pressurization, and recycle pressurization. These three types of dissolved air flotation are illustrated by flow diagrams in Figure 3–24. In the total pressurization system the entire wastewater stream is injected with air and pressurized and held in a retention tank before entering the flotation cell. The flow is direct, and all recycled effluent is repressurized. In partial pressurization only part of the wastewater stream is pressurized, and the remainder of the flow bypasses the air dissolution system and enters the separator directly. Recycling serves to protect the pump during periods of low flow, but it does hydraulically load the separator. Partial pressurization requires a smaller pump and a smaller pressurization system. In recycle pressurization, clarified effluent is recycled for the purpose of adding air and then is injected into the raw wastewater. Approximately 20–50% of the effluent is pressurized in this system. The recycle flow is blended with the raw water flow in the flotation cell or in an inlet manifold.

Important parameters in the design of a flotation system are hydraulic loading rate, including recycle, concentration of suspended solids con-

Table 3-12

Summary of Typical Dissolved Air Flotation Performance

Investigator and Location	Coagulant and Dose (MG/L)	Overflow Rate (GPM/SF)	Detention Time (MIN)	BOD$_5$ Influent (mg/l)	BOD$_5$ Effluent (mg/l)	BOD$_5$ % Removed	Suspended Solids Influent (mg/l)	Suspended Solids Effluent (mg/l)	Suspended Solids % Removed
Parker[26] Stockton, Ca.	Alum: 225 mg/l acid added to pH 6.4	2.7[a]	17[a]	46	5	89	104	20	81
Ort[27] Lubbock, Texas	Lime:[c] 150	NA	12[b]	280-450	0-3	>99	240-360	0-50	>79
Komline-Sanderson[28] El Dorado, Arkansas	Alum: 200 mg/l	4.0[c]	8[c]	93	<3	>97	450	36	92
Bare[29] Logan, Utah	Alum: 300 mg/l	1.3-2.4[d]	NA	NA	NA	NA	100	4	96
Stone et al.[30] Sunnyvale, Ca.	Alum: 175 mg/l acid added to pH 6.0 to 6.3	2.0[e]	11[e]	NA	NA	NA	150	30	80

Source: From Ref. 26; Courtesy of Center for Research In Water Resources, The University of Texas at Austin.

[a] Including 33% pressurized (35-60 psig) recycle.
[b] Including 30% pressurized (50 psig) recycle.
[c] Including 100% pressurized recycle.
[d] Including 25% pressurized (45 psig) recycle.
[e] Including 27% pressurized (55-70 psig) influent.

FIGURE 3-24
Types of dissolved air flotation systems.
Courtesy of Center for Research in Water
Resources, The University of Texas at Austin
(Ref. 31).

FULL FLOW PRESSURIZATION

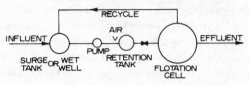

PARTIAL PRESSURIZATION

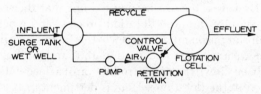

RECYCLE PRESSURIZATION

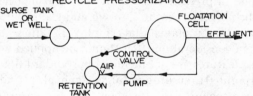

tained within the flow, coagulant dosage, and the air-to-solids ratio required to effect efficient removal. Pilot-plant studies by Stone et al.,[30]
Bare,[29] and Snider[31] have shown the maximum hydraulic loading rate to
range between 81.5 and 101.8 l/min·m² (2 and 2.5 gpm/ft²). A most efficient
air-to-solids ratio was found to be 0.019:1.0 by Bare.[29] Solids concentrations
during Bare's studies were 125 mg/l. Experimental results with the removal
of algae indicate that lower hydraulic rates and air-to-solids ratios than
those recommended by the manufacturers of industrial equipment should
be utilized when attempting to remove algae.

In combined sedimentation flotation pilot-plant studies at Windhoek,
Southwest Africa van Vuuren and van Duuren[32] reported effective
hydraulic loading rates to range between 11.2 and 30.5 l/min·m²
(0.275–0.75 gpm/ft²), with flotation provided by the naturally dissolved
gases. Because air was not added, air-solids ratios were not reported. They
also noted that it was necessary to use from 125 to 175 mg/l of aluminum
sulfate to flocculate the effluent containing from 25 to 40 mg/l of algae.
Subsequent reports (1965) on a total flotation system by van Vuuren[33]

stated that a dose of 400 mg/l of aluminum sulfate was required to flocculate a 110 mg/l algal suspension sufficiently to obtain a removal that was satisfactory for consumptive reuse of the water. Based on data provided by Parker et al.,[34] Stone et al.,[30] Bare,[29] and Snider,[31] it appears that a much lower dose of alum would be required to produce a satisfactory effluent to meet present discharge standards.

Dissolved air flotation with the application of coagulants performs essentially the same function as coagulation-flocculation-sedimentation, except that a much smaller system is required with the flotation device. Flotation will occur in shallow tanks with hydraulic residence times of 7–20 min, compared with hours in deep sedimentation tanks. Overflow rates of 81.5–101.8 l/min·m² (2–2.5 gpm/ft²) can be employed with flotation; whereas a value of less than 40.7 l/min·m² (1 gpm/ft²) is recommended with sedimentation. However, it must be pointed out that the sedimentation process is much simpler than the flotation process, and when applied to small systems, consideration must be given to this factor. The flotation process does not require a separate flocculation unit, and this has definite advantages. It has been shown that the introduction of a flocculation step after chemical addition in the flotation system is detrimental. It is best to add alum at the point of pressure release where its mixing occurs and a good dispersion of the chemicals occurs. Brown and Caldwell[26] have designed two tertiary treatment plants that employ flotation, and they have developed design considerations that should be applied when employing flotation. These features are not included in standard flotation units and should be incorporated to ensure good algae removal.[26] In addition to incorporating various mechanical improvements, the Brown and Caldwell study recommended that the tank surface be protected from excessive wind currents to prevent float movement to one side of the tank. They found that the relatively light float is easily moved across the water surface by wind action. It was also recommended that the flotation tank be covered in rainy climates to prevent the breakdown of the floc by rain. Another alternative proposed has been to store the wastewater in stabilization ponds during the rainy season and then operate the flotation process at a higher rate during dry weather.

Alum-algae sludge has been returned to the wastewater stabilization ponds for over 3 years at Sunnyvale, California with no apparent detrimental effect.[35] Neither sludge banks, floating mats of material, nor increased SS concentrations in the pond effluent have been observed. Return of the float to the pond system is an alternative, at least for a few years. Most estimates of a period of time that sludge can be returned range from 10 to 20 years.

Sludge disposal from a dissolved air flotation system can impose considerable difficulties. Alum-algae sludge is very difficult to dewater and discard. Centrifugation and vacuum filtration of unconditioned-algae-alum sludge have produced marginal results. Indications are that lime coagulation may prove to be as effective as alum and produce a sludge more easily dewatered.

Brown and Caldwell[36] evaluated heat treatment of alum-algae sludges using the Porteous, Zimpro low-oxidation, and the Zimpro high-oxidation processes and found relatively inefficient results. The Purifa process using chlorine to stabilize the sludge produces a sludge dewaterable on sand beds or in a lagoon. However, the high cost of chlorine eliminates this alternative. If algae are killed before entering an anaerobic digestor, volatile matter destruction and dewatering results are reasonable. But as with the other sludge treatment and disposal processes, additional operations and costs are incurred, and the option of dissolved air flotation loses its competitive position.

Controlled Discharge

Controlled discharge is defined as limiting the discharge from a lagoon system to those periods when the effluent quality will satisfy existing discharge requirements. The usual practice is to prevent discharge from the lagoon during the winter period and during the spring and fall overturn period and algal bloom periods. Many states currently do not permit lagoon discharges during winter months.

Pierce[37] has reported on the quality of lagoon effluent obtained from 49 lagoon installations in Michigan that practice controlled discharge. Of these 49 lagoon systems, 27 have two cells, 19 have three cells, 2 have four cells, and 1 has five cells. Discharge from these systems is generally limited to late spring and early fall. However, several of the systems discharged at various times throughout the year. The period of discharge varied from fewer than 5 days to more than 31 days. The lagoons were emptied to a minimum depth of approximately 0.46 m (18 in.) during each controlled discharge to provide storage capacity for the nondischarge periods.

During the discharge period, the lagoon effluent was monitored for BOD_5, SS, and FC. The effluent BOD_5 and SS concentrations measured during the study are illustrated in terms of probability of occurrence in Figure 3-25. All values are arranged in order of magnitude and plotted on normal probability paper with concentration (mg/l) plotted against the probability that the value would not be exceeded under similar conditions. The plot compares the performance of two-cell lagoon systems versus three-or-more-cell lagoon systems. The results of Figure 3-25 are summarized in Table 3-13.

The results of the study indicated that the most probable effluent BOD_5 concentration for controlled discharge systems was 17 mg/l for the two-cell lagoon systems and 14 mg/l for three-or-more cell lagoon systems. There

FIGURE 3-25
Comparison of effluent quality two-cell systems versus three or
more with long period storage before discharge, Michigan
(Ref. 37).

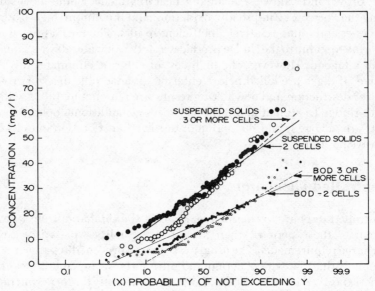

was a 90% probability that the effluent BOD_5 concentration from both
two-cell and three-or-more cell lagoon systems would not exceed 27 mg/l.
This value was slightly less than the 30 mg/l BOD_5 Federal Secondary Treatment Standard.

The most probable effluent SS concentration was found to be 30 mg/l

Table 3-13
Effluent Quality Resulting from Controlled Discharge
Operation of 49 Michigan Lagoon Installations

PERCENT PROBABILITY OF OCCURRENCE	EFFLUENT BOD_5 CONCENTRATION (MG/L)		EFFLUENT SUSPENDED SOLIDS CONCENTRATION (MG/L)	
	Two Cells	Three or More Cells	Two Cells	Three or More Cells
50% probability (i.e., most probable)	17	14	30	27
90% probability (i.e., will not be exceeded 9 out of 10 samples)	27	27	27	47

SOURCE: From Ref. 37.

for two-cell lagoon systems and 27 mg/l for three-or-more cell lagoon systems. However, the 90% probability levels for effluent SS concentrations were 46 mg/l for two-cell lagoon systems and 47 mg/l for three-or-more cell lagoon systems.

The results of the study also indicate that the FC levels were generally less than 200/100 ml, although this standard was exceeded on several occasions when chlorination was not employed.

A similar study of controlled discharge lagoon systems was conducted in Minnesota.[37] The discharge practices of the 39 installations studied were similar to those employed in Michigan. The results of that study from the fall discharge period indicated that the effluent BOD_5 concentrations for 36 of 39 installations sampled were less than 25 mg/l, and the effluent SS concentrations were less than 30 mg/l. In addition, effluent FC concentrations were measured at 17 of the lagoon installations studied. All the installations reported effluent FC concentrations of less than 200/100 ml.

During the spring discharge period 49 municipal lagoon installations were monitored. Effluent BOD_5 concentrations exceeded 30 mg/l at only three installations, while the maximum effluent BOD_5 concentration reported was only 39 mg/l. However, effluent SS concentrations ranged from 7 mg/l to 128 mg/l, with 16 of the 49 installations reporting effluent SS concentrations greater than 30 mg/l. Only 3 of the 45 installations monitored for effluent FC concentrations exceeded 200/100 ml.

An evaluation of the four facultative lagoon systems described earlier[38-41] indicates that all these systems could satisfy the 30 mg/l BOD_5 federal secondary treatment standard throughout the entire year with the adoption of controlled discharge operations. There were only 4 of the 12 months studied during which the Peterborough site did not produce a final effluent BOD_5 and SS concentration of less than 30 mg/l. Because this system practiced chlorination, the geometric mean FC concentration never exceeded 200/100 ml. However, monthly average effluent SS concentrations were often greater than 30 mg/l. Thus controlled discharge at these two sites would be possible to implement, but would require a very limited discharge period and large storage volumes.

The Corinne site is an excellent example of where controlled discharge can be implemented. The monthly average effluent BOD_5 concentration of this system never exceeded 30 mg/l, and during only 3 of the 13 months studied did the monthly average effluent SS concentration exceed 20 mg/l. In addition, FC concentrations were never greater than 200/100 ml.

The controlled discharge of lagoon effluent is a simple, economical, and practical method of achieving a high degree of treatment. Experience indicates that routine monitoring of the lagoon effluent is necessary to determine the proper discharge period. However, these discharge periods may extend throughout the major portion of the year. It will be necessary to increase the storage capacity of certain lagoon systems that employ controlled discharge. However, many lagoon systems already have additional freeboard and storage capacity which could be utilized without significant modification.

Total Containment

In areas with inexpensive land and high evaporation rates, total containment lagoons are a suitable alternative for wastewater disposal (see Figure 1–1). Fluctuating water levels can produce odor problems, but limited experience indicates that normal isolation practices for lagoons are adequate. Salt buildup may eventually produce a limited growth which could result in poor degradation of the organics. Because of the limited operating experience, it is difficult to predict what problems might occur. Economically total containment is competitive when applicable.

In-Pond Removal of Particulate Matter

There are several factors to be considered for in-pond removal of particulate matter:

1. The subsequent decay of settled matter and degradation by microorganisms to produce dissolved BOD_5, which would then have an effect on the receiving water.
2. The possibility that settled material will not remain settled.
3. The lack of positive control of effluent particulate matter.
4. The problem of eventually filling in the oxidation pond.
5. The possibility that anaerobic reactions within the settled material will produce malodors.

At first glance it seems that some of these problems could be resolved by rather simple changes in operation. For example, in ponds that are in series, possibly all the settled material could be removed from the bottom of the last pond and transferred to the anaerobic pond or primary pond in which biological degradation is encouraged and malodors are not such a problem. Positive control could be achieved by adding coagulants to the final pond to ensure that settling takes place. For example, chemicals such as lime, ferric chloride, and alum might be used in this manner. Filling in a pond is not necessarily as much of a problem as one might think unless chemicals are to be added. Generally, ponds that are used for complete containment have a life expectancy of about 20 years.[42] In areas short of land, filling could be a problem, but it might be possible to dredge and remove the solid material after 10–20 years have elapsed and to restore the pond to its initial status. In areas in which land is available and cost is not prohibitive, filling would not be a problem.

Specific in-pond mechanisms of particulate removal that will be considered include complete containment, biological disks, baffles or raceways, chemical additions for precipitation, autoflocculation, and biological harvesting. These mechanisms are discussed in the following sections.

In-Pond Treatment with Chlorination

Caldwell[43] reported the use of chlorine to kill algae and clarify the effluent from a four-cells-in-a-series oxidation pond in California. A 13.5-hr detention time chlorine contact pond followed the four-cell oxidation pond system, and when 12 mg/l of chlorine were added, all the algae were killed. The BOD_5 concentrations were reduced from 45 to 25 mg/l, SS concentrations were reduced from 110 mg/l to 40 mg/l, and turbidity was reduced from 70 to 40 units. In a later study, Dinges and Rust[44] reported similar results. Chem Pure, Inc.[45] has a proprietary system that uses chlorine to kill algae, followed by settling to remove the algae. This process is essentially the same as the procedure used by Caldwell mentioned above, but it has a few proprietary twists.

Recent studies at Utah State University on the chlorination of lagoon effluent indicate that the concern expressed in earlier articles by Echelberger et al.[46] and Hom[47] is not confirmed and that the destruction of algae and the lysis of cells with high doses of chlorine occur only when free residual chlorine is available.[48,49] These studies have shown relatively little chemical oxygen demand (COD) released to the effluent by the chlorination of algae laden waters with chlorine dosages adquate for disinfection. Dosages of chlorine as high as 30 mg/l with 63 mg/l of SS have produced very little change in the COD of lagoon effluents.

The majority of the changes in COD values showed an increase with free chlorine residual available, but at residual concentrations less than 2 mg/l there appears to be no consistent pattern. With a reasonable degree of control, disinfection or killing of the algae to improve settling should produce an effluent meeting the requirement of 30 mg/l of BOD_5 and SS concentrations.

Biological Harvesting

The use of biological systems to harvest algae from wastewater stabilization pond effluents appears reasonable and would have tremendous economic advantage if the processes were controllable. Most of the studies which have been conducted have focused on the production of high concentrations of algae rather than attempting to develop design and operating procedures. However, there are currently several studies being conducted throughout the United States which may eventually lead to a well-designed and easily operated and controlled biological system for producing high-quality effluents. Duffer[50] has presented an effective summary of projects being conducted throughout the world in which the principal objective is the removal of algae from lagoon effluents. This reference is still current and should be consulted if additional information is desired.

Some of the most promising work directed toward developing a biological harvesting system was conducted at the Woods Hole Oceanographic Institute in Woods Hole, Massachusetts.[51] Experiments have not been successful using prototype processes. Several growth systems involving marine phytoplankton, oysters, deposit feeders, and seaweed were combined in series to treat secondary effluent diluted with filtered seawater. A large-scale multispecies food chain concept has been in operation at Woods Hole for several years.[52]

Fish culture in enriched waters is a very old practice of treating wastewater, employing principally fish of the carp family. The silver carp is the fish most frequently used to consume phytoplankton. Studies in the United States are being conducted to determine its effectiveness as a means of consuming algae from wastewater stabilization pond effluents.[50] Other types of fish have been employed successfully in removing algae from wastewater stabilization ponds. But control of the system still remains a principal problem. This is particularly true when the possibility of introducing toxicants to the lagoon system occurs.

Two vascular plants, water hyacinths and alligator weeds, have been used to polish wastewater stabilization pond effluents on laboratory- and full-scale systems, and excellent results have been obtained.[53,54] An engineering assessment of water hyacinth systems is presented in Chapter 8.

Several "natural" recycling systems are being evaluated that incorporate lagoon systems or other holding and treatment systems which are followed by marshes and ponds or meadows.[55] The City of Arcata, California is evaluating a wastewater treatment facility that includes existing primary treatment, lagoons, and a freshwater marsh operated in conjunction with ocean ranching. The first released salmon returned to the Arcata lagoons in the fall of 1977.[56]

Serfling and Alsten[57] reported results from an integrated, controlled environment aquaculture treatment system using five different but relatively well-proven technologies. The Solar AquaCell Process uses the basic multicell lagoon process, floating aquatic macrophytes, greenhouse-covered ponds, and high surface area fixed-film substrates to provide a habitat for the bio-film and associated microorganisms and invertebrate detritivore community, and it also has a dual aeration and solar heat exchange system. A section view of the Solar AquaCell is shown in Figure

FIGURE 3-26
Section view of Solar AquaCell lagoon system (Ref. 57).

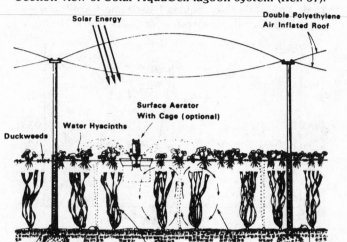

3-26. Two demonstration facilities have been evaluated at Solana Beach and Cardiff, California. The Solana facility was a pilot-scale unit consisting of a series of four tanks averaging 5.7 m³ (1500 gal) each, and 1.83 m × 2.44 m × 1.31 m (6 ft × 8 ft × 4.3 ft) deep in size. The average flow rate through the system was 5.7 m³ (1500 gal) per day to achieve a total theoretical hydraulic residence time of 4 days. To overcome the inconsistent flow of wastewater to the system at Solana, another pilot plant was constructed at Cardiff consisting of three cells in series, each 2.44 m × 3.96 m × 1.88 m (8 ft × 13 ft × 5.5 ft) deep, with a capacity of approximately 15.1 m³ (4000 gal) in each cell. Design flow rates were 11.4–15.1 m³ (3000–4000 gal) per day, providing mean theoretical hydraulic residence times from 2 to 4 days. Pretreatment to the Cardiff system consisted of a two-stage anaerobic tank 4.54 m³ (1200 gal) in capacity in stage 1 and 9.08 m³ (2400 gal) in stage 2.

BOD$_5$ and dissolved oxygen concentrations in the effluents from the Solana and Cardiff AquaCells are shown in Figure 3–27. SS and nitrogen-forms concentrations from the AquaCells are shown in Figures 3–28 and 3–29. Coliform reduction by the two pilot plants is shown in Figure 3–30.

The City of Hercules, California has constructed a 1325 m³/day (0.35 mgd) Solar AquaCell Process, but the system has just started operation. The Hercules facility is shown in Figure 3–31. Some difficulties have been encountered, and costs are reportedly equivalent to actived sludge treatment.

FIGURE 3-27

BOD$_5$ and dissolved oxygen concentration in relation
to detention time for Solana Beach and Cardiff Solar
AquaCell demonstration facilities (Ref. 57).

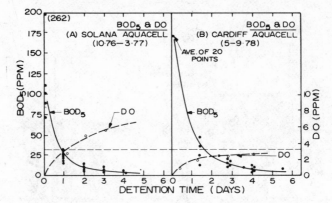

More information should become available on this process in the near
future.

Results from carefully controlled experiments such as those being conducted are required before it will become feasible to recommend biological
harvesting as an alternative in the control of algae from small wastewater
stabilization systems. In addition to the lack of data substantiating the
reliability and performance of such systems, there are no cost data available
on which economic comparisons could be made. To date, it appears to be
unfeasible to incorporate biological harvesting into small wastewater

FIGURE 3-28

SS treatment performance for Solana Beach and
Cardiff solar AquaCell demonstration facilities
(Ref. 57).

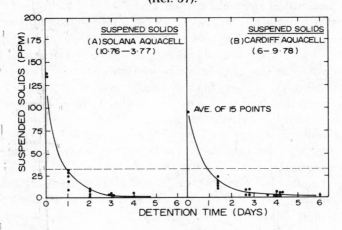

FIGURE 3-29
Treatment and removal of ammonia, nitrite, nitrate,
and total Kjeldahl nitrogen, expressed as mg/l
nitrogen, for the Solana Beach and Cardiff AquaCell
facilities (Ref. 57).

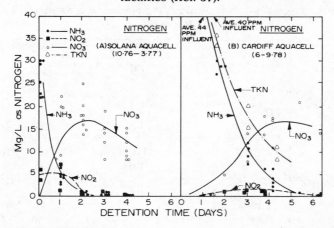

stabilization pond systems, but it is an idea that holds some promise and should be considered as more research results become available.

Intermittent Discharge Lagoons with Chemical Addition

A series of reports distributed by the Canadian government[58,59,60] has reported success with the treatment of intermittent discharge lagoons by adding various coagulants from a motorboat. Excellent quality effluents are produced, and the costs are relatively inexpensive. The cost of in-pond

FIGURE 3-30
Average total coliform reduction in
relation to detention time (Ref. 57).

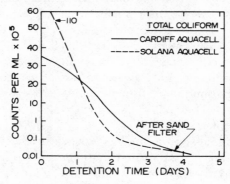

FIGURE 3-31
Plan view of the proposed 2.0 mgd Solar AquaCell Lagoon
Treatment Facility for the City of Hercules, California. Each
AquaCell will be 2.0 acres (6 acres total). The 0.35 mgd treatment
phase currently under construction consists of a 1.5 acre
AquaCell system with anaerobic, facultative, and aerobic stages
(Ref. 57).

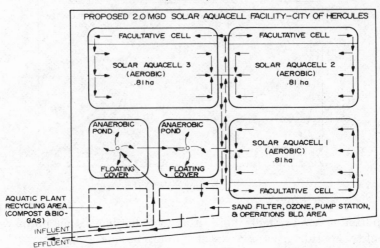

treatment and the long detention times required must be balanced against
the alternatives available to an engineer. Man-hour requirements for the
full-scale batch treatment systems employed in Canada are summarized in
Table 3–14.

In addition to the usual design considerations applied to intermittent
discharge lagoons, the following physical design requirements were recommended by Graham and Hunsinger:[58]

1. A roadway to the edge of each cell with a turnabout area sufficient
   to carry 45 metric tons (50 tons) in early spring and late fall or a piping system to deliver the chemical to each cell and a road adequate
   to get the boats to the lagoon edge.
2. A boat ramp and a small dock installed in each cell.
3. Separate feed and outlet facilities to allow diversion of raw sewage
   during treatment and draw-down in multiple cell installations for
   maintaining optimum effluent quality.
4. A low-level outlet pipe in the lagoon to allow complete drainage of
   the cell contents.
5. A discharge pipe from the lagoon of sufficient size and design to
   allow drainage of the treated area over a 5–10 day period.
6. In new large installations, a number of medium-sized cells of 4–6 ha

Table 3-14
Labor Requirements for Full-Scale Batch Treatments

	MAN-HOURS PER ACRE	MAN-HOURS PER MILLION GALLONS	MAN-HOURS FOR SETUP AND CLEANUP PER APPLICATION
Alum, liquid	2	1.6	16
Ferric chloride			
Liquid	1.5	1.2	16
Powder	13	9.6	16
Lime,			
Dry chemical method	24	17.7	125
Haliburton method	1.7	1.4	16

SOURCE: From Ref. 58.

1 hectare = 2.471 acres.

1 million gallons = 3785 m^3.

(10–15 acres) would be better suited to this type of treatment than one or two large cells. These medium-sized cells could be treated individually and drawn down over a relatively short period of time, thus maintaining optimum water quality in the effluent.

Using the typical wastewater stabilization pond design required in cold climates with 120 days retention time for cold weather storage, three chemical treatments per year would be required. Designing a system for a 1136 m^3 (0.3 mgd) operation would result in approximately 136,000 m^3 of stored wastewater per treatment. Applying alum at a rate of 150 mg/l would result in 20,400 kilograms (44,900 lb) of alum required per treatment, assuming that the hydraulic design would control the sizing of the storage ponds and neglecting evaporation. The installation would require approximately 8.1 ha (20 acres) of lagoon surface with a depth of 1.83 m (6 ft). Assuming that a relatively small boat and supply system would be adequate to distribute and mix the chemicals with the lagoon water, a capital investment of approximately $33,000 (1977 U.S. $) would be required to obtain the tank trucks, storage facilities, boat and motor to carry out the operation. If we amortize the equipment for a useful life of 10 years and assume 7% interest, it would cost $4700 per year. Liquid alum costs approximately $105/ton (equivalent dry), and using 68 tons annually would cost $7140. Approximately 1.36 × 10^5 m^3 (36 mg) of wastewater would be treated before each discharge. Using the labor requirements shown in Table 3–14 of 1.6 man-hours/mg and 16 man-hours for setup and cleanup per application results in a total labor requirement of 221 man-hours/year. At labor costs of $7.50/hr, the cost would be $1658. Adding all the above costs, exclusive of the capital cost for the lagoon system, results in an annual cost of $13,500 or $0.0325/m³ ($123/mg) of wastewater treated ($0.12/1000 gals).

The above costs do not include storage facilities for the alum and the additional design requirements to accommodate the alum-handling equipment and the boats. However, even if the estimated costs are doubled, it is apparent that intermittent discharge with chemical treatment is a suitable alternative where applicable.

In addition to the cost advantages outlined above, batch chemical treatment of intermittent discharge lagoons can produce an effluent containing less than 1 mg/l of total phosphorus. SS and BOD$_5$ concentrations of less than 20 mg/l can be produced consistently, and only occasionally does a bloom occur during draw down of the lagoon. Rapid draw down would overcome this disadvantage. Sludge buildup is insignificant and would require years of operation before cleaning would be required.

The EPA is supporting a study of chemical addition to polish lagoon effluents at St. Charles, South Carolina, and results are promising. The project has been in operation since January 1977, and the report should be forthcoming.

Autoflocculation and Phase Isolation

Autoflocculation of algae has been observed during some studies.[61-64] *Chlorella* was the predominant alga occurring in most of the cultures. Laboratory-scale continuous experiments with mixtures of activated sludge and algae have produced large bacteria-algae flocs with good settling characteristics.[64,65]

Floating algae blankets have been reported in some cases in the presence of chemical coagulants.[66,32] The phenomenon may be caused by the entrapment of gas bubbles produced during metabolism or by the fact that in a particular physiological state the algae have a neutral buoyancy. In a 11,355-l/hr (3000-gph) pilot plant (combined flocculation and sedimentation), a floating algal blanket occurred with alum doses of 125–170 mg/l. About 50% of the algae removed was skimmed from the surface.[32]

Because of the infrequent occurrence of conditions necessary for autoflocculation, it is not a usable alternative to remove algae from wastewater stabilization ponds. However, phase isolation experiments to remove algae from lagoon effluents is based on this concept, and some success has been reported.[67] Other work currently in progress should be described soon.[68]

Baffles

The encouragement of attached microbial growth in oxidation ponds is an apparent practical solution for maintaining biological populations while still obtaining the treatment desired. Although baffles are considered useful primarily to ensure good mixing and to eliminate the problem of short-circuiting, they behave similarly to the biological disks, in that they provide a substrate on which bacteria, algae, and other microorganisms can grow.[69,70] In general, attached growth surpasses suspended growth if sufficient surface area is available. In anaerobic or facultative ponds with baffling or biological disks, the microbiological community consists of a gradient of algae to photosynthetic, chromogenic bacteria and, finally, to nonphotosynthetic, nonchromogenic bacteria.[69,70] In these baffle experiments, the presence of attached growth to the baffles has been the reason for the higher efficiency of treatment than that in the nonbaffled systems.

Whereas an attached growth system has the advantage of requiring little maintenance in terms of the biological operations, its initial cost, subsequent treatment requirement, and unproven capability seem to preclude serious consideration at this time.

Rock Filter

The rock filter[71-74] is essentially a porous rock wall or rock embankment located at the end of a lagoon system through which the lagoon effluent is allowed to flow. As the lagoon effluent flows through the void space within the rock filter, SS settle out in the voids and onto the rock surfaces. The accumulated SS are then biologically degraded. This biological degradation process is generally aerobic during warm months and anaerobic during cooler periods.

Laboratory-scale studies on rock filters have been conducted by Martin[75] and O'Brien et al.[76] In addition, pilot-scale rock filter studies have been conducted by Martin and Weller.[77] However, the principal full-scale rock filter research has been conducted by O'Brien[71,72,73] and Williamson and Swanson.[74]

Eudora, Kansas

O'Brien's work[71,72,73] was conducted at Eudora, Kansas during 1974–1975. The study was conducted on both a large rock filter and a small rock filter (Table 3–15). A cross section of each filter is shown in Figure 3–32. The hydraulic loading of the large rock filter ranged from 2977.8 $1/m^3$· day (22.3 gal/ft^3·day) to 535.6 $1/m^3$·day (4.0 gal/ft^3·day) and on the small rock filter from 62.0 $1/m^3$·day (0.5 gal/ft^3·day) to 2187.9 $1/m^3$·day (16.4 gal/ft^3· day). After approximately 11 months of operation, the small rock filter ceased to function because of the growth of a dense algal mat on the downstream side of the filter face.

The BOD_5 of the large rock filter is shown in Figure 3–33. In general, the rock filter did not receive BOD_5 loadings that exceeded 30 mg/l.

Table 3-15
Size Gradation of the Rock Used in the Large
and Small Rock Filters of Eudora, Kansas

Sieve Opening	% Weight Retained	
(cm)	Large Rock	Small Rock
5.08	7.4	
3.81	28.8	
2.54	52.0	13.4
1.91	10.4	33.1
1.27	1.3	29.0
0.95	0.1	10.4
0.67		3.2
0.47		0.9
Porosity	0.44	0.44

Source: From Ref. 72.

However, rock filter effluent BOD_5 concentrations were generally less than 20 mg/l. During the winter months of February, March, and April influent BOD_5 concentrations to the large rock filter did exceed 30 mg/l, and during these months the rock filter effluent BOD_5 concentrations were generally between 25 and 30 mg/l.

The SS performance of the large rock filter is illustrated in Figure 3–34. In general, influent SS concentrations were 50–60 mg/l. Rock filter effluent SS concentrations were generally less than 30 mg/l during the warm summer months, when biological activity with the rock filter is high. However, during the cooler fall, winter, and spring months, rock filter effluent SS concentrations were generally between 40 and 50 mg/l.

O'Brien[72] concluded that lagoon effluents polished by a rock filter could produce an effluent with a BOD_5 concentration of less than 30 mg/l

FIGURE 3-32
Cross section of rock filter at Eudora, Kansas (Ref. 72).

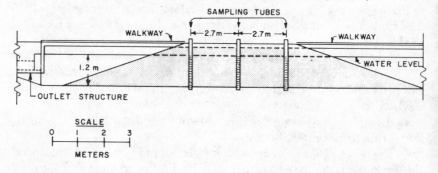

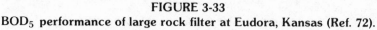

FIGURE 3-33
BOD₅ performance of large rock filter at Eudora, Kansas (Ref. 72).

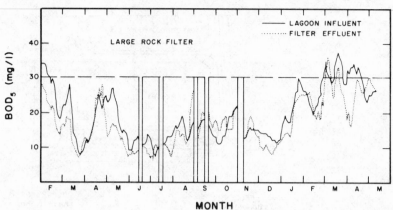

and a SS concentration of less than 30 mg/l during the warm summer months of the year. O'Brien [72] recommended that rock filters should be constructed with rocks greater than 2.54 cm (1 in.) and less than 12.70 cm (5 in.), with most of the rock being approximately 5.1 cm (2 in.) in diameter. Recommended hydraulic loading rates range from 400.9 l/m³·day (3 gal/ft³·day) during cold weather periods to 1203 l/m³·day (9 gal/ft³·day) during warm-weather periods.

California, Missouri

Lane-Rochelle Engineers, Inc. designed a rock filter for California, Missouri [72] in June 1974 to upgrade an existing lagoon. This was placed

FIGURE 3-34
SS performance of large rock filter at Eudora, Kansas (Ref. 72).

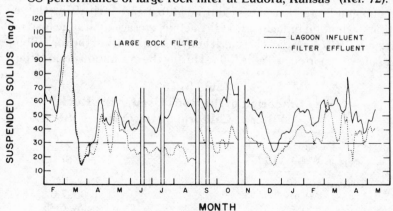

FIGURE 3-35
Rock filter installation at California, Missouri (Ref. 72).

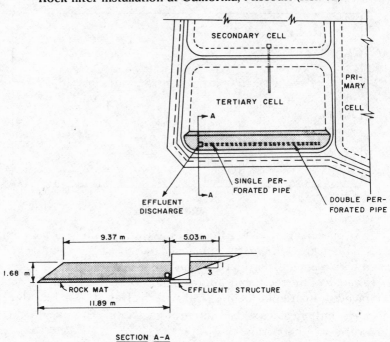

SECTION A-A

along one side of a tertiary lagoon, as illustrated in Figure 3–35. The rock filter was designed for a hydraulic loading rate of 400.9 $l/m^3 \cdot day$ (3.0 $gal/ft^3 \cdot day$). The gradation of rock employed in the filter is reported in Table 3–16. The filter was constructed by city employees in August of 1974.

A performance evaluation of the California, Missouri rock filter was conducted by the Surveillance and Analysis Division, Region 7, U.S. Environmental Protection Agency during March and April 1975. The results of that evaluation indicated that the actual average hydraulic load on the filter was 250.7 $l/m^3 \cdot day$ (1.9 $gal/ft^3 \cdot day$). A summary of the rock filter performance is reported in Table 3–17. During the evaluation period, the rock

Table 3-16
**Size Gradation of the Rock Used in the Rock
Filter at California, Missouri**

Passing Screen by Weight (%)	Screen Size (cm)
85-100	7.62-12.70
0-15	6.35-7.62
0-5	below 6.35

Source: From Ref. 72.

Table 3-17
Performance of Rock Filter at California, Missouri

Date	BOD$_5$ (MG/L)		SS (MG/L)		Dissolved Oxygen (MG/L)	
	Influent	Effluent	Influent	Effluent	Influent	Effluent
March 5, 1975	15	8	35	24	19	6.6
March 13, 1975	14	6	64	26	16.8	7.9
March 19, 1975	19	7	80	24	—	12.9
March 26, 1975	25	13	94	20	12.9	5.2
April 2, 1975	30	15	74	16	13.3	4.6
Average	21	12	69	22	16	7.4

SOURCE: From Ref. 72.

FIGURE 3-36
SS concentrations in lagoon effluent applied to the
California, Missouri rock filter and the rock filter
effluent (Ref. 78).

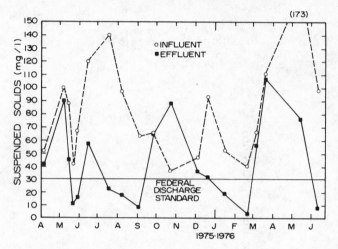

filter average effluent BOD$_5$ concentration was 12 mg/l; however, the average rock filter influent BOD$_5$ concentration was only 21 mg/l. The rock filter average influent SS concentration was 69 mg/l, while the rock filter average effluent SS concentration was 22 mg/l.

A routine monitoring program for the California, Missouri rock filter was initiated by the Missouri Department of Natural Resources[78] in March 1975, and the results of this work are summarized in Figure 3-36. All solids determinations were made on grab samples. The performance of the rock filter was sporadic and failed to meet the federal discharge standard of 30 mg/l of SS in the effluent on 11 of the 19 sampling dates.

Veneta, Oregon

A full-scale rock filter was constructed at Veneta, Oregon in 1975, and an evaluation of the performance was initiated at the beginning of operation.[74] A schematic flow diagram of the Veneta facility is shown in Figure 3-37. The system receives wastewater from a population of 2200, and no industrial wastes are discharged to the sewer system. A plan and cross section of the rock filter are shown in Figure 3-38. Effluent from the lagoon system is discharged into a 0.3-m (1.0-ft) square channel at the bottom of the filter. Trench grating covers the influent channel and improves the distribution of the wastewater. The wastewater flows toward an effluent channel around the periphery of the filter. Water from all sides is collected in a common channel, where the effluent is chlorinated and discharged to the Long Tom

FIGURE 3-37
Schematic flow diagram and aerial photograph of the wastewater treatment system at Veneta, Oregon (Ref. 74).

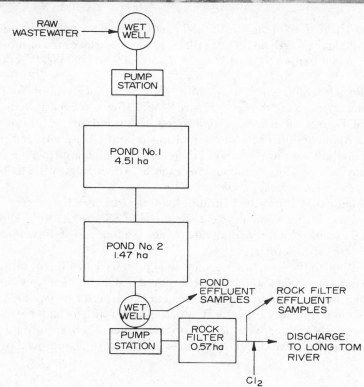

FIGURE 3-38
Rock filter located at Veneta, Oregon
(Ref. 74).

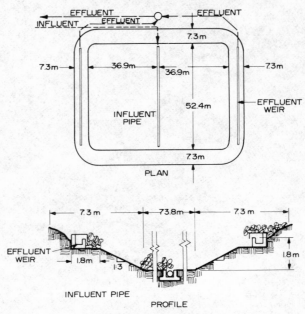

River. The surface of the rock is 0.30 m (1 ft) above the water elevation to prevent algae growth on the rock filter. Effective surface area and volume of the rock filter is 5400 m² (58,000 ft²) and 8200 m³ (290,000 ft³), respectively.

Rock filter influent and effluent BOD_5 and SS concentrations for seven sampling periods are shown in Figure 3–39. Effluent concentrations of SS and BOD_5 met the federal secondary effluent standards on all sampling dates. Ammonia-nitrogen, organic nitrogen, and nitrate-nitrogen in the rock filter effluents are shown in Figure 3–40. Ammonia-nitrogen concentrations increased during the summer months. This increase was attributed to increased biological degradation rates.

Figure 3–41 shows the hydraulic loading rate versus the percentage of SS removal at the Veneta system. Removal of SS appears to be directly related to the hydraulic loading rate, which is a function of the net settling velocity in the rock filter.

Although the Veneta, Oregon system produced an effluent satisfying the federal secondary standards, it was designed hydraulically essentially the same as the California, Missouri system which performed sporadically. The major differences are the upflow of the lagoon effluent and the 0.3 m of rock above the surface of the liquid in the rock filter. There are no data to indicate that these two design differences account for the improved perfor-

FIGURE 3-39
Influent and effluent BOD$_5$ and TSS over study period
(Ref. 74).

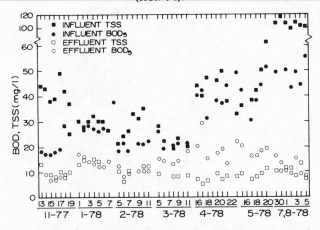

mance; however, both features appear logical and should be incorporated into a design. Other rock filters are being built, and anyone anticipating the use of a rock filter should inquire about these later experiences.

Coagulation-Flocculation

Coagulation followed by sedimentation has been applied extensively for the removal of suspended and colloidal materials from water. Lime, alum, and

FIGURE 3-40
Weekly averages of NH$_4^+$-N, Org-N and NO$_3^-$-N for
rock filter (Ref. 74).

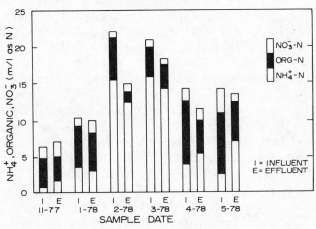

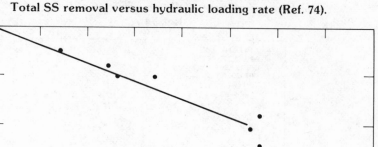

FIGURE 3-41
Total SS removal versus hydraulic loading rate (Ref. 74).

ferric salts are the most commonly used coagulating agents. Floc formation is sensitive to parameters such as pH, alkalinity, turbidity, and temperature. Most of these variables have been studied, and their effects on the removal of turbidity of water supplies have been evaluated. In the case of the chemical treatment of wastewater stabilization pond effluents, however, the data are not comprehensive.

Shindala and Stewart[65] investigated chemical treatment of stabilization pond effluents as a post-treatment process to remove the algae and to improve the quality of the effluent. They found that the optimum dosage for best removal of the parameters studied was 75–100 mg/l of alum. When this dosage was used, the removal of phosphate was 90% and the COD was 70%.

Tenney[79] has shown that at a pH range of 2–4, algal flocculation was effective when a constant concentration of a cationic polyelectrolyte (10 mg/l of C-31) was used. Golueke and Oswald[61] conducted a series of experiments to investigate the relation of hydrogen ion concentrations to algal flocculation. In this study, only H_2SO_4 was used, and only to lower the pH. Golueke and Oswald found that flocculation was most extensive at a pH value of 3, which agrees with the results reported by Tenney. They obtained algal removals of about 80–90%. Algal removal efficiencies were not affected in the pH range of 6–10 by cationic polyelectrolytes.

The California Department of Water Resources[80] reported that of 60 polyelectrolytes tested, 17 compounds were effective in coagulation of algae and were economically competitive when they were compared with mineral coagulation used alone. Generally, less than 10 mg/l of the polyelectrolytes was required for effective coagulation. A daily addition of 1 mg/l of ferric chloride to the algal growth pond resulted in significant reductions in the required dosage of both organic and inorganic coagulants.

McGarry[81] has studied the coagulation of algae in stabilization pond effluents. He reported the results of a complete factorially designed experiment using the common jar test. Tests were performed to determine the economic feasibility of using polyelectrolytes as primary coagulants alone or in combination with alum. He also investigated some of the independent variables that affected the flocculation process, such as concentration of alum, flocculation turbulence, concentration of polyelectrolyte, pH after the addition of coagulants, chemical dispersal conditions, and high rate oxidation pond suspension characteristics. Alum was found to be effective for coagulation of algae from high rate oxidation pond effluents, and the polyelectrolytes used did not reduce the overall costs of algal removal. The minimum cost per unit algal removal was obtained with alum alone (75–100 mg/l). The most significant effects occurred with alum and polyelectrolyte concentrations. The time of polyelectrolyte addition alone had no significant effect. The more important interactions occurred between alum and polyelectrolyte, alum and polyelectrolyte concentrations, time of polyelectrolyte addition and alum concentration, and time of polyelectrolyte addition and polyelectrolyte concentration.

Al-Layla and Middlebrooks[82] evaluated the effects of temperature on algae removal using coagulation-flocculation-sedimentation. Algae removal at a given alum dosage decreased as the temperature increased. Maximum algae removal generally occurred at an alum dosage of approximately 300 mg/l at 10°C. At higher temperatures alum dosages as high as 600 mg/l did not produce removals equivalent to the results obtained at 10°C with 300 mg/l of alum. Settling time required to achieve good removals, flocculation time, organic carbon removal, total phosphorus removal, and turbidity removal were found to vary adversely as the temperature of the wastewater increased.

Dryden and Stern[83] and Parker[26] have reported on the performance and operating costs of a coagulation-flocculation system followed by sedimentation, filtration, and chlorination, with discharge to recreational lakes. This system probably has the longest operating record of any coagulation-flocculation system treating wastewater stabilization pond effluent. The SS concentrations applied to the plant have ranged from about 120 to 175 mg/l, and the plant has produced an effluent with a turbidity of less than 1 Jackson turbidity unit (JTU) most of the time. Aluminum sulfate dosages have ranged from 200 to 360 mg/l. The design capacity is 1893 m³/day (0.5 mgd), and the plant was constructed in 1970 at a cost of

$243,000. Operating and maintenance costs for 1973–1974 were $0.08/m³ ($304/mg). Because of seasonal flow variations, operations and maintenance costs ranged from $0.053 to $0.21/m³ ($200/mg to $800/mg).

Coagulation-flocculation is not easily controlled and requires expert operating personnel at all times. A large volume of sludge is produced, and this introduces an additional operating problem that would very likely be ignored in a small community that is accustomed to a minimum of operation and maintenance of a wastewater lagoon. Therefore, coagulation-flocculation seems infeasible for application in small communities.

Normal Granular Media Filtration

Granular media filtration (rapid sand filters) has proven very successful as a means of liquid-solids separation. The simple design and operation process makes it applicable to wastewater streams containing up to 200 mg/l suspended solids. Automation based on easily measured parameters results in minimum operation and maintenance costs. However, when regular granular media filtration has been applied to the removal of algae from wastewater stabilization effluents, very poor results have been obtained. Good efficiencies can be obtained when chemicals are added prior to filtration or when the wastewater is treated by coagulation and flocculation prior to filtration. Table 3–18 contains a summary of the results with direct granular media filtration.

Diatomateous earth filtration is capable of producing a high-quality effluent when treating wastewater stabilization pond water, but the filter cycles are generally less than 3 hr. This results in excessive usage of backwash water and diatomateous earth, which leads to very high cost and eliminates this method of filtration as an alternative for polishing wastewater stabilization pond effluents.

Microstraining

Microstrainers have been used to remove algae from water in reservoirs before water treatment, to remove solids from industrial wastes, and to polish activated sludge effluents.[89-95] The major problems with the process are incomplete solids removal and difficulty in handling solids fluctuations. These problems may be partially overcome by varying the speed of rotation. In general, drum rotation should be at the slowest rate possible that is consistent with the throughput and should provide an acceptable head differential across the fabric. The controlled variability of drum rotational speed is an important feature of the process, and the speed may be automatically increased or decreased according to the differential head best suited to the circumstances involved.

Table 3 - 18
Summary of Direct Filtration with Rapid Sand Filters

Investigator	Coagulant Aid and Dose (mg/l)	Filter Loading (gpm/sf)	Filter Depth (ft)	Sand Size (mm)	Finding
Borchardt and O'Melia[a,84]	none	0.2-2	2	$d_{50} = 0.32$	removal declines to 21-45% after 15 hr
Davis and Borchardt[85]	Fe: 7 mg/l	2.1	2	$d_{50} = 0.40$	50% algae removal
	none	0.49	NA	$d_{50} = 0.75$	22% algae removal
	none	0.49	NA	$d_{50} = 0.29$	34% algae removal
	none	1.9	NA	$d_{50} = 0.75$	10% algae removal
	none	1.9	NA	$d_{50} = 0.29$	2% algae removal
Foess and Borchardt[86]	Fe	NA	NA	$d_{50} = 0.75$	45% algae removal
	none	2.0	.2	$d_{50} = 0.71$	pH 2.5 90% algae removal pH 8.9 14% removal
Lynam et al. [87]	none	1.1	0.92	$d_{50} = 0.55$	62% SS removal
Kormanik and Cravens[88]	none	—	—	—	11-45% SS removal

Source: From Ref. 26, courtesy of Center For Research In Water Resources, the University of Texas at Austin.
[a] Lab culture of algae.

Pilot plant studies with wastewater-grown algae were conducted at flow rates of 189–379 l/m (50–100 gpm).[61] Only a small amount of algae was removed, even when filter aid was added. Other investigators, however, have found substantial removal of algal cells and other SS in both pilot-plant and full-scale operations. Scriven[92] reported successful removal of heavy phytoplankton loads from Lake Ontario water before treatment by rapid sand filtration. Installation of the 30,280 m³/day (8 mgd) microstraining facility reduced filter backwashes by 75%. A 7570 m³/day (2 mgd) microstraining facility has been constructed for removal of microorganisms from lake water before the water enters a filter plant in Denver.[93] The process has reduced microorganism concentrations by an average of over 90%. Berry[89] reported on the operation of 15 microstrainer plants in Canada and concluded that they are effective and economical for the removal of algae. In a series of nine investigations over a period of years, the average reduction of the plankton was 89%. Microstraining has also been effective as a tertiary treatment process, and BOD_5 and SS reductions of 50 to over 90% have been observed.[94,95] During studies using microstraining on effluents from combined sewer overflows, up to 98% SS reduction was obtained.[96]

Although microstraining has been reasonably successful in pretreating water supplies prior to filtration, very little success has been reported in removing algae from wastewater stabilization pond effluents until recently. Among others, Golueke and Oswald,[61] Dryden and Stern,[83] Lynam et al.,[87] and the California Department of Water Resources[80] have evaluated the microstraining process as a means of removing algae from wastewater stabilization pond effluents and have reported very little success. The size, approximately 23 μm, of the microstraining fabric has been the principal reason for failure, because algae in general are small and pass through the fabric. In addition, binding has created a considerable problem, resulting in very high usages of backwash water.

Kormanik and Cravens[88] have conducted 2 years of full-scale field studies using 1 μm polyester fabric for the removal of algae. As shown in Tables 3–19 to 3–22, the new fabric produces an excellent effluent. Table 3–19 compares the algae removal results obtained with microscreens and dissolved air flotation (DAF) units, and Table 3–20 compares microscreens and pressure filters. As mentioned above, DAF units produce an excellent-quality effluent, but a normal granular media filter removes very little algae.

Based on the results shown in Tables 3–19 to 3–22, the microscreen manufacturers are promoting the use of the 1 μm screen with the return of the filtered algae to the pond. Short-term experience indicates that the return of filtered algae does not cause problems; however, the potential exists for the filtered material to accumulate and cause overloading of the screen. The effect of solids recylce through the lagoon system should be monitored in the newly constructed microscreen systems. Multicell lagoons

Table 3-19
Summary of Side-by-Side Microscreens versus DAF
Continuous 24-Hr Runs

Dissolved Air Flotation Operating Conditions	Pond Effluent (mg/l) T-BOD/TSS	Effluent (mg/L)	
		DAF T-BOD/TSS	Microscreen T-BOD/TSS
100 mg/l alum No flocculation 25% Recycle Overflow rate 2 gpm/ft^2	29/37	9/12	23/4
100 mg/l alum No flocculation 25% recycle 3 gpm/ft^2	35/47	13/23	23/19
100 mg/l alum 11 min flocculation 25% recycle 3.4 gpm/ft^2	47/61	—/23	4/4
145 mg/l alum 10.5-min flocculation 19% recycle 3.2 gpm/ft^2	—	16/20	3/17
100 mg/l alum 19-min flocculation 25% recycle 2 gpm/ft^2	42/58	15/20	33/20
100 mg/l alum 3 mg/l polymer 7.5-min flocculation 20% recycle 3.6 gpm/ft^2	44/34 soluble BOD 18	9/4 soluble BOD 4	29/10 soluble BOD 4

Source: From Ref. 88.

would be less likely to develop an accumulation problem if the filtered solids were returned to the primary cell. A separate small cell for collection of the screen wash water would probably not eliminate the economic advantages of microscreens if this became necessary.

Centrifugation

Centrifugation has been found to be an effective process for the dewatering of industrial and domestic sludges.[97] From 50 to 95% of the influent solids may be removed by the process, and the solids concentration in the sludge

Table 3-20
Summary of Side-by-Side Microscreen versus
Pressure Filter Continuous 24-Hr Test Runs

Run Number	Microscreen Influent Source	Hydraulic Loading (GPM/FT^2) Microscreen	Hydraulic Loading (GPM/FT^2) Pressure Filter	SS (MG/L) Microscreen and Pressure Filter Influent	SS (MG/L) Microscreen Effluent	SS (MG/L) Pressure Filter Effluent
1	Polishing pond effluent	2.40	2.06	29	7	21
2	Polishing pond	2.90	1.93	27	14	24
3	Polishing pond	1.65	1.93	19	4	13
4	Polishing pond	1.72	1.93	40	4	22
5	Polishing pond	2.50	—	28	10	24

SOURCE: From Ref. 88

cake is normally in the range of 15 to 40%. Centrifuge operation is characterized by liquid throughput, solids throughput, speed of rotation, and pool depth.

Pilot-plant experiments conducted on wastewater-grown algae resulted in about 77% removal at feed rates of about 3 gpm (11.4 l/min).[61] Solids concentration in the "cake" was about 10%, and the effluent contained about 260 mg/l of SS. Centrifugation was found to be an effective means for dewatering algal sludge.

The major disadvantage of centrifugation is the temperamental nature of the equipment. Abrasive solids in the water can cause rapid deterioration of the scroll. Operating problems associated with the relatively

Table 3-21
Comparison of Lagoon and Microscreen Effluents
BOD/SS Ratios versus EPA Lagoon Standards

Site	Actual Pond Discharge[a] (Microscreen Influent)	Ratio	Required for EPA Standard	Microscreen Effluent Actually Achieved
1	43/51	0.84	30/36	12/19
2	38/100	0.38	30/79	23/21
3	61/81	0.75	30/40	19/22
4	44/83	0.53	30/57	14/11
5	50/72	0.70	30/43	28/9
6	30/59	0.51	30/59	18/15

SOURCE: From Ref. 88.

[a] 30 mg/l BOD currently required for lagoon effluent discharges less than 2 mgd.

Table 3-22
Summary of the Average Microscreen Influent and
Effluent Total SS and T-BOD$_5$ (mg/l)

Run Number	Lagoon Effluent (Microscreen Influent) TSS (mg/l)/T-BOD$_5$	Microscreen Effluent TSS (mg/l)/T-BOD$_5$
6	69/72	9/9
7	58/30	15/15
8	44/45	12/14
9	126/38	19/23
10	44/—	13/—
11	64/32	22/16
12	26/—	6/—

Source: From Ref. 88.

sophisticated equipment take up a considerable amount of a skilled operator's time. Because one of the advantages of stabilization ponds is the low operating expense, it would seem to be impractical to couple these advantages with a unit process that has a major disadvantage of high operating costs.

References

1. Environmental Protection Agency. 1975. *Process Design for Suspended Solids Removal.* Technology Transfer, EPA–625/1–75–003a, Washington,D.C., January 1975.

2. Environmental Protection Agency. 1973. *Upgrading Lagoons.* Technology Transfer, Washington, D.C., August 1973.

3. Environmental Protection Agency. 1974. *Process Design Manual for Upgrading Existing Wastewater Treatment Plants.* Technology Transfer, Washington, D.C., October 1974.

4. Environmental Protection Agency. 1981. *Process Design Manual for Land Treatment of Municipal Wastewaters,* ERIC, US EPA Cincinnati, Ohio. EPA 625/1–77–008.

5. Reed, S. C., et al. 1979. *Cost of Land Treatment Systems,* EPA 430/9-75-003, MCD–10, Sept. 1979 revision, OWPO, US EPA, Washington, D.C.

6. Reed, S. C. 1979. *Health Aspects of Land Treatment,* June 1979 Land Treatment Technology Transfer Seminars, ERIC, US EPA, Cincinnati, Ohio.

7. Hinrichs, D. J., et al. 1980. *Assessment of Current Information on Overland Flow Treatment of Municipal Wastewater,* EPA 430/9–80–002, MCD–66, OWPO, US EPA, Washington, D.C.

200   **Wastewater Stabilization**

8. Jewell, W. J., B. Seabrook. 1979. *A History of Land Application as a Treatment Alternative*, EPA 430/9-79-012, OWPO, US EPA, Washington, D.C.

9. Thomas, R. E. 1974. Upgrading Lagoon Treatment with Land Application. In *Upgrading Wastewater Stabilization Ponds to Meet New Discharge Standards*, PRWG 159-1, Utah Water Research Laboratory, Utah State University, Logan, Utah, November

10. Peters, R. E., C. R. Lee, D. J. Bates. 1980. *Field Investigations of Overland Flow Treatment of Municipal Lagoon Effluent*, USAE WES Technical Report, USAE WES, Vicksburg, Mississippi.

11. Martel, C. J., et al. 1980. *Rational Design of Overland Flow Systems*. In Proceedings Am. Soc. Civil Engineers National Conference of Environmental Engineering, July 8-10, 1980, New York, New York.

12. Spraham, V. 1976. Tertiary Treatment and Advanced Waste Water Treatment-Sheet Flow Over Grassland. In *Draft Project Report. City of Davis-Algae Removal Facilities*, prepared by Brown and Caldwell, Walnut Creek, California, November 1976.

13. Brown and Caldwell. 1972. *Report on Pilot Flotation Studies at the Main Water Quality Control Plant*. Prepared for the City of Stockton, California, February 1972.

14. Hill, D. W., J. H. Reynolds, D. S. Filip, and E. J. Middlebrooks. 1977. *Series Intermittent Sand Filtration of Wastewater Lagoon Effluents*. PRWR 159-1, Utah Water Research Laboratory, Utah State University, Logan, Utah.

15. Harris, S. E., D. S. Filip, J. H. Reynolds, and E. J. Middlebrooks. 1978. *Separation of Algal Cells from Wastewater Lagoon Effluents. Vol. I: Intermittent Sand Filtration to Upgrade Waste Stabilization Lagoon Effluent*. EPA-600/2-78-033. Municipal Environmental Research Laboratory, U.S. Environmental Protection Agency, Cincinnati, Ohio.

16. Marshall, G. R., and E. J. Middlebrooks. 1974. *Intermittent Sand Filtration to Upgrade Existing Wastewater Treatment Facilities*. PRJEW 115-2, Utah Water Research Laboratory, Utah State University, Logan, Utah. 80 p.

17. Reynolds, J. H., S. E. Harris, D. W. Hill, D. S. Filip, and E. J. Middlebrooks. 1974. *Single and Multi-Stage Intermittent Sand Filtration to Upgrade Lagoon Effluents—Preliminary Report, Presented at EPA Technology Transfer Seminar on Wastewater Lagoons*, November 19-20, Boise, Idaho.

18. Harris, S. E., J. H. Reynolds, D. W. Hill, D. S. Filip, and E. J. Middlebrooks. 1975. *Intermittent Sand Filtration for Upgrading Waste Stabilization Pond Effluents*. Presented at 48th Water Pollution Control Federation Conference, Miami, Florida (October 5-10, 1975).

19. Reynolds, J. H., S. E. Harris, D. W. Hill, D. S. Filip, and E. J. Middlebrooks. 1974. *Intermittent Sand Filtration to Upgrade Lagoon Effluents*. Preliminary Report, pp. 71-88. PRWG 159-1. Utah Water Research Laboratory, Utah State University, Logan, Utah.

20. Messinger, S. S., J. H. Reynolds, and E. J. Middlebrooks. 1977. *Anaerobic Lagoon—Intermittent Sand Filter System for the Treatment of Dairy Parlor Wastes*. Agricultural Experiment Station, Utah State University, Logan, Utah.

21. Bishop, R. P., S. S. Messinger, E. J. Middlebrooks, and J. H. Reynolds. 1976.

Treating Aerated and Anaerobic Lagoon Effluents with Intermittent Sand Filtration. *31st Purdue Industrial Waste Conference*, May 4–6, Purdue University, West Lafayette, Indiana.

22. Tupyi, B., D. S. Filip, J. H. Reynolds, and E. J. Middlebrooks. 1979. *Separation of Algal Cells from Wastewater Lagoon Effluents. Vol. II: Effect of Sand Size on the Performance of Intermittent Sand Filters.* EPA–600/2–79–152. Municipal Environmental Research Laboratory, U.S. Environmental Protection Agency, Cincinnati, Ohio.

23. Fair, G. M., J. C. Geyer, and D. A. Okun. 1968. *Water and Wastewater Engineering*, Vol. II. Wiley, New York, New York.

24. Reynolds, J. H., S. E. Harris, D. W. Hill, D. S. Filip, and E. J. Middlebrooks. 1975. Intermittent Sand Filtration for Upgrading Waste Stabilization Ponds. In *Ponds as a Wastewater Treatment Alternative*. Water Resources Symposium No. 9. July 22–24. Center for Research in Water Resources. University of Texas, Austin, Texas.

25. Russell, J. S., E. J. Middlebrooks, and J. H. Reynolds. 1980. *Wastewater Stabilization Lagoon-Intermittent Sand Filter System.* EPA–600/2–80–032, Municipal Environmental Research Laboratory, U.S. Environmental Protection Agency, Cincinnati, Ohio.

26. Parker, D. S. 1976. Performance of Alternative Algae Removal Systems. In *Ponds as a Wastewater Treatment Alternative*, edited by E. F. Gloyna, J. F. Malina, Jr., and E. M. Davis, Center for Research in Water Resources, College of Engineering, The University of Texas at Austin.

27. Ort, J. E. 1972. Lubbock WRAPS It Up. *Water Wastes Eng.*, **9**, 9, September.

28. Komline-Sanderson Engineering Corporation. 1972. *Algae Removal Application of Dissolved Air Flotation.* Peapack, New Jersey, August 1972.

29. Bare, W. F. R. 1971. Algae Removal from Waste Stabilization Lagoon Effluents Utilizing Dissolved Air Flotation. M.S. thesis, Utah State University, Logan.

30. Stone, R. W., D. S. Parker, and J. A. Cotteral. 1975. Upgrading Lagoon Effluent to Meet Best Practicable Treatment. *J. Water Poll. Cont. Fed.* **46**, 8, 2019–2042.

31. Snider, E. F., Jr. 1976. Algae Removal by Air Flotation. In *Ponds as a Wastewater Treatment Alternative*, edited by E. F. Gloyna, J. F. Malina, Jr., and E. M. Davis, Center for Research in Water Resources, College of Engineering, The University of Texas at Austin.

32. van Vuuren, L. R. J., and F. A. van Duuren. 1965. Removal of Algae from Wastewater Maturation Pond Effluent. *J. Water Poll. Contr. Fed.*, **37**, 1256.

33. van Vuuren, L.R.J., P. G. J. Miiring, M. R. Henzen, and F. F. Kolbe. 1965. The Flotation of Algae in Water Reclamation. *Inter. J. Air Water Poll.*, **9**, 12, 823.

34. Parker, D. S., J. B. Tyler, and T. J. Dosh. 1973. Algae Removal Improves Pond Effluent. *Water Wastes Eng.*, **10**, 1, January.

35. Farnham, Helen. 1981. Personal communication, Sunnyvale, California Wastewater Treatment Plant, Sunnyvale, California.

36. Brown and Caldwell. 1976. Draft Project Report, City of Davis-Algae Removal Facilities. Walnut Creek, California, November 1976.

37. Pierce, D. M. 1974. Performance of Raw Waste Stabilization Lagoons in Michigan with Long Period Storage Before Discharge. In *Upgrading Wastewater Stabilization Ponds to Meet New Discharge Standards*. PRWG151. Utah Water Research Laboratory, Utah State University, Logan, Utah.

38. Bowen, S. P. 1977. *Performance Evaluation of Existing Lagoons, Peterborough, New Hampshire*. EPA–600/2–77–085. Technology Series, Municipal Environmental Research Laboratory, U.S. Environmental Protection Agency, Cincinnati, Ohio.

39. Hill, D. O., and A. Shindala. 1977. *Performance Evaluation of Kilmichael Lagoon*. EPA–600/2–77–109. EPA Technology Series, Municipal Environmental Research Laboratory, U.S. Environmental Protection Agency, Cincinnati, Ohio.

40. McKinney, R. E. 1977. *Performance Evaluation of an Existing Lagoon System at Eudora, Kansas*. EPA–600/2–77–167. Technology Series, Municipal Environmental Research Laboratory, U.S. Environmental Protection Agency, Cincinnati, Ohio.

41. Reynolds, J. H., R. E. Swiss, C. A. Macko, and E. J. Middlebrooks. 1977. *Performance Evaluation of an Existing Seven Cell Lagoon System*. EPA–600/2–77–086. Technology Series, Municipal Environmental Research Laboratory, U.S. Environmental Protection Agency, Cincinnati, Ohio.

42. Oswald, W. J. 1968. Advances in Anaerobic Pond Systems Design. In *Advances in Water Quality Improvement*. Edited by E. F. Gloyna and W. W. Eckenfelder, Jr. University of Texas Press, Austin. p. 409.

43. Caldwell, D. H. 1946. Sewage Oxidation Ponds—Performance, Operation and Design. *Sewage Works J.* 18, 3, 433–458.

44. Dinges, R., and A. Rust. 1969. Experimental Chlorination of Stabilization Pond Effluent. *Public Works*, 100, 3, 98–101.

45. Chem Pure, Inc. 1972. Algae Destruction Chamber Upgrades Sewage Lagoon Effluents. *Civil Eng.* 42, 11, 106.

46. Echelberger, W. F., J. L. Pavoni, P. C. Singer, and M. W. Tenney. 1971. Disinfection of Algal Laden Waters. *J. Sanit. Eng. Div. ASCE*, 97, SA5, 721–730.

47. Hom L. W. 1972. Kinetics of Chlorine Disinfection in an Ecosystem. *J. Sanit. Eng. Div. ASCE*, 98, SA1, 183–194.

48. Wight, J. L. 1976. *A Field Study: The Chlorination of Lagoon Effluents*. M.S. thesis, Utah State University, Logan, Utah.

49. Johnson, B. A. 1976. *A Mathematical Model for Optimizing Chlorination of Waste Stabilization Lagoon Effluent*. PhD. dissertation, Utah State University, Logan, Utah.

50. Duffer, W. R. 1974. Lagoon Effluent Solids Control by Biological Harvesting. In *Upgrading Wastewater Stabilization Ponds to Meet New Discharge Standards*, PRWG159–1, Utah Water Research Laboratory, Utah State University, Logan, Utah.

51. Ryther, J. H. 1979. Treated Sewage Effluent as a Nutrient Source for Marine Polyculture. In *Agriculture Systems for Wastewater Treatment*. EPA–

430/9–80–006. Office of Water Program Operations, U.S. Environmental Protection Agency, Washington, D.C.

52. Goldman, J. C. 1976. Personal communication, Woods Hole Oceanographic Institute.

53. Wolverton, B. C., R. M. Barlow, and R. C. McDonald. 1975. Application of Vascular Aquatic Plants for Pollution Removal, Energy and Food Production in a Biological System. *National Space Technology Laboratories*, TM–X–72726, Bay St. Louis, Mississippi, May 12, 1975.

54. Wolverton, B. C., R. C. McDonald, and J. Gordon. 1975. Water Hyacinths and Alligator Weeds for Final Filtration of Sewage. *National Space Technology Laboratories*, TM–X–72724, Bay St. Louis, Mississippi, May 21.

55. Small, M. M. 1977. *National Sewage Recycling Systems, BNL 50630*. Brookhaven National Laboratory, U.S. Energy Research and Development Administration, Upton, New York.

56. Klopp, R. F. 1977. Personal communication, Director of Public Works, City of Arcata, California.

57. Serfling, S. A., and C. Alsten. 1978. An Integrated, Controlled Environment Aquaculture Lagoon Process for Secondary or Advanced Wastewater Treatment. In Proceedings of a Conference Held August 23–25, 1978, at Utah State University, Logan, Utah, entitled *Performance and Upgrading of Wastewater Stabilization Ponds*.

58. Graham, H. J., and R. B. Hunsinger. Undated. *Phosphorus Removal in Seasonal Retention Lagoons by Batch Chemical Precipitation*. Project No. 71–1–13, Wastewater Technology Centre, Environment Canada, Burlington, Ontario.

59. Pollutech Pollution Advisory Services, Ltd. Undated. *Nutrient Control in Sewage Lagoons*. Project No. 72–5–12, Wastewater Technology Centre, Environment Canada, Burlington, Ontario.

60. Pollutech Pollution Advisory Services, Ltd. 1975. *Nutrient Control in Sewage Lagoons, Volume II*. Project No. 72–5–12, Wastewater Technology Centre, Environment Canada, Burlington, Ontario, October 1975.

61. Golueke, C. G., and W. J. Oswald. 1965. Harvesting and Processing Sewage-Grown Planktonic Algae. *J. Water Poll. Cont. Fed.*, 37, 4, 471–498.

62. McGriff, E. C., and R. E. McKinney. 1971. Activated Algae: A Nutrient Removal Process. *Water Sewage Works*, 118, 337.

63. McKinney, R. E., et al. 1971. Ahead: Activated Algae? *Water Wastes Eng.*, 8, 51.

64. Hill, D. W., J. H. Reynolds, D. S. Filip, and E. J. Middlebrooks. 1977. *Series Intermittent Sand Filtration of Wastewater Lagoon Effluents*. PRWG 159–1, Utah Water Research Laboratory, Utah State University, Logan, Utah.

65. Hill, D. O., and A. Shindala. 1976. *Performance Evaluation of Kilmichael Lagoon*. Unpublished Report for Contract No. 68–03–2061. Environmental Protection Agency, Cincinnati, Ohio.

66. Shindala, A., and J. W. Stewart. 1971. Chemical Coagulation of Effluents from Municipal Waste Stabilization Ponds. *Water Sew. Works*, 118, 4, 100–103.

67. Koopman, B. L., J. R. Benemann, and W. J. Oswald. 1978. Pond Isolation and Phase Isolation For Control of Suspended Solids and Concentration in Sewage Oxidation Pond Effluents. In: Proceedings of a Conference Held August 23–25, 1978, at Utah State University, Logan, Utah, entitled *Performance and Upgrading of Wastewater Stabilization Ponds.*

68. McGriff, E. C. 1979. Personal communication.

69. Reynolds, J. H. 1971. *The Effects of Selected Baffle Configurations on the Operation and Performance of Model Waste Stabilization Ponds.* M.S. thesis, Utah State University, Logan, Utah.

70. Nielson, S. B. 1973. *Loading and Baffle Effects on Performance of Model Waste Stabilization Ponds.* M.S. thesis, Utah State University, Logan, Utah.

71. O'Brien, W. J. 1974. Polishing Lagoon Effluents with Submerged Rock Filters. *Upgrading Wastewater Stabilization Ponds to Meet New Discharge Standards.* Report PRWG 159-1, Utah Water Research Laboratory, Utah State University, Logan.

72. O'Brien, W. J. 1975. Algal Removal by Rock Filtration. *Transactions, Twenty-fifth Annual Conference on Sanitary Engineering,* The University of Kansas, Lawrence, Kansas.

73. O'Brien, W. J. 1975. Algal Removal by Rock Filtration. In: *Ponds as a Wastewater Treatment Alternative.* Water Resources Symposium No. 9, July 22–24. Center for Research in Water Resources, University of Texas, Austin, Texas.

74. Williamson, K. J., and G. R. Swanson. 1978. Field Evaluation of Rock Filters for Removal of Algae from Lagoon Effluents. In Proceedings of a Conference Held August 23–25, 1978, at Utah State University, Logan, Utah, entitled *Performance and Upgrading of Wastewater Stabilization Ponds.*

75. Parker, D. S. 1976. Performance of Alternative Algae Removal Systems. In *Ponds as a Wastewater Treatment Alternative,* edited by E. F. Gloyna, J. F. Malina, Jr., and E. M. Davis, Center for Research in Water Resources, College of Engineering, The University of Texas at Austin.

76. O'Brien, W. J., R. E. McKinney, M. D. Turvey, and D. M. Martin. 1973. Two Methods for Algae Removal from Oxidation Pond Effluents. *Water Sewage Works,* **120**, 3, 66.

77. Parker, D. S., J. B. Tyler, and T. J. Dosh. 1973. Algae Removal Improves Pond Effluent. *Water Wastes Eng.,* **10**, 1, January.

78. Forester, T. H. 1977. Personal communication. Missouri Department of Natural Resources, Jefferson City, Missouri.

79. Tenney, M. W. 1968. Algal Flocculation with Aluminum Sulfate and Polyelectrolytes. *Appl. Microbiol.,* **18**, 6, 965.

80. California Department of Water Resources. 1971. *Removal of Nitrate by an Algal System; Bioengineering Aspects of Agricultural Drainage.* San Joaquin Valley, California. Sacramento, California.

81. McGarry, M. G. 1970. Algal Flocculation with Aluminum Sulfate and Polyelectrolytes. *J. Water Poll. Cont. Fed.,* **42**, Part 2, 5, R191.

82. Al-Layla, M. A., and E. J. Middlebrooks. 1975. Effect of Temperature on Algal

Removal From Wastewater Stabilization Ponds by Alum Coagulation. *Water Res.* **9**, 873–879.

83. Dryden, F. D., and G. Stern. 1968. Renovated Wastewater Creates Recreational Lake. *Environmental Science and Technology*, **2**, 4, 266–278.

84. Borchardt, J. A., and C. R. O'Melia. 1961. Sand Filtration of Algae Suspensions. *J. Amer. Water Works Ass.*, **53**, 12, 1493–1502.

85. Davis, E., and J. A. Borchardt. 1966. Sand Filtration of Particulate Matter. *J. Sanit. Eng. Div. ASCE*, **92**, SA5, 47–60.

86. Foess, G. W., and J. A. Borchardt. 1969. Electrokinetic Phenomenon in the Filtration of Algae Suspension. *J. Amer. Water Works Ass.*, **61**, 7, 333–337.

87. Lynam, G., G. Ettelt, and T. McAloon. 1969. Tertiary Treatment of Metro Chicago by Means of Rapid Land Filtration and Microstrainers. *J. Water Poll. Cont. Fed.*, **41**, 2, 247–279.

88. Kormanik, R. A., and J. B. Cravens. 1978. Microscreening and Other Physical-Chemical Techniques for Algae Removal. In Proceedings of a Conference Held August 23–25, 1978, at Utah State University, Logan, Utah, entitled *Performance and Upgrading of Wastewater Stabilization Ponds*.

89. Berry, A. E. 1961. Removal of Algae by Microfilters. *J. Amer. Water Works Assn.*, **53**, 1503.

90. Eckenfelder, W. W., Jr. 1966. *Industrial Water Pollution Control.* McGraw-Hill, New York, New York.

91. Evans, G. R., and P. L. Boucher. 1962. Microstraining—Description and Application. *Water Sewage Works*, **109**, R-184.

92. Scriven, J. 1960. Microstraining Removes Algae and Cuts Filter Backwashing. *Water Works Eng.*, **113**, 6, 554–555.

93. Turre, G. J., and G. R. Evans. 1959. Use of Microstrainer at Denver, Colorado. *J. Amer. Water Works Ass.*, **51**, 354.

94. Diaper, E. W. J. 1969. Tertiary Treatment by Microstraining. *Water Wastes Eng.*, **116**, 202.

95. Bodien, D. G., and R. L. Stenburg. 1969. Microscreening Effectively Polishes Activated Sludge Effluent. *Water Wastes Eng.*, **3**, 9, 74.

96. Federal Water Quality Administration. 1970. *Microstraining and Disinfection of Combined Sewer Overflows.* Water Poll. Control Res. Ser. 11023 EVO 06/70, FWQA, U.S. Dept. of the Interior, Washington, D.C.

97. Dugan, G. L., et al. 1970. Photosynthetic Reclamation of Agricultural Solid and Liquid Wastes. *SERL Rept. No. 70-1*, University of California, Berkeley.

CHAPTER
4

The Hydraulics of
Waste Stabilization Ponds

THE TREATMENT EFFICIENCY of waste stabilization ponds depends on the biological factors such as the type of waste and organic loading. However, the biological activity in a pond is influenced by the environmental conditions of temperature, wind, sunlight, and humidity and the physical factors of pond geometry and the hydraulic flow patterns. The flow patterns within stabilization ponds are affected by the shape of the pond or lagoon, the presence of dead spaces, the existence of density differences, and the positioning of inlets and outlets. These hydraulic flow characteristics obviously have an effect on the dispersion of the waste material as well as the average detention time and, ultimately, on the organic (BOD_5) and pathogenic organism removal efficiency of the treatment process.

In this chapter the effects of these hydraulic flow characteristics on the treatment efficiency are evaluated. The information presented draws strongly on the work of Mangelson[1] and George.[2] The approach used is to obtain information from the age distribution functions of the fluid particles within continuous-flow process vessels or tanks. Since the biological reactions that occur in waste stabilization ponds have been found to closely follow a first-order chemical reaction, they can be coupled with the age distribution functions to give a quantitative measure of treatment efficiency.

This method of analysis in this chapter applies whether the age distribution functions are mathematically determined, established by dye injection studies on hydraulic models or full-sized ponds, or synthesized from functions associated with similar ponds. Age distribution functions for a wide variety of pond dimensions, inlet-outlet configurations, and baffling schemes are presented and discussed in this chapter. Evaluations of treatment efficiency using the age distribution functions are demonstrated, and a computer program is provided to assist the designer in this task.

The data used to generate the age distribution functions presented in this chapter were obtained from dye studies performed on laboratory models verified with prototype ponds. The prototype ponds used for model-verification purposes were operated by the City of Logan, Utah. The model ponds were constructed in the Utah Water Research Laboratory.

Background

Previous approaches to the design of waste stabilization ponds have been primarily empirical, with such parameters as depth, detention time, physical shape of pond, and BOD_5 reduction derived from experience. Advances in biological oxidation, photosynthetic phenomena, and algology make a more theoretical approach both possible and feasible from the biological standpoint.

The parameters which must be established for the design of stabilization ponds include detention period, hydraulic loading, depth, recirculation, mixing, pond size and shape, and inlet and outlet systems. Because there are many different methods for determining these parmeters and design methods for the three types of ponds (aerobic, anaerobic, and facultative), a detailed listing of these is not made here. Those parameters, which are all determined in essentially the same way, are briefly mentioned.

1. *Detention period (theoretical)*—The volume of the pond divided by the flow rate into the pond.
2. *Hydraulic loading*—The depth of the pond divided by the theoretical detention time.
3. Depth.

The determination of the pond depth is somewhat arbitrary, depending on the pond type. Most of the other parameters have been determined essentially from past experience with sewage ponds.

Thirumurthi[3] has made an attempt to determine these design parameters from an experimental model and to incorporate these into a design equation based on chemical reactor design methods. The goal was to develop design formulas based on sound scientific and mathematical principles related to chemical engineering unit operations and reactor design concepts.

Present design concepts tend to neglect the shape of the treatment pond, the existence of dead spaces, short-circuiting, density differences, and the inlet and outlet flow patterns. These hydraulic flow characteristics will have an obvious effect on molecular and turbulent diffusion as well as the detention time, and hence on BOD_5 removal efficiency. Because a quantitative evaluation of hydraulic flow characteristics depends on some chemical reactor concepts, those concepts must be developed first.

Chemical Reactor Design

In recent years some researchers[3,4] have tried to develop a method for designing algal waste stabilization ponds based on existing chemical engineering practice for reactor design. The discussion of these approaches is made later. First, some of the important chemical engineering terms are defined.

1. *Chemical Reactor*—A chemical reactor is a vessel in which a chemical reaction takes place. There are three general types of reactors—namely, the batch, the steady flow, and the unsteady flow. A steady flow reactor is one in which the influent and effluent flow rates remain constant with time, and as a result it is mathematically easier to work with than the unsteady flow reactor. Also, the steady flow reactor is more realistic than a batch reactor from the viewpoint of a sanitary engineer, because the biological processes of stabilization pond operation are continuous rather than batch. These two reasons point out the desirability of using steady flow reactor principles in the design of waste stabilization ponds.

2. *Types of Fluid Flow in a Vessel*—When a fluid passes through a reactor, tank, or pond, a number of possible patterns of flow could exist, depending on the entrance and exit arrangements, short-circuiting flow rate, velocity, volume of tank, and fluid properties. The two general categories of flow are: (a) ideal flow and (b) nonideal flow.

(a) *Ideal Flow*—This flow consists of two types: The first is plug flow (piston, slug, tubular, or nonmix flow), which is characterized by the orderly flow of fluid through the tank with no mixing and no element of fluid overtaking any other element. Consequently, there is no velocity gradient and no diffusion or dispersion. The residence or detention time of all fluid elements is the same. The second is completely mixed (total back mix, or stirred tank) flow by a uniform composition in the tank. Any fluid element has an equal chance of being found at the outlet.

(b) *Nonideal Flow*—An actual process vessel (e.g., an aeration tank or a waste stabilization pond) is obviously far from ideal. In an actual situation stagnant pockets or dead space, short-circuiting, and dispersion will occur to create nonideal-type flows characterized by channeling, recycling, eddying, and the like. Since the flow in a tank or pond is in reality nonideal, the problem should be approached using nonideal flow terms.

3. *Stimulus-Response Methods of Characterizing Flow*—A tracer or stimulus is applied at the inlet to a tank or vessel, and the response is measured as a function of time at the outlet.

4. *Open and Closed Vessels*—In a closed vessel the fluid moves in and out by bulk flow alone. Plug flow exists in the entering and leaving streams. In an open vessel neither the entering nor the leaving fluid streams satisfy the plug flow requirements of the closed vessel. A waste stabilization pond is a closed vessel.

5. *Mean Residence Time of Fluid (Theoretical)*—The theoretical detention time, $\bar{t}$, is defined as the volume of the vessel divided by the flow rate through the vessel.

6. *Reduced or Dimensionless Time*—The dimensionless time, θ, is defined as time, t, divided by the theoretical detention time, $\bar{t}$.

Age Distribution Functions

Danckwerts[5] first introduced the general idea of age distribution functions. These functions give information about the general behavior of the fluid that resides a certain time in a closed vessel. They do not give information about point-to-point change of the variables and do not yield complete information about the behavior of the fluid in the vessel.

Some of the most useful age distribution functions follow.

F-Diagrams. If the incoming fluid flowing into a vessel is suddenly changed from a white to a red color, then $F(\theta)$ is the fraction of red material that occurs at the outlet as a function of dimensionless time θ. A plot of $F(\theta)$ versus θ is called the *F*-diagram. (The *F*-diagram, in reality, is a dimensionless mass diagram.) Figure 4–1 shows *F*-diagrams for some representative types of flow.

FIGURE 4-1
The *F*-diagram.

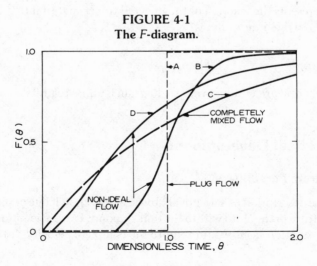

Since there is always some longitudinal mixing with real fluids because of viscous effects and molecular or eddy diffusion, the perfect piston or plug flow (curve A, Figure 4–1) will never occur. Curve B in Figure 4–1 shows a representative F-diagram for flow with some longitudinal mixing. Curve C is a diagram for complete mixing given by the equation

$$F(\theta) = 1 - e^{-\theta} \tag{4-1}$$

Curve D is the F-diagram where there is considerable "dead water." The term "dead water" or "stagnant pockets" means that some fraction of the fluid is trapped in corner eddies and spends much more than the average length of time in the vessel.

The shape of an F-diagram depends on the relative times taken by various portions of an incoming fluid element to flow through the vessel, that is, on the distribution of particle residence times.

C-Diagrams. C-diagrams are dimensionless plots similar to F-diagrams where $C(\theta) = c(\theta)/c_0$. The term $c(\theta)$ is the tracer concentration at the outlet as a function of time, $c_0 = M_T/V$ and θ is defined the same as for F-diagrams. The term M_T is the mass of tracer instantaneously dropped into the inlet of a pond of volume V. The same information as that obtained with an F-diagram can be obtained with a C-diagram. The term $C(\theta)$ is the dimensionless concentration of that tracer at the outlet as a function of dimensionless time; $C(\theta)$ is also a measure of time of residence in the vessel of each fraction of the injected dye. Figure 4–2 shows some typical C-diagrams for the systems whose F-diagrams are given in Figure 4–1.

The relationship between F-diagrams and C-diagrams can be shown as

$$C = \frac{c}{c_0} = \frac{dF}{d\theta} \tag{4-2}$$

$F(\theta)$ represents the fraction of the injected tracer material that has left the vessel up to the time θ, that is,

$$F(\theta) = \int_0^\theta \frac{c}{c_0} \, d\theta \tag{4-3}$$

Also, the total area under every C-diagram is equal to unity.

Theoretical Development

C-Diagram Parameters

A precise description of a nonideal flow requires knowledge of the complete flow pattern of the fluid within the tank or pond. It is difficult to determine this information, as it would require the complete velocity distribution of

FIGURE 4-2
The C-diagram.

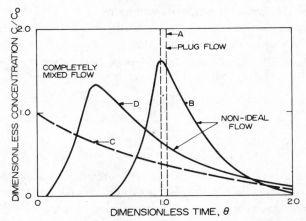

DIMENSIONLESS TIME, θ

the fluid within the vessel. Therefore, an alternate approach has been used, requiring knowledge only of how long different elements of the fluid remain in the vessel, that is, the distribution of residence times. This information is easy to obtain experimentally and to interpret. Although it cannot completely define the nonideal flow pattern within the pond, this approach yields sufficient information in many cases to determine the conversion of waste in the pond.

The experimental technique used for finding this desired distribution of residence times of fluid in the pond is a stimulus-response technique using tracer material in the flowing fluid. The known quantity of tracer is injected into the inflow (stimulus), and the concentration in the outflow (response) is measured as a function of time.

The treatment is limited to steady-state flow with one entering and one exiting stream. From the concentration versus time curve at the outlet, C- and F-diagrams are determined.

From the C-diagrams and F-diagrams several parameters can be defined that give some measure of the extent and effectiveness of mixing in the vessel. These parameters are used to check quantitatively the effectiveness of various factors such as inlet and outlet positions, geometry, and fluid properties of the pond. The C-diagrams is also used in determining the expected conversion of the reactant for selected experiments.

Mean Residence Time. The mean dimensionless residence time $\bar{\theta}_c$ is a measure of the average time the tracer slug spends in the vessel.

$$\bar{\theta}_c(\theta_0) = \frac{\int_0^{\theta_c} \frac{c}{c_0} \, \theta \, d\theta}{\int_0^{\theta_0} \frac{c}{c_0} \, d\theta} \qquad (4\text{--}4)$$

The value of $\bar{\theta}_c(\theta_0)$ is the distance from the origin to the centroid of that portion of the C-diagram between the origin and θ_0. The value of θ_0 is arbitrarily taken as 2, because after two detention times c/c_0 on the C-diagram has generally been observed to be small, and data taken beyond this point are near the limit of readability of the tracer-sensing instruments. Furthermore, if all parameters are constructed using data for $0 < \theta < 2$, the comparison of the results is meaningful. For purposes of this work, Equation 4–4 becomes

$$\bar{\theta}_c = \frac{\int_0^2 \frac{c}{c_0}\, \theta\, d\theta}{\int_0^2 \frac{c}{c_0}\, d\theta} \qquad (4\text{–}5)$$

Dead Space. In any flow vessel there are generally regions in which mixing is less active than desirable. Generally, this occurs in corners of the vessel. These regions of poor mixing are called dead spaces if the fluid moving through these spaces takes 5–10 times as long to pass through the vessel as does the main flow. The amount of dead space in a flow vessel is reflected by the long tail on the C-curve.

If the flow through the vessel has a minimum of dead space, then the mean residence time $\bar{t}_c$ will approach the detention time and $\bar{\theta}_c$ will approach 1.0. If there are substantial dead water regions in the flow, then a large portion of the tracer will leave the vessel before $\theta = 1$. Consequently, the centroid of the C-diagram will shift toward the origin and make $\bar{\theta}_c < 1.0$. To define dead space the following expression is used:

$$\overline{V}_d = \frac{V_d}{V} = 1 - \bar{\theta}_c [\mathrm{F}]_{\theta = 2} \qquad (4\text{–}6)$$

where V_d/V is defined as the dead-space parameter $\overline{V}_d$.

Deviation from Plug Flow. It is recognized that the ideal situation for flow through a waste stabilization pond is typified by plug flow. If the incoming waste mixes vertically and horizontally just enough to assure continuous treatment and then moves through the pond as a plug, then the optimum condition—called plug flow—has been realized. That is, all the fluid will have remained in the pond long enough for the desired degree of treatment, and none will have remained longer than necessary. This condition leads to a pond of minimum volume for a given waste load and hence, minimum cost.

The deviation of a given flow from plug flow can be considered a measure of pond treatment efficiency. It is the waste that leaves the pond before one detention time that contributes to poor treatment. Both the amount of waste leaving too soon and also the amount of time the waste lacks of remaining in the pond for one detention time are important. Consequently, the deviation from plug flow parameter $\bar{\theta}_{pf}$ is defined, as shown in

FIGURE 4-3
The plug flow deviation parameter.

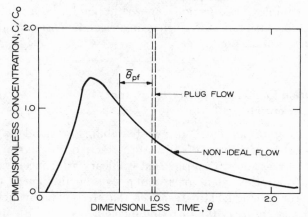

Figure 4–3, as the distance from the line $\theta = 1$ to the centroid of the area under the C-curve from $0 \le \theta \le 1$. The equation used to calculate $\bar{\theta}_{pf}$ is

$$\bar{\theta}_{pf} = \frac{\int_0^1 (1 - \theta) \dfrac{c}{c_0} \, d\theta}{\int_0^1 \dfrac{c}{c_0} \, d\theta} \tag{4-7}$$

Each experiment was evaluated with these parameters to determine the most efficient design from a hydraulic standpoint. For the best hydraulic performance that will result in maximum biological performance the dimensionless mean residence time $\bar{\theta}_c$ should approach 1.0. As $\bar{\theta}_c$ increases for given hydraulic conditions, the biological conversion increases, because the average detention time of the fluid particles would be greater.

Dead space has an obvious effect on conversion. Quantitatively, as the dead space parameter $\bar{V}_d$ increases, the effective flow area decreases, which results in a decreased detention time of the wastewater and poorer biological performance.

The plug flow deviation parameter, θ_{pf}, decreases in value as the hydraulic efficiency increases. Since maximum hydraulic efficiency and maximum conversion occur when $\theta_c = 1.0$, the same will hold true when $\bar{\theta}_{pf} = 0$.

The flow situation that would result in an optimal detention time (and a minimum value for dead space and for the plug flow deviation parameter) would be plug or one-dimensional flow. However, in an actual pond it is important that the incoming flow mixes well at the outset to effectuate intimate contact of the waste material with the microorganisms in the pond. For maximum conversion, the incoming wastewater should be diffused in such a way as to effectuate complete local vertical and lateral mixing and then travel more or less as a plug toward the outlet. This type of flow would utilize the complete cross-sectional area of the pond, which would then give maximum conversion for the given geometrical dimensions and hydraulic characteristics of the pond.

Quantitative Evaluation of the Treatment Efficiency Parameters

Conversion Equations. In a waste stabilization pond the major concern is the extent of reaction of the waste material during its stay within the pond. For a reaction with a rate that is linear with concentration (first-order reaction) the extent of reaction can be predicted solely from knowledge of the length of time each reactant element has spent in the reactor. The exact nature of the surrounding elements is of little importance. The distribution of residence times (*C*-diagram) gives information on how long various elements of fluid spend in a reactor, but not on the detailed exchange of matter within and between the elements. Because of this, the distribution of residence times in conjunction with the reaction equations yields sufficient information for the prediction of the average concentration in the reactor effluent.

A waste stabilization pond is a biological reactor that closely corresponds to a first-order (linear) chemical reactor, thereby supporting the use of chemical reactor design principles. To determine the conversion of a chemical or biological reactor using the distribution of residence times, the following equation[6] can be used:

$$
\begin{array}{l}
\text{mean} \\
\text{concentration} \\
\text{of reactant} \\
\text{leaving the} \\
\text{reactor} \\
\text{unreacted}
\end{array}
=
\begin{array}{l}
\Sigma \\
\text{all elements} \\
\text{of exit} \\
\text{stream}
\end{array}
\left[
\begin{array}{l}
\text{concentration} \\
\text{of reactant} \\
\text{remaining in} \\
\text{an element of} \\
\text{age between } t \\
\text{and } t + dt
\end{array}
\times
\begin{array}{l}
\text{fraction of} \\
\text{exit stream} \\
\text{which consists} \\
\text{of elements of} \\
\text{age between} \\
t \text{ and } t + dt
\end{array}
\right]
\tag{4-8}
$$

For a steady flow this equation shows that if a sample were taken of the entire exit stream at some time and the extent of conversion of each waste or reactant element were determined and summed for all the elements in the exit stream or sample, the resulting quantity would be the mean concentration of the waste or reactant leaving the reactor unreacted. This same information can be obtained in a more convenient way by the following equation:

$$
N_\infty = \int_0^\infty N \, \frac{c}{c_0} \, d\theta
\tag{4-9}
$$

in which

N = concentration of reactant as a function of time

N_∞ = mean concentration of reactant leaving the reactor in an unreacted state

c/c_0 = residence time distribution of fluid particles

For this analysis c/c_0 is used as an indication of the residence time of a non-conservative reactant. Equation 4–9 assumes a slug injection of reactant or pollutant and thereby requires a summation over time to account for the conversion of all the pollutant elements injected into the reactor. The first-order equation is

$$\frac{dN}{dt} = -KN \qquad (4\text{--}10)$$

in which K = first-order reaction coefficient. When $t = 0$, $N = N_0$ with N_0 = initial concentration of reactant. Integration of Equation 4–10 yields

$$N = N_0 e^{-Kt\theta} \qquad (4\text{--}11)$$

which is the well-known equation that gives the remaining BOD_5 of a waste as a function of N_0, K, and time. Incorporating Equation 4–11 into Equation 4–9 gives

$$\frac{\overline{N}_\infty}{N_0} = \int_0^\infty e^{-Kt\theta} \left[\frac{c}{c_0}\right] d\theta \qquad (4\text{--}12)$$

Equation 4–12 is the general equation for the conversion of a reactant in a chemical or biological reactor. The extent to which Equation 4–12 accurately predicts the fraction of unreacted material leaving a reactor is dependent on (1) how closely the reactant or waste material follows a first-order reaction, (2) the selection of a correct value for the first-order reaction coefficient, and (3) the accuracy of the expression for the residence time distribution of fluid particles in the reactor.

In using experimentally determined c/c_0 functions, the data are generally unusable beyond $\theta = 2.0$, because the concentration becomes too small to measure. Consequently, in determining the extent of reaction of a waste, Equation 4–12 can be modified to one in which experimental c/c_0 functions are used:

$$\frac{\overline{N}_2}{N_0} = \int_0^2 e^{-Kt\theta} \left[\frac{c}{c_0}\right] d\theta \qquad (4\text{--}13)$$

Equation 4–13 gives the fraction of material that has not undergone reaction at $\theta = 2.0$. The extent of reaction (treatment) efficiency as a percentage is

$$\% \text{ treatment efficiency} = 100 \left[1 - \frac{\overline{N}_2}{N_0}\right] \qquad (4\text{--}14)$$

Values of $\overline{N}_2/N_0$ and expected treatment efficiencies have been determined for a number of pond configurations. These results are presented later in the chapter.

Experimental Studies

Prototype Studies

The purpose of the prototype experimental studies was to determine the characteristics of the circulation patterns in existing waste stabilization ponds and to collect data on several of the existing ponds in the Logan, Utah city pond system (Figure 4–4). The circulation patterns were determined by introducing a dye into the inlet flow and determining how rapidly and efficiently the flow disperses to all areas of the pond. By sampling at the exit, time of travel through the pond was established to determine if the requirements on the time of retention were being met and collect data for constructing the $C(\theta)$ curves. The data collected were also used in the laboratory model verification.

Experiments were performed on ponds A–1, A–2, D, and E, which are shown in Figure 4–4, a plan view of the Logan pond system. Figure 4–4 also shows where samples were taken, in addition to the outlet samples of each pond. The grid size was chosen to give a good representation of the pond geometry, yet small enough (15 grid points) to allow the samples to be taken in a reasonable length of time.

The tracing dye, rhodamine WT, was in the form of a 20% solution.

FIGURE 4-4
The Logan City waste stabilization pond system.

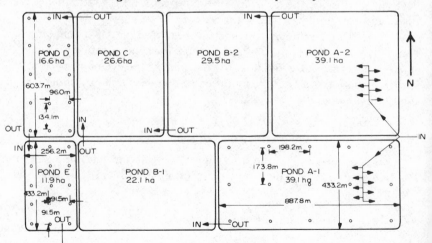

This dye was chosen because it is relatively nonabsorbent and inexpensive. A Turner Model 110 fluorometer was calibrated to measure the dye concentrations.

The results shown in Figure 4–5 are typical of the seven tests completed on the prototype ponds. The results in Figure 4–5 happen to apply to ponds A–1 and A–2 where stratified flow exists. It is clear from this figure that substantial short-circuiting is resulting, because peak concentrations are occurring at times considerably less than 1.0. Additional C-diagrams and more detailed discussion are presented by Mangelson[1] and Mangelson and Watters.[7]

Some general conclusions and observations can be drawn from the prototype experiments.

1. In every case, dye arrived at the pond outlet in less than 10% of the theoretical travel time through the pond. Observations suggest that this effect can be attributed to:

(a) *Density Currents*—Heavy (or light) inflow retains its identity and flows through the pond in a narrow stream, mixing very little. The result is that the pond is effectively reduced in size by a large amount.

(b) *Strong Inlet Jets*—The relatively high-velocity flows jet into the pond and retain enough momentum to reach the vicinity of the outlet in a short time.

(c) *Velocity Profiles*—Because the high-velocity flow regions transport dye at two or more times the average rate of flow, initial dye concentrations occur at the outlet more quickly.

FIGURE 4-5

Normalized concentration versus time curves at outlets, ponds A-1 and A-2.

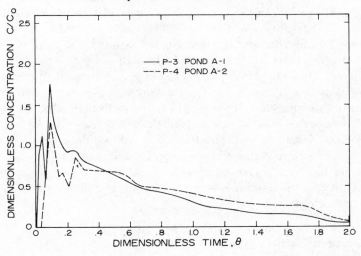

(d) *Dispersion*—Dispersion can string out the dye stream considerably, causing dye to reach the outlet sooner.

2. Mixing of dye throughout the pond occurred within a day or so, and the long, relatively uniform tail on the C-diagram became established. This result suggests that diffusion and dispersion are active enough in the ponds to play a major role in detention time. Plug flow was never closely approached in any of the experiments.

3. Circulatory flows were observed to be superposed on the general flow-through hydraulic pattern. These phenomena caused the multiple-peak C-diagrams commonly found in the prototype ponds (see Figure 4–5). These circulatory flows were generally caused either by strong inlet jets (circulation in the horizontal plane) or wind-generated circulation (in the vertical plane).

It was of interest to duplicate these phenomena in the laboratory. Once convinced that the laboratory model could adequately approximate prototype behavior (verification), design could proceed based on laboratory-model studies.

Laboratory-Model Studies

A hydraulic model of the prototype ponds was constructed in the Utah Water Research Laboratory at Utah State University. The model pond was 40 × 20 ft and 3.5 ft deep (see Figure 4–6). The facility was designed so that tracer concentration could be continuously monitored at the outlet. The inflow apparatus was configured so that flow rate could be continuously monitored and various mixtures of saltwater could be injected (for density-stratification experiments).

The model pond was equipped with a motorized carriage that spanned the width to enable samples to be taken at any location in the pond. An extra wall was constructed, along with baffles and a number of inlet and outlet devices. The inlet devices were constructed so as to break up or diffuse the incoming flow to nullify the effects of jets on the flow patterns. This effect was accomplished by diffusers made out of plastic pipe and graded gravel. The outflow was mixed to give a good representative sample before being pumped through the fluorometer. The raw data from the strip charts were extracted at equal intervals of time, and the concentration of the tracer was determined from temperature-dependent calibration charts. The data were processed by computer so that C- and F-diagrams could be drawn and other pertinent parameters computed.

Following are descriptions of the different groups of experiments that were performed:

1. Several identical experimental tests were made to test reproducibility of results.

FIGURE 4-6
Experimental apparatus and model pond.

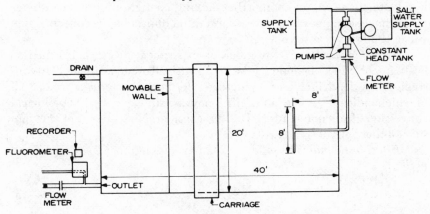

2. Verification experiments to test model reliability and its capability of duplicating prototype behavior were performed.
3. The effect of variable Reynolds numbers on the C-diagram and other treatment efficiency parameters was evaluated.
4. The effect of depth variation on flow patterns was tested.
5. Variations in number, type, and position of inlets and outlets were evaluated.
6. Length-to-width (L/W) ratios were varied, and different baffling schemes were tested as means of improving treatment efficiency.
7. Density-stratified experiments were attempted to determine the influence of density difference on C-diagrams and other pertinent parameters.

Experimental Reproducibility. A number of identical experiments were performed on the hydraulic model to determine the extent of the reproducibility of the results, that is, an idea of the reliability of the model data. The C-diagrams for these sets of experiments exhibit variability, particularly between $\theta = 0.0$ and $\theta = 1.0$; however, the parameter variation ($\bar{\theta}_c, \bar{\theta}_{pf}$, etc.) *is not as extreme as the* C-diagrams seem to indicate. If the lack of exact reproducibility is assumed to be of the order of 10%, then the variability about the dimensionless mean, $\bar{\theta}_c$, plug flow deviation parameter, $\bar{\theta}_{pf}$, and dead space, $\bar{V}_d$, are within the realm of experimental reproducibility. It apparently takes more pronounced variations in the C-diagram to affect these parameters greatly.

A number of factors have an influence on reproducibility and are important to consider in determining the reliability of the data. They are residual eddies resulting from filling the model, errors in measuring discharge and concentration of dye and depth, and errors in instrument calibration. The one problem encountered over which there is no control is the random eddying of the turbulent motion. The scale of turbulence in the horizontal plane is large enough that the eddy configuration at the instance of dye injection appears to have a substantial impact on the concentration versus time curves at the outlet.

Effect of Depth. A number of experiments were performed to evaluate the effect of changing the depth with all other variables held constant. As the depth increases, $\overline{V}_d$ increases, $\overline{\theta}_{pf}$ increases, and $\overline{\theta}_c$ decreases, all of which indicate less hydraulic efficiency and subsequently less biological conversion of organic matter. There is considerable scatter, but the data suggest that a slight trend exists.

Effect of Reynolds Number. The Reynolds number was defined and used as

$$R_e = \frac{V d_0}{\nu} = \frac{Q}{W\nu} \tag{4-15}$$

in which

$\quad Q$ = flow rate
$\quad W$ = width of pond
$\quad d_0$ = depth of pond
$\quad \nu$ = kinematic viscosity

The data show a minimum variation in $\overline{V}_d$, $\overline{\theta}_{pf}$, $\overline{\theta}_c$ with Reynolds number. Dead space and $\overline{\theta}_{pf}$ show a slight decrease as R_e increases, which means greater efficiency of operation. At higher Reynolds numbers, this would be expected, because of greater turbulence and larger diffusivity coefficients which result in greater mixing and interchange of the fluid particles.

Effects of Inlets and Outlets. Two types of outlet configurations were tried, along with a number of inlet variations. Figure 4-7 shows the inlet and outlet configurations that were used. Groups D and E of Table 4-1 give the results of these experiments. Figures 4-8 and 4-9 illustrate the C-diagrams for the experiments of groups D and E, respectively. From these figures and from Table 4-1 it appears that O-5 is the best design of these groups for maximum hydraulic efficiency. It has the highest dimensionless mean retention time and the lowest values for $\overline{\theta}_{pf}$ and dead space which indicate maximum efficiency.

Length-to-width (L/W) Ratio Effects. The model was modified to determine the effects of L/W ratio changes on the hydraulic efficiency of ponds (see Figure 4-7). The results of this G-series of experiments are tabulated in Table 4-1 as group F with the corresponding C-diagrams given in Figure 4-10. If we refer to the tabulated results for experiments G-2, G-3, G-4, and G-5, we can see that an increase in the hydraulic efficiency occurs for an increase in the L/W ratio. As the L/W ratio increases, the conditions for plug flow are closer to being satisfied, and as an end result, the efficiency of the treatment increases.

FIGURE 4-7
Inlet and outlet configurations employed.

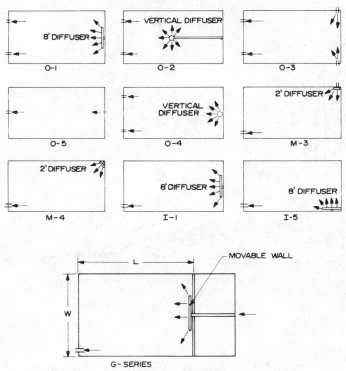

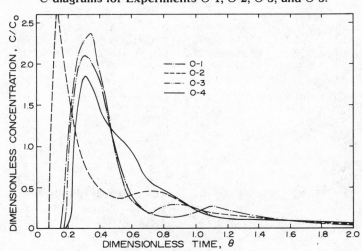

FIGURE 4-8
C-diagrams for Experiments O-1, O-2, O-3, and O-5.

FIGURE 4-9
C-diagrams for Experiments M-3, M-4, O-4, and I-5.

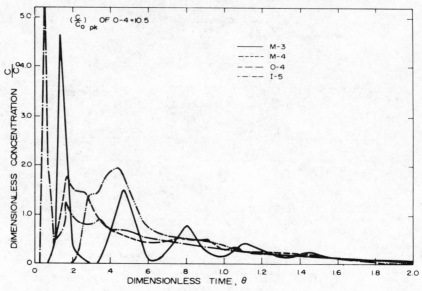

Table 4-1
Model Pond Data and Experimental Results
for Constant Density Experiments

Test No.	Depth (FT)	Q (GPM)	Renolds No.	$\bar{t}$, (MIN)	$\bar{\theta}_c$	$\bar{\theta}_{pf}$	$\overline{V}_d$	L/W	F ($\theta = 2.0$
Group D									
O-1	1.5	72.0	561	145.4	0.552	0.602	0.534	2.0	0.844
O-2	1.5	72.0	561	145.4	0.517	0.599	0.614	2.0	0.744
O-3	1.5	73.9	521	121.5	0.548	0.572	0.574	2.0	0.778
O-5	1.5	74.0	535	121.3	0.625	0.500	0.492	2.0	0.814
Group E									
O-4	1.75	77	608	136.0	0.456	0.668	0.570	2.0	0.944
M-3	1.5	80	606	112.2	0.641	0.576	0.493	2.0	0.791
M-4	1.5	80	606	112.2	0.632	0.555	0.479	2.0	0.824
I-5	1.5	80	632	112.2	0.621	0.552	0.449	2.0	0.887
Group F									
G-3	1.5	80	586	42.1	0.583	0.596	0.508	0.375	0.844
G-5	1.5	80	586	70.1	0.561	0.613	0.569	0.625	0.769
G-4	1.5	80	578	84.2	0.579	0.616	0.571	0.750	0.740
G-2	1.5	80	578	56.1	0.505	0.656	0.579	1.0	0.834
I-1	1.5	80	632	112.2	0.548	0.503	0.545	2.0	0.831

FIGURE 4-10
C-diagrams for experiments with different L/W ratios.

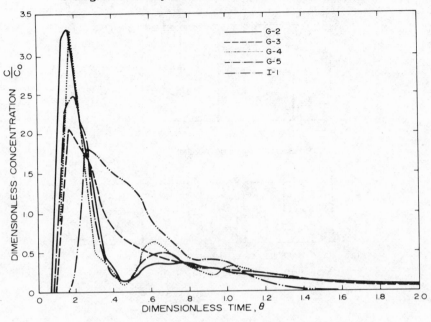

Baffling Effects. There were 17 tests performed using different arrangements of evenly spaced baffles to cause the flow to meander toward the exit. The dimensionless parameter used to express baffle length is the length of the baffles kW divided by the width of the pond W (see Figure 4–11a). The k-value for each of three groups of tests is 0.50, 0.70, and 0.90, respectively.

Four different lateral baffle spacings were used. The lateral spacing of baffles is defined by a dimensionless parameter obtained by dividing the length L of the pond by the baffle spacing S (see Figure 4–11a). Each test can now be identified by a number. For example, test 9.7 means that $k = 0.7$ (baffles extend across 70% of the width of the pond) and the spacing is $L/S = 9$ (there are four baffles extending from each wall of the pond). In Figure 4–11a, test 5.5 is depicted.

The C-diagrams for the most efficient baffle configurations are shown in Figures 4–12, 4–13, and 4–14 with the results of the set of experiments given in Table 4–2. It is clear from the figures that the peak concentration is occurring closer to 1.0 when the flow guidance is optimized.

One surprising result occurs. When baffles are extended too far across the pond (90%), higher velocity jets are created in the narrow corridors. The jets tend to form "jet streams" which enhance transport to the exit and

FIGURE 4-11
Baffling schemes used in model tests.

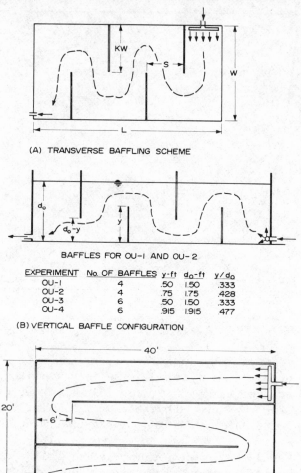

(A) TRANSVERSE BAFFLING SCHEME

BAFFLES FOR OU-1 AND OU-2

EXPERIMENT	No. OF BAFFLES	y-ft	d_0-ft	y/d_0
OU-1	4	.50	1.50	.333
OU-2	4	.75	1.75	.428
OU-3	6	.50	1.50	.333
OU-4	6	.915	1.915	.477

(B) VERTICAL BAFFLE CONFIGURATION

PLAN VIEW

(C) LONGITUDINAL BAFFLE CONFIGURATION FOR TEST C-3

reduce efficiency. As a consequence, the intermediate patterns of baffling seem to give the best results (eight baffles extending 70% across pond).

Tests were also run on over-and-under baffling (see Figure 4–11*b* and test series *OU* in Table 4–2) and a longitudinal baffling scheme (see Figure 4–11*c* and test *C*–3 in Table 4–2). The over-and-under schemes did not seem as effective as the horizontal baffling, whereas the longitudinal baffling seemed equally as effective.

FIGURE 4-12
C-diagrams for experiments 3.5, 5.5, 7.5, and 9.5.

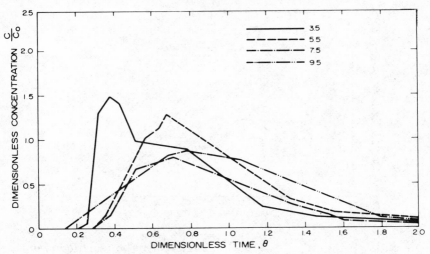

Stratified Flow Experiments

Tests were run in which the incoming flow was both more dense and less dense than the receiving pond. When the inflow is more dense than the receiving body, the inflow "dives" to the bottom and moves toward the outlet. Conversely, a less-dense inflow "floats" to the top and moves toward

FIGURE 4-13
C-diagrams for Experiments 3.7, 5.7, 7.7, and 9.7.

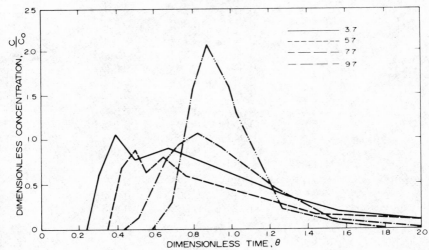

FIGURE 4-14
C-diagrams for Experiments 3.9, 5.9, 7.9, and 9.9.

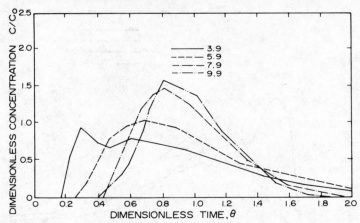

the outlet. In the former case the flow is retarded by bottom friction and the friction of the overlying fluid. In the latter case only the underlying fluid provides any frictional resistance. Hence, short-circuiting and poor performance are more likely for a less-dense inflowing fluid. Management of inlet

Table 4-2
Baffled Pond Experimental Results

Test No.	$\bar{\theta}_c$	$\bar{\theta}_{pf}$	F ($\theta = 2.0$)
3.5	0.724	0.416	0.881
5.5	0.870	0.309	0.850
7.5	0.855	0.303	0.630
9.5	0.920	0.314	0.866
3.7	0.836	0.374	0.794
5.7	0.864	0.286	0.827
7.7	0.932	0.212	0.651
9.7	0.964	0.136	0.782
3.9	0.837	0.398	0.884
5.9	0.936	0.296	0.954
7.9	0.947	0.216	0.985
9.9	0.947	0.188	0.952
No baffles	0.650	0.550	
OU-1	0.849	0.325	0.828
OU-2	0.832	0.364	0.802
OU-3	0.813	0.355	0.823
OU-4	0.768	0.395	0.824
C-3	0.931	0.250	0.870

and outlet positions is quite important in maximizing treatment in density-stratified situations.

Treatment Efficiency

In the design of a waste stabilization pond, it would be desirable to be able to predict the degree of conversion of the waste material to stable end products during its residence in the pond. Equation 4–12 or Equation 4–13 can be used for this purpose, providing the age distribution function, c/c_0, the theoretical detention time, $\bar{t}$, and the first-order reaction coefficient, K, are known. Since all pollutant remaining in the pond longer than two detention times is ignored in computing $\overline{N_2}/N_0$, *the quantity* $(1 - \overline{N_2}/N_0)$ represents the fraction of waste material that has undergone reaction.

Equation 4–13 consists of two parts: (1) the residence time distribution of the fluid particles c/c_0 as a function of θ, and (2) the biological first-order reaction equation, $N/N_0 = e^{-K\bar{t}\theta}$ as a function of K, $\bar{t}$, and θ. To evaluate the effect of the hydraulic design variations on the treatment efficiency for a given pond, K and $\bar{t}$, must be held constant. Since the extent of reaction of a waste is highly dependent on $\bar{t}$, for a given K, the detention time used in Equation 4–13 must be the same for all pond design alternatives for a meaningful comparison (constant volume). The c/c_0 data as a function of θ are then sufficient to determine the overall treatment efficiency for each pond design. By comparing the treatment efficiencies for these designs, it will be possible to determine the most efficient designs from the hydraulic standpoint.

Table 4–3 gives $\overline{N_2}/N_0$ for some of the experiments discussed earlier.

Table 4-3
Treatment Efficiency for Selected Experiments

Test No.	$\bar{t}$ (DAYS)	$\overline{N_2}/N_0$					% Treatment Efficiency
		$K = 0.20$	$K = 0.25$	$K = 0.30$	$K = 0.35$	$K = 0.40$	$K = 0.40$
0-2	5.0	0.495	0.452	0.415	0.383	0.355	64
0-3	5.0	0.477	0.428	0.386	0.348	0.316	68
0-5	5.0	0.463	0.408	0.361	0.320	0.285	71
0-4	5.0	0.618	0.575	0.539	0.507	0.479	52
M-4	5.0	0.480	0.429	0.386	0.349	0.318	68
I-5	5.0	0.486	0.424	0.372	0.307	0.289	71
G-2	5.0	0.547	0.503	0.464	0.430	0.401	60
G-3	5.0	0.517	0.461	0.418	0.381	0.349	65
Plug Flow	5.0	0.368	0.286	0.223	0.173	0.135	86

The theoretical detention time used to determine $\overline{N}_2/N_0$ was the same for all the experiments listed in Table 4–3. This specification makes it possible to compare the pond designs on the same basis, even though each experiment actually had a different theoretical detention time. The values of K in day^{-1} used in Equation 4–13 were based on the average values from Fair et al.[8] Also included in Table 4–3 are the values of $\overline{N}_2/N_0$ for plug flow to be used as a comparison with the treatment efficiencies of the other experiments.

The model pond results for treatment efficiency do indicate the importance of proper hydraulic design of waste stabilization ponds. The advantage of large L/W ratios is further substantiated by the values of efficiency for the baffled experiments in Table 4–4. Virtually all the experiments show 80% removal, with some as high as 89% for $K = 0.40$. In contrast, the percentages in Table 4–3 rarely reach 70%. This points to the desirability of using baffled ponds to increase effectively the L/W ratio. It should be noted that because all the dye in the experimental tests is not accounted for, $\overline{N}_2/N_0$ will be smaller than it should be (Equation 4–13) and efficiency values will be larger than they should be (Equation 4–14). This is why tests 7.7 and 9.7 in Table 4–3 show better efficiencies than the ideal 86% for plug flow. Consequently, the efficiency values shown in Tables 4–3 and 4–4 should be viewed more as relative levels of performance rather than absolute values.

Table 4-4
Treatment Efficiency for Baffled Experiments

Test No.	$\bar{t}$ (DAYS)	$\bar{\theta}_{pf}$	% Treatment Efficiency $K = 0.30$	$K = 0.40$
3.5	5.0	0.416	66	74
5.5	5.0	0.309	74	82
7.5	5.0	0.303	82	87
9.5	5.0	0.314	75	82
3.7	5.0	0.374	74	81
5.7	5.0	0.286	73	83
7.7	5.0	0.212	83	89
9.7	5.0	0.136	81	88
3.9	5.0	0.398	70	78
5.9	5.0	0.296	73	81
7.9	5.0	0.216	74	83
9.9	5.0	0.188	74	83
C-3	5.0	0.250	76	84
OU-1	5.0	0.325	74	81
OU-2	5.0	0.364	73	81
OU-3	5.0	0.355	72	80
OU-4	5.0	0.395	70	78

Wind Effects

The wind causes internal circulation patterns to form that can have a marked effect on the residence time of the waste in the pond. The wind can enhance mixing within the pond, and it can create strong enough internal currents to break down density stratification in shallow ponds. All these effects are of interest to the designer of a waste stabilization pond.

Nonstratified Flow.   Wind shear on the water surface causes a recirculating flow to rapidly build up in the pond. A surface current develops in the same direction as the wind flow at a magnitude of about 3% of the wind velocity. A return flow along the pond bottom occurs in the direction opposing the wind. It is easy to see how pond short-circuiting can occur if inlet and outlet structures are in line with prevailing wind directions. The strength of wind-generated surface currents, when compared with typical pond flow velocities, are seen to be substantial contributors to wastewater mass movement.

Rather than getting involved in difficult evaluations of longitudinal mixing and diffusion, it seems best to align the inlet-outlet axis of the pond along a direction approximately perpendicular to the prevailing wind direction. In this configuration, the wind will enhance mixing and probably not seriously reduce pond treatment efficiency. When this cannot be done or when wind direction is variable, a judicious use of baffling may be required to direct the flow properly.

Stratified Flow.   The effect of wind on stratified flow is very similar to that on nonstratified flow. If the pond is stratified into two layers, wind-generated currents are set up in both layers. The result is two counter-rotating circulatory flows in the pond. Again, the basic strategy to preventing short-circuiting is to align the inlet-outlet axis perpendicular to the prevailing wind direction.

However, there is one feature in stratified flow that it was not required to address in nonstratified flow. Breakdown of stratification can occur if wind becomes strong enough and persistent enough. The phenomenon is referred to as critical mixing.

Critical Mixing.   There are basically two phenomena that cause mixing between the layers of a stratified flow. The first kind of diffusion occurs when an exchange of fluid between the layers causes a gradual smoothing of the density gradient at the interface. This mixing is generally caused by low levels of turbulence in the upper layer that causes small waves to travel along the interface between the layers. Eddies are formed at the interface, and the two layers are mixed near the interface. As the level of turbulence increases, the wave activity at the interface approaches that of breaking waves. Eddies are shed more often, and the waves grow larger. A point is reached at which any increase in the turbulence and velocity of the upper layer causes the phenomena to change from a turbulent diffusion-type exchange into the second mode of mixing.

This second mode of mixing is characterized by the net transfer of fluid from the denser lower layer to the upper layer. This transfer process causes complete mixing to begin at the upwind end of the channel. The mixed front then advances downwind almost as a single progressive wave.

Initial stages of wind buildup are characterized by the diffusion-type mixing between the layers. At higher velocities the mixing tends to be characterized by the complete mixing of the layers. The distinction between the two modes lies in the stability of the interface. If there is laminar-type flow at the interface, diffusion will be the source of mixing. The exchange-type mixing is obtained when a rough turbulent flow exists at the interface.

Tests for critical mixing were conducted by George,[2] and the results are shown in Figure 4–15. The dimensionless density ratio $\Delta\rho/\rho$ shown on the

FIGURE 4-15
Critical mixing Reynolds number versus density difference and layer thickness.

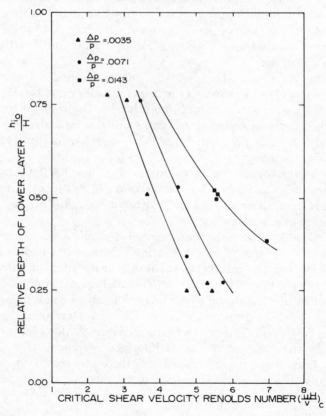

graph is the difference in density between the two layers divided by the lower layer density. The term h_{i_o}/H is the depth of the lower layer divided by the total depth. The critical shear velocity Reynolds number $(u^*H/\nu)_c$ employs the wind-generated surface shear τ_s to compute u^* from

$$u^* = \sqrt{\frac{\tau_s}{\rho}} \qquad (4\text{--}16)$$

For a given two-layer pond with the densities of the layers known, Figure 4–15 enables the designer to establish the wind surface shear velocity that would cause complete mixing. The difficulty is in relating surface shear τ_s to wind velocity. For accurate evaluation, one should address recent literature on the subject. For approximation purposes, the equation of Hidy and Plate[9] may be used:

$$\tau_s = 4 \times 10^{-7} U_a{}^3 \qquad (4\text{--}17)$$

where U_a is the wind velocity in fps and τ_s is the surface shear in lb/ft^2.

Design Recommendations

The design recommendations below relate only to the hydraulics of waste stabilization ponds. Recommendations should be implemented only if they are compatible with the biological requirements of the design. It is assumed that the discharge through the ponds and the pond depth have already been established.

Number, Shape, and Size of Ponds

The number of ponds in series is generally determined by the extent of treatment desired and the confidence of the designer in his or her ability to control detention times in the ponds. If ponds were designed so that more efficient treatment were assured, fewer ponds in series would be required.

Rectangular ponds are most commonly used because of simplicity of construction and ease of clustering a series of ponds. Circular, pie-shaped, or kidney-shaped ponds hold no inherent advantage over well-designed rectangular ponds. Data and results shown in this chapter should be directly applied only to rectangular ponds. Care should be used in any extrapolation of results to other shapes of ponds.

The size of the pond is intimately related to the desired amount of treatment in the pond. Tables 4–3 and 4–4 provide estimates of treatment efficiency for a variety of possible configurations. Depending on the configuration selected, treatment efficiencies can range from the 80% range (very good) down to the 50% range (very poor).

A pond configured to give an 80% efficiency, if made 25% or so larger, might give 90% or better treatment. Note that the relationship between increase in volume and increase in efficiency is not linear, because of the exponential nature of the first-order reaction equation. A technique that may be employed to estimate treatment efficiency for a proposed design is discussed at the end of this section.

Inlet and Outlet Configuration

Figure 4–7 provides a selection of inlet and outlet configurations and Table 4–3 illustrates what effect the variations in type and placement have on treatment efficiency. To get the greatest efficiency, the outlet should be located as far from the inlet as possible. The inflow should be introduced at as low a velocity as possible and preferably by means of a long diffuser. The outlet(s) (and inlets) should be placed so that the flow through the pond is symmetric. Using dual outlets such as those of tests O–1, O–3, and O–5 is a good way to accomplish this.

Baffling

Baffling schemes shown in Figure 4–11 and the corresponding treatment efficiencies given in Table 4–4 demonstrate that better treatment is obtained when the flow is guided more carefully through the pond. In addition to treatment efficiency, economics and aesthetics play an important role in deciding whether baffling is desirable.

Because there is little horizontal force on baffling except that caused by wave action, the baffle structure need not be particularly strong. It may also be placed below the pond surface to help overcome aesthetic objections. A typical type of baffle to consider might be a submerged fence attached to posts driven into the pond bottom and covered with a heavy plastic flexible membrane.

In general, the more baffling used, the better the flow guidance and treatment efficiency. The lateral spacing and length of the baffle should be specified so that the cross-sectional area of the flow is as close to a constant as possible. For example, in Figure 4–11c the longitudinal channel width is 6.7 ft. The baffle is terminated so that a 6-ft opening remains around the end of the baffle. For this configuration there will not be large velocity changes in the flow. Test 7.7 in Table 4–4 supports this approach, because it shows the best treatment efficiency and has a baffle spacing of 5.7 ft and a baffle end opening of 6 ft.

Baffling has an additional virtue. The spiral flow induced when flow occurs around the end of the baffles enhances mixing and tends to break up or prevent any stratification or tendency to stratify.

Wind Effects

As mentioned earlier, wind generates a circulatory flow in bodies of water. To minimize wind-caused short-circuiting, align the pond inlet-outlet axis perpendicular to the prevailing wind direction.

If for some reason the inlet-outlet axis cannot be oriented properly, baffling can be used to control, to some extent, the wind-induced circulation. It should be kept in mind that in a constant-depth pond, the surface current is in the direction of the wind, and the return flow is in the upwind direction along the bottom.

Stratified Ponds

Ponds that are stratified because of temperature differences between the inflow and the pond tend to behave differently in winter and summer. In summer the inflow is generally colder than the pond, so it sinks to the pond bottom and flows toward the outlet. In the winter the reverse is generally true, and the inflow rises to the surface and flows toward the outlet.

A likely consequence of this behavior is that the effective volume of the pond is reduced to that of the stratified inflow layer (density current). The result can be a drastic decrease in detention time and an unacceptable level of treatment.

One strategy to employ is to use selective pond outlets positioned vertically so that outflow is drawn from the layer with density different from that of the inflow. For example, under summer conditions the inflow will occur along the pond bottom. Hence the outlets should draw from water near the pond surface. This concept has not been tested, but it seems likely to improve performance over "full-depth" outlet structures.

Another approach is to premix the inflow with pond water while in the pipe or diffuser system, thereby decreasing the density difference. This could be accomplished by regularly constricting the submerged inflow diffuser pipe and locating openings in the pipe at the constrictions. The low pressure at the pipe constrictions would draw in pond water and mix it with the inflow to alter the density. However, clogging of openings with solid material could be a problem.

Estimating Treatment Efficiency

The treatment efficiencies shown in Tables 4–3 and 4–4 are for a very limited range of situations. It would be most useful if there were a technique for investigating efficiencies for various K-values and theoretical detention times $\bar{t}$. If, in addition, it were possible to synthesize C-diagrams for proposed designs and use them to compute treatment efficiencies, then a useful design aid would be available.

A variety of pond configurations have been presented in this chapter, and the C-diagrams for each have been included. It may be possible for a designer to sketch a reasonably accurate C-diagram for a proposed design by referring to similar configurations in this chapter or in Mangelson.[1] The C-diagram is dimensionless, and model-prototype tests have demonstrated that scale effects are not significant.[1] The C-diagram can represent a geometrically similar pond of any size if the depth scale is ignored. In other words, its form depends strongly on pond shape, inlet-outlet type, and location and baffling. It is not affected a great deal by depth.

Because the area under the C-diagram is theoretically 1.0, the syn-

FIGURE 4-16
Computer program for computing treatment efficiency from C-diagrams.

```
 1 -- C*************************************************************************************
 2 -- C    PROGRAM FOR COMPUTING WASTE STABILIZATION POND TREATMENT EFFICIENCY
 3 -- C          NUMERICAL INTEGRATIONS PERFORMED USING TRAPEZOIDAL RULE
 4 -- C*************************************************************************************
 5 -- C
 6 -- C TITLE( ) = JOB DESCRIPTION CARD (80 CHARACTERS MAXIMUM)
 7 -- C NDATA    = NUMBER OF C-ZERO VALUES TO BE INPUT
 8 -- C TBAR     = THEORETICAL DETENTION TIME - DAYS
 9 -- C K        = FIRST-ORDER REACTION COEFFICIENT - /DAYS
10 -- C CZERO( ) = C-ZERO VALUES IN SEQUENCE. FIRST ENTRY SHOULD BE 0.0.
11 -- C THETA( ) = THETA VALUES CORRESPONDING TO ABOVE C-ZERO VALUES
12 -- C
13 --         DIMENSION CZERO(100),THETA(100),E(100),TITLE(20)
14 --         REAL K
15 -- C
16 -- C------------------------------------------------------------------------------
17 --         READ(5,100) (TITLE(I),I=1,20)
18 --     100 FORMAT(20A4)
19 -- C------------------------------------------------------------------------------
20 --         WRITE(6,200) (TITLE(I),I=1,20)
21 --     200 FORMAT(1H1,5X,20A4/)
22 -- C------------------------------------------------------------------------------
23 --         READ(5,101) NDATA,TBAR,K
24 --     101 FORMAT(I10,2F10.0)
25 -- C------------------------------------------------------------------------------
26 --         WRITE(6,201) TBAR,K
27 --     201 FORMAT(10X,'DETENTION TIME =',F7.2,' DAYS'/10X,'FIRST-ORDER REACTI
28 --        $ON COEFFICIENT =',F5.2,' /DAY'/)
29 --         TK=-K*TBAR
30 -- C------------------------------------------------------------------------------
31 --         READ(5,102) (THETA(I),CZERO(I),I=1,NDATA)
32 --     102 FORMAT(8F10.0)
33 -- C------------------------------------------------------------------------------
34 --         WRITE(6,202)
35 --     202 FORMAT(15X,'THETA    CZERO'/15X,'-----    -----')
36 --         DO 20 I=1,NDATA
37 --      20 WRITE(6,203) THETA(I),CZERO(I)
38 --     203 FORMAT(14X,F6.2,3X,F6.2)
39 -- C
40 --         DO 10 I=1,NDATA
41 --      10 E(I)=EXP(TK*THETA(I))
42 --         CSUM=0.0
43 --         THSUM=0.0
44 --         DO 30 I=2,NDATA
45 --         CSUM=CSUM+0.5*(CZERO(I-1)+CZERO(I))*(THETA(I)-THETA(I-1))
46 --      30 THSUM=THSUM+0.5*(E(I-1)*CZERO(I-1)+E(I)*CZERO(I))*(THETA(I)-
47 --        $THETA(I-1))
48 --         EFF=100.*(1.0-THSUM)
49 --         WRITE(6,204) CSUM,THSUM,EFF
50 --     204 FORMAT(//5X,'AREA UNDER CZERO CURVE =',F5.3/
51 --        $5X,'FRACTION OF WASTE UNTREATED =',F5.3/
52 --        $5X,'WASTE TREATMENT EFFICIENCY =',F5.1,' PERCENT')
53 --         END
```

FIGURE 4-17
Computed printout of analysis of treatment
efficiency for test O-5.

```
DETENTION TIME =     5.00 DAYS
FIRST-ORDER REACTION COEFFICIENT = 0.40 /DAY

       THETA     CZERO
       -----     -----
       0.18      0.00
       0.22      0.12
       0.27      1.66
       0.31      1.86
       0.43      1.33
       0.60      0.95
       0.70      0.59
       1.00      0.22
       1.20      0.11
       1.30      0.12
       1.40      0.10
       1.60      0.09
       1.80      0.09
       2.00      0.05


   AREA UNDER CZERO CURVE =0.807
   FRACTION OF WASTE UNTREATED =0.288
   WASTE TREATMENT EFFICIENCY = 71.2 PERCENT
```

thesized diagram may need some adjustment from the initial trial. If the curve is extended out to $\theta = 2$ or 3, then the area under the curves should be 0.90 or better (see F-values in Tables 4–1 and 4–2).

A computer program is provided in Figure 4–16 to compute the area under the C-diagram and the treatment efficiency. The COMMENT cards in the program describe how to prepare the input data. Figure 4–17 shows the printout from the computer program when test O–5 is analyzed. The input data were scaled from Figure 4–8, and the results compared with those in Table 4–3. Using the basic C-diagram sketched by the designer, variations in detention time and first-order reaction coefficient can be related to treatment efficiency.

Ideally, a mathematical model of flow through waste stabilization ponds is needed that would provide velocity fields and the age distribution functions required to construct the C-diagram. Such a successful model would add a degree of confidence to the validity of C-diagrams for proposed designs. At the present time, hydrodynamic models still require large computers and expensive computer runs to generate the required information. Even then, there is still a lack of confidence in their ability to represent the flow processes in the pond accurately.

References

1. Mangelson, Kenneth A. 1971. Hydraulics of Waste Stabilization Ponds and Its Influence on Treatment Efficiency. Ph.D. dissertation, Utah State University, Logan, Utah.
2. George, Robert L. 1973. Two-dimensional Wind-generated Flow Patterns, Dif-

fusion and Mixing in a Shallow Stratified Pond. Ph.D. dissertation, Utah State University, Logan, Utah.

3. Thirumurthi, D. 1969. Design Principles of Waste Stabilization Ponds. *ASCE San. Eng. J.*, **95**(SA2), 311–330.

4. Oswald, W. J. 1963. Fundamental Factors in Stabilization Pond Design. *Proceedings of the Third Conference on Biological Waste Treatment*, Pergamon Press, Oxford.

5. Danckwerts, P. V. 1953. Continuous Flow Systems. *Chem. Eng. Sci.* **1**:1–13.

6. Levenspiel, O., and K. B. Bischoff. 1963. Patterns of Flow in Chemical Process Vessels. *Advances in Chemical Engineering*, Vol. 4, Academic Press, New York, pp. 95–198.

7. Mangelson, Kenneth, A., and G. Z. Watters. 1972. Treatment Efficiency of Waste Stabilization Ponds. *ASCE San. Eng. J.*, **98**(SA2), April.

8. Fair, G. M., J. C. Geyer, and D. A. Okun. 1968. *Water and Wastewater Engineering*. Wiley, New York.

9. Hidy, G. M., and E. J. Plate. 1966. Wind Action on Waters Standing in a Laboratory Channel. *J. Fluid Mech*, **26**(Part 4), 651–687.

CHAPTER 5

Lagoon Lining and Sealing

THE NEED FOR a well-sealed wastewater stabilization lagoon has become part of modern lagoon design, construction, and maintenance. The primary motive for sealing wastewater stabilization lagoons is to prevent seepage. Seepage has two effects on lagoon performance. First, seepage affects the treatment capabilities of the lagoon by causing fluctuations in the lagoon water depth. For consistent and sufficient treatment, lagoons require a constant water depth at a specified design depth. Second, seepage can cause pollution of groundwater, which can have serious effects on groundwater uses.

Many types of lagoon liners exist, but all can be classified into three major categories: (1) synthetic and rubber liners, (2) earthen and cement liners, and (3) natural and chemical treatment sealers. Within each category also exists a wide variety of application characteristics.

Choosing the appropriate lining for a specific lagoon is a critical issue in lagoon design and in the improvement of seepage control. The criteria for lining a lagoon are highly dependent on the specific geographical location, on climate, on local hydrogeology, and on materials found in the lagoon wastewater. Detailed information is available from the manufacturers, in a book by Kays,[1] and a report by Middlebrooks et al.[2]

Sealing and Seepage Rates

The most interesting and complex techniques of lagoon sealing, either separately or in combination, are natural lagoon sealing and chemical treatment sealing.[3,4] Natural lagoon sealing has been found to occur when the settled solids form a bottom layer that physically clogs soil pores.

Chemical treatment has changed the chemical nature of the bottom soil to ensure sealing.

Infiltration characteristics of anaerobic lagoons were studied in New Zealand.[5] Certail soil additives were added (bentonite, sodium carbonate, sodium triphosphate) to 12 pilot lagoons with varying pond depth, soil types, and compacted bottom soil thickness. It was found that chemical sealing was effective for soils with a minimum clay content of 8 % and a silt content of 10 %. Effectiveness increased with clay and silt content.

Four different soil columns were placed at the bottom of an animal wastewater pond to study physical and chemical properties of soil and sealing of wastewater ponds.[6] It was discovered that the initial sealing that occurred at the top 5.1 cm (2 in.) of the soil columns was caused by the trapping of suspended matter in the soil pores. This was followed by a secondary mechanism of microbial growth that completely sealed off the soil from water movement.

A similar study was performed in Arizona.[7] The double mechanism of physical and biological sealing was also found to occur. The sealing of the lagoon was also induced by the use of an organic polymer united with bentonite clay. This additive could have been applied with the pond full or empty, although the treatment was more effective when the pond was empty.

An experiment was performed by Matthew and Harms[8] in South Dakota in an effort to relate the sodium adsorption ratio (SAR) of the *in situ* soil to the sealing mechanism of wastewater stabilization ponds. No definite quantitative conclusions were formed. The general observation was made that the equilibrium permeability ratio decreases by a factor of 10 as SAR varies from 10 to 80. For 7 out of 10 soil samples, the following were concluded: (1) SAR did affect permeability of soils studied; (2) as the SAR increased, the probability that the pond would seal naturally also increased; and (3) soils with higher liquid limits would probably be less affected by the SAR.

Polymeric sealants have been used to seal both filled and unfilled ponds.[9] Unfilled ponds have been sealed by admixing a blend of bentonite and the polymer directly into the soil lining. Filled ponds have been sealed by spraying the fluid surface with alternate slurries of the polymer and bentonite. It has been recommended that the spraying take place in three subsequent layers: (1) polymer, (2) bentonite, and (3) polymer. The efficiency of the sealant has been found to be significantly affected by the composition of the impounded water. Most importantly, calcium ions in the water exchanged with sodium ions in the bentonite to cause failure of the compacted bentonite linings.

Natural sealing of lagoons has been found to occur from three mechanisms: (1) physical clogging of soil pores by settled solids, (2) chemical clogging of soil pores by ionic exchange, and (3) biological and organic clogging caused by microbial growth at the pond lining. The dominant mechanism of the three has been shown to depend on the composition of the wastewater being treated. Davis et al.[10] found that for liquid dairy waste the biological clogging mechanism predominated. In their San Diego County study site located on sandy loam, the infiltration rate of a virgin pond was measured. A clean water infiltration rate for the pond was 122 cm (48 in.) per day. After two weeks of manure water, infiltration averaged 5.8 cm (2.3 in.) per day; after 4 months 0.51 cm (0.2 in.) per day.

A study performed in southern California[11] indicated similar results. After waste material was placed in the unlined pond in an alluvial silty soil, the seepage rate was reduced. The initial 11.2 cm (4.4 in.) per day seepage rate dropped to 0.56 cm (0.22 in.) per day after 3 months, and after 6 months to 0.30 cm (0.12 in.) per day.

Stander et al.[12] presented a summary of information (Table 5-1) on measured seepage rates in wastewater stabilization ponds. Seepage is a function of so many variables it is impossible to anticipate or predict rates without extensive soils tests. The importance of controlling seepage to protect groundwater dictates that careful evaluations be conducted before construction of lagoons to determine the need for linings and the acceptable types.

The Minnesota Pollution Control Agency[13] initiated an intensive study to evaluate the effects of stabilization pond seepage from five municipal systems. The five communities were selected for study on the basis of geologic setting, age of the system, and past operating history of the wastewater stabilization pond. The selected ponds were representative of the major geomorphic regions in the state, and the age of the systems ranged from 3 to 17 years.

Estimates of seepage were calculated by two independent methods for each of the five pond systems. Water balances were calculated by taking the difference between the recorded inflows and outflows, and pond seepage was determined by conducting in-place field permeability tests of the bottom soils at each location. Good correlation was obtained with both techniques.

Field permeability tests indicated that the sealing ability of the sludge blanket was insignificant in locations in which impermeable soils were used in the construction process. In the case of more permeable soils, it appeared that the sludge reduced the permeability of the bottom soils from a background level of 2.54×10^{-4} or 10^{-5} cm (10^{-4} or 10^{-5} in.) per sec to the order of 2.54×10^{-6} cm (10^{-6} in.) per sec. At all five systems evaluated, the stabilization pond was in contact with the local groundwater table. Local groundwater fluctuations had a significant impact on seepage rates. The reduced groundwater gradient resulted in a reduction of seepage losses at three of the sites. The contact with the groundwater possibly explains the reduction in seepage rates with time observed in many ponds. In the past

Table 5-1
Reported Seepage Rates from Pond Systems

Literature Source	Initial Rates				
	Initial Seepage Rate		Hydraulic Load gals/ acre·day	Seepage Rate as % of Hydraulic Load	Settling-in Period
	in./day	gal/ acre·day			
California SWPCB (Nos. 15, 18)[14,15]	8.8	199,000	316,000	63	9 months
Neal and Hopkins[a,16]	5.5	127,850	141,970	90	1 year
Voights[b,17]	—	—	—	—	Average over 5 years (1951-55)
Shaw[18]	6	136,000	54,000	Exceeded inflow rate	± 1 Year
Windhoek municipal maturation ponds:[c] Nos. 5	0.16	3,500	776,000	0.45	Over period June 14-
6	0.17	3,900	1,190,000	0.32	22 1967 after all
7	0.015	300	712,000	0.046	ponds in
8	0.06	1,400	715,000		full op-
9	0.23	5,300	455,000		eration

this reduction in seepage rates has been attributed to a sludge buildup, but perhaps the increase in contact with groundwater accounts for this reduction. In an area underlain by permeable material where little groundwater mounding occurs, there is probably little influence on seepage rates. The buildup of sludge on the bottom of a pond appears to enhance the ability of the system to treat seepage water. Sludge accumulation apparently increases the cation exchange capacity of the bottom soils.

Groundwater samples obtained from monitoring wells did not show any appreciable increases in nitrogen, phosphorus, or fecal coliform over the background levels after 17 years of operation. The groundwater down gradient from the waste stabilization pond showed an increase in soluble salts as great as 20 times over background levels. Concentrations from 25 mg/l of chlorides to 527 mg/l were observed.

A comparison of observed seepage rates for various types of liner

Table 5-1 (*Cont.*)
Reported Seepage Rates from Pond Systems

EVENTUAL RATES				GEOLOGY OF POND BASE	PLACE
Eventual Seepage Rate		*Hydraulic Load gals/ acre·day*	*Seepage Rate as % of Hydraulic Load*		
in./day	*gal/ acre·day*				
0.35	7,940	±380,000	2.1	Desert soil (sandy soil)	Mojave, California
0.61	13,810	47,323	29.2	Sand and gravel	Kearney, Nebraska
0.34	7,660	9,160	84	Sandy soil	Filer City, Michigan
0.3	6,800	50,000	13.6	Clay loam and shale	Pretoria
—	—	—	—	Mica and schist	Windhoek, S.W.A.
—	—	—	—	Mica and schist	
—	—	—	—	Mica	
—	—	—	—	Mica and schist	
—	—	—	—	Mica and schist with side wall seepage to river	

SOURCE: From Ref. 12. Courtesy of Ann Arbor Science Publishers, Inc., Ann Arbor, Michigan.

[a] Evaporation and rainfall effects were apparently not corrected for. Seepage losses were also influenced by a high water table at times.

[b] These lagoons were constructed in sandy soil with the express purpose of seeping away Paper Mill NSSC liquor.

[c] Ponds constructed for the express purpose of water reclamation.

material is presented in Table 5-2.[1] If an impermeable liner is required, it appears that one of the synthetic materials must be used. Protection of the synthetic liners is essential if impermeability is specified.

Design and Construction Practice

The presentation of recommended design and construction procedures is divided into two categories: (1) bentonite, asphalt, and soil cement liners

Table 5-2
Seepage Rate Comparisons[a]

Material	Thickness (in.)	Minimum Expected Seepage Rate at 20 ft of Water Depth after 1 yr of Service (in./day)
Open sand and gravel		96
Loose earth		48
Loose earth plus chemical treatment*		12
Loose earth plus bentonite*		10
Earth in cut		12
Soil cement (continuously wetted)	4	4
Gunite	1.5	3
Asphalt concrete	4	1.5
Unreinforced concrete	4	1.5
Compacted earth	36	0.3
Exposed prefabricated asphalt panels	0.5	0.03
Exposed synthetic membranes	0.045	0.001

Source: From Ref. 1, courtesy of John Wiley & Sons, Inc., New York, New York.

[a]The data are based on actual installation experience. The chemical and bentonite (*) treatments depend on pretreatment seepage rates, and in the table loose earth values are assumed.

and (2) thin membrane liners. This division was selected because of the major differences between the application techniques. There is some similarity between the application of asphalt panels and the elastomer liners, and of necessity there will be some repetition in these two major subdivisions. A partial listing of the trade names and sources of common lining materials is presented in Table 5-3.

Regardless of the type of material selected as a liner, there are many common design, specification, and construction practices. A summary of the common effective design practices in cut-and-fill reservoirs is given in Table 5-4. Most of these practices are common-sense items and would appear to not require mentioning. Unfortunately, experience has shown these items to be the most commonly ignored practices.

A lining material must be selected with the type of waste to be contained in mind. Kays[1] has developed a lining selection guide chart (Table 5-5) for various types of wastes and the common types of lining materials. The chart should be used only as a guide, and before selecting one of the materials, a careful evaluation of the waste and the proposed liner must be conducted.

Bentonite, Asphalt, and Soil Cement. The application of bentonite, asphalt, and soil cement as lining materials for lagoons and reservoirs has a long history.[1] The following summary includes consideration of the method of using the materials, resultant costs, and evaluations of durability and ef-

Table 5-3
Trade Names and Sources of Common Lining Materials

Trade Name	Production Description	Manufacturer
Aqua Sav	Butyl rubber	Plymouth Rubber Canton, Mass.
Armor last	Reinforced neoprene and Hypalon	Cooley, Inc. Pawtucket, R.I.
Armorshell	PVC-nylon laminates	Cooley, Inc. Pawtucket, R.I.
Armortite	PVC-coated fabrics	Cooley, Inc. Pawtucket, R.I.
Arrowhead	Bentonite	Dresser Minerals Houston, Tex.
Biostate Liner	Biologically stable PVC	Goodyear Tire & Rubber Co. Akron, Ohio
Careymat	Prefabricated asphalt panels	Phillip Carey Co. Cincinnati, Ohio
CPE (resin)	Chlorinated PE resin	Dow Chemical Co. Midland, Mich.
Coverlight	Reinforced butyl and Hypalon	Reeves Brothers, Inc. New York, N.Y.
Driliner	Butyl rubber	Goodyear Tire & Rubber Co. Akron, Ohio
EPDM (resin)	Ethylene propylene diene monomer resins	U.S. Rubber Co. New York, N.Y.
Flexseal	Hypalon and reinforced Hypalon	B. F. Goodrich Co. Akron, Ohio
Geon (resin)	PVC resin	B. F. Goodrich Co. Akron, Ohio
Griffolyn 45	Reinforced Hypalon	Griffolyn Co., Inc. Houston, Tex.
Griffolyn E	Reinforced PVC	Griffolyn Co., Inc. Houston, Tex.
Griffolyn V	Reinforced PVC, oil resistant	Griffolyn Co., Inc. Houston, Tex.
Hydroliner	Butyl rubber	Goodyear Tire & Rubber Co. Akron, Ohio
Hydromat	Prefabricated asphalt panels	W. R. Meadows, Inc. Elgin, Ill.
Hypalon (resin)	Chlorosulfonated PE resin	E. I. Du Pont Co. Wilmington, Del.
Ibex	Bentonite	Chas. Pfizer & Co. New York, N.Y.

(continued)

Table 5-3 (*cont.*)
Trade Names and Sources of Common Lining Materials

TRADE NAME	PRODUCTION DESCRIPTION	MANUFACTURER
Koroseal	PVC films	B. F. Goodrich Co. Akron, Ohio
Kreene	PVC films	Union Carbide & Chemical Co. New York, N.Y.
Meadowmat	Prefabricated asphalt panels with PVC Core	W. R. Meadows, Inc. Elgin, Ill.
National Baroid	Bentonite	National Lead Co. Houston, Tex.
Nordel (resin)	Ethylene propylene diene monomer resin	E. I. Du Pont Co. Wilmington, Del.
Panelcraft	Prefabricated asphalt panels	Envoy-APOC Long Beach, Calif.
Paraqual	EPDM and butyl	Aldan Rubber Co. Philadelphia, Pa.
Petromat	Polypropylene woven fabric (Base fabric-spray linings)	Phillips Petroleum Co. Bartlesville, Okla.
Pliobond	PVC adhesive	Goodyear Tire & Rubber Co. Akron, Ohio
Polyliner	PVC-CPE, alloy film	Goodyear Tire & Rubber Co.
Red Top	Bentonite	Wilbur Ellis Co. Fresno, Calif.
Royal Seal	EPDM and butyl	U.S. Rubber Co. Mishawaka, Ind.
SS-13	Waterborne dispersion	Lauratan Corp. Anaheim, Calif.
Sure Seal	Butyl, EPDM, neoprene, and Hypalon, plain and reinforced	Carlisle Corp. Carlisle, Pa.
Vinaliner	PVC	Goodyear Tire & Rubber Co. Akron, Ohio
Vinyl Clad	PVC, reinforced	Sun Chemical Co. Paterson, N.J.
Visqueen	PE resin	Ethyl Corp. Baton Rouge, La.
Volclay	Bentonite	American Colloid Co. Skokie, Ill.
Water Seal	Bentonite	Wyo-Ben Products Billings, Mont.

(*continued*)

Table 5-3 (*cont.*)
Trade Names and Sources of Common Lining Materials

Materials	Manufacturers	Locations
Bentonite	American Colloid Co.	Skokie, Ill.
	Archer-Daniels-Midland	Minneapolis, Minn.
	Ashland Chemical Co.	Cleveland, Ohio
	Chas. Pfizer & Co.	New York, N.Y.
	Dresser Minerals	Houston, Tex.
	National Lead Co.	Houston, Tex.
	Wilbur Ellis Co.	Fresno, Calif.
	Wyo-Ben Products, Inc.	Billings, Mont.
Butyl and EPDM	Carlisle Corp.	Carlisle, Pa.
	Goodyear Tire & Rubber Co.	Akron, Ohio
Butyl and EPDM, reinforced	Aldan Rubber Co.	Philadelphia, Pa.
	Carlisle Corp.	Carlisle, Pa.
	Plymouth Rubber Co.	Canton, Mass.
	Reeves Brothers, Inc.	New York, N.Y.
CPE, reinforced	Goodyear Tire & Rubber Co.	Akron, Ohio
Hypalon	Burke Rubber Co.	San Jose, Calif.
	B. F. Goodrich Co.	Akron, Ohio
Hypalon, reinforced	Burke Rubber Co.	San Jose, Calif.
	Carlisle Corp.	Carlisle, Pa.
	B. F. Goodrich Co.	Akron, Ohio
	Plymouth Rubber Co.	Canton, Mass.
	J. P. Stevens Co.	New York, N.Y.
EPDM	See "Butyl and EPDM"	
EPDM, reinforced	See "Butyl and EPDM, reinforced"	
Neoprene	Carlisle Corp.	Carlisle, Pa.
	Firestone Tire & Rubber Co.	Akron, Ohio
	B. F. Goodrich Co.	Akron, Ohio
	Goodyear Tire & Rubber Co.	Akron, Ohio
Neoprene, reinforced	Carlisle Corp.	Carlisle, Pa.
	B. F. Goodrich Co.	Akron, Ohio
	Firestone Tire & Rubber Co.	Akron, Ohio
	Plymouth Rubber Co.	Canton, Mass.
	Reeves Brothers, Inc.	New York, N.Y.
PE	Monsanto Chemical Co.	St. Louis, Mo.
	Union Carbide, Inc.	New York, N.Y.
	Ethyl Corp.	Baton Rouge, La.
PE, reinforced	Griffolyn Co., Inc.	Houston, Tex.
PVC	Firestone Tire & Rubber Co.	Akron, Ohio

(*continued*)

Table 5-3 (*cont.*)
Trade Names and Sources of Common Lining Materials

MATERIALS	MANUFACTURERS	LOCATIONS
	B. F. Goodrich Co.	Akron, Ohio
	Goodyear Tire & Rubber Co.	Akron, Ohio
	Pantasote Co.	New York, N.Y.
	Stauffer Chemical Co.	New York, N.Y.
	Union Carbide, Inc.	New York, N.Y.
PVC, reinforced	Firestone Tire & Rubber Co.	Akron, Ohio
	B. F. Goodrich Co.	Akron, Ohio
	Goodyear Tire & Rubber Co.	Akron, Ohio
	Reeves Brothers, Inc.	New York, N.Y.
	Cooley, Inc.	Pawtucket, R.I.
	Sun Chemical Co.	Paterson, N.J.
Prefabricated asphalt panels	Envoy-APOC	Long Beach, Calif.
	Gulf Seal, Inc.	Houston, Tex.
	W. R. Meadows, Inc.	Elgin, Ill.
	Phillip Carey Co.	Cincinnati, Ohio
3110	E. I. Du Pont Co.	Louisville, Ky.

SOURCE: From Ref. 1, courtesy of John Wiley & Sons, Inc., New York, New York.

fectiveness in limiting seepage. The cost analysis is, necessarily, somewhat arbitrary, since this cost depends primarily on the availability of the materials. A summary of state standards developed or being developed to control the application of these types of materials are presented in a report by Middlebrooks et al.[2]

Bentonite is a sodium-type montmorillonite clay and exhibits a high degree of swelling, imperviousness, and low stability in the presence of water. Different ways in which bentonite may be used to line lagoons are listed below.

1. A suspension of bentonite in water (with a bentonite concentration approximately 0.5% of the water weight) is placed over the area to be lined, and the bentonite settles to the bottom, forming a thin blanket.

2. The same procedure as (1), except frequent harrowing of the surface produces a uniform soil bentonite mixture on the surface of the soil. The amount of bentonite used in this procedure is approximately 4.89 kg/m^2 (1 lb/ft^2) of soil.

3. A gravel bed approximately 6 in. deep is first prepared and the bentonite application performed as in (1). The bentonite will settle through the gravel layer and seal the void spaces.

4. Bentonite is spread as a membrane 1 or 2 in. thick and covered with

Table 5-4
Summary of Effective Design Practice for Placing
Lining in Cut-and-Fill Reservoirs

1. Lining must be placed in a stable structure.
2. Facility design and inspection should be the responsibility of professionals with backgrounds in liner applications and experienced in geotechnical engineering.
3. A continuous underdrain to operate at atmospheric pressure is recommended.
4. A leakage tolerance should be included in the specifications. The East Bay Water Company of Oakland, California, developed the following formula for leakage tolerance which has been modified by inserting more stringent factors in the denominator, 100, 200, etc.

$$Q = \frac{A\sqrt{H}}{80}$$

where

   Q = maximum permissible leakage tolerance, gal/min
   A = lining area, 1000 ft^2
   H = maximum water depth
   $Q \geqq 1.0$

5. Continuous, thin, impermeable type linings should be placed on a smooth surface of concrete, earth, Gunite, or asphalt concrete.
6. Except for asphalt panels, all field joints should be made perpendicular to the toe of the slope. Joints of Hypalon formulations and 3110 materials can run in any direction, but generally joints run perpendicular to the toe of the slope.
7. Formal or informal anchors may be used at the top of the slope. (See details in Figures 5-1 through 5-5.)
8. Inlet and outlet structures must be sealed properly. (See details in Figures 5-6 through 5-10.)
9. All lining punctures and cracks in the support structure should be sealed. (See details in Figures 5-4 and 5-12.)
10. Emergency discharge quick-release devices should be provided in large reservoirs (20-30 million gal).
11. Wind problems with exposed thin membrane liners can be controlled by installing vents built into the lining. See details in Figure 5-13.
12. Adequate protective fencing must be installed to control vandalism.

an 20.3–30.5 cm (8–12 in.) blanket of earth and gravel to protect the membrane. A mixture of earth and gravel is more satisfactory than soil alone, because of the stability factor and resistance to erosion.

5. Bentonite is mixed with sand at approximately 1:8 volume ratio. The mixture is placed in a layer of approximately 5.08–10.16 cm (approximately 2–4 in. in thickness) on the reservoir bottom and covered with a protective cover. This method takes about 14.68 kg/m² (3 lb/ft²) of bentonite. [18]

In methods 4 and 5 above, certain construction practices are recommended. They are as follows:

Table 5-5
Lining Selection Guide Chart[a,b]

Substance						Type of Lining					
	PE	Hypalon	PVC	Butyl Rubber	Neoprene	Asphalt Panels	Asphalt Concrete	Concrete	Steel	CPE	3110
Water	OK	OK	OK	OK	OK	OK	OK	OK	CP	OK	OK
Animal oils	OK[c]	OK	ST	OK	OK	Q	Q	NR	OK	OK	OK
Petroleum oils (no aromatics)	OK[c]	Q	NR	NR	SW	NR	NR	OK	OK	OK	OK
Domestic sewage	OK	OK	OK	OK	OK	OK	OK	OK	OK	OK	OK
Salt solutions	OK	OK	OK	OK	OK	OK	Q	NR	NR	OK	OK
Base solutions	OK	OK	OK	OK	OK	OK	OK	Q	OK	OK	OK
Mild acids	OK	OK	OK	OK	OK	OK	OK	NR	NR	OK	OK
Oxidizing acids	NR	NR	NR	NR	Q	NR	NR	NR	NR	NR	NR
Brine	OK	OK	OK	OK	OK	OK	OK	Q	NR	OK	NR
Petroleum oils (aromatics)	Q	NR	NR	NR	NR	NR	NR	OK	OK	NR	NR

SOURCE: From Ref. 1, courtesy of John Wiley & Sons, Inc., New York, New York.

[a] OK = generally satisfactory, Q = questionable, NR = not recommended, ST = stiffens, SW = swells, CP = cathodic protection suggested.

[b] It is recommended that immersion tests be run on any lining being considered for use in an environment in which a question exists concerning its longevity. Consult the lining manufacturer or an experienced testing laboratory when in doubt.

[c] Must be a one-piece lining.

248

1. The section must be overexcavated (30.5 cm or more) with drag lines or graders.
2. Side slopes should probably be not steeper than two to one.
3. Subgrade surface should be dragged to remove large rocks and sharp angles. Normally, two passes with adequate equipment are sufficient to smooth the subgrade.
4. Subgrade should be rolled with a smooth steel roller.
5. The subgrade should be sprinkled to eliminate dust problems.
6. A membrane of bentonite or soil bentonite should then be placed.
7. The protective cover should contain sand and small gravel, in addition to cohesive, fine-grained material so that it will be erosion resistant and stable.

The performance of bentonite linings is greatly affected by the quality of the bentonite. Some bentonite deposits may contain quantities of sand, silt, and clay impurities. Wyoming-type bentonite, which is a high swelling sodium montmorillonite clay, has been found to be very satisfactory. Fine-ground bentonite is generally more suitable for the lining than pit-run bentonite. If the bentonite is finer than a No. 30 sieve, it may be used without specifying size gradation, but if the material is coarser than the No. 30 sieve, it should be well graded. Bentonite should usually contain a moisture content of less than 20%. This is especially important for thin membranes. Some disturbance and possible cracking of the membrane may take place during the first year after construction, because of settlement of the subgrade on saturation. A proper maintenance program, especially at the end of the first year, is necessary.[20]

Asphalt linings may be buried or on the surface and may be composed of asphalt or a prefabricated asphalt. Some possibilities are as follows.

1. *Asphalt Membrane*—an asphalt membrane is produced by spraying asphalt at high temperatures. This lining may be either on the surface or buried. A large amount of special equipment is needed for installation. Useful lives of 18 years or greater have been observed when these membranes are carefully applied and covered with an adequate layer of fine-grained soil.
2. *Asphaltic Membrane Macadam*—This is similar to the asphaltic membrane, but it is covered with a thin layer of gravel penetrated with hot-blown asphalt cement.
3. *Buried Asphaltic Membrane*—This is similar to item 1, except a gravel-sand cover is applied over the asphaltic membrane. This cover is usually more expensive than cover in 2 and less effective in discouraging plant growth.
4. *Builtup Linings*—These include several different types of materials. One type could be a fiberglass matting, which is applied over a sprayed asphalt layer and then also sprayed or broomed with a seal coat of

asphalt or clay. A 286 gr (10 oz) jute burlap has also been used as the interior layer between two hot sprayed asphalt layers. In this case the total asphalt application should be about 11.3 l/m^2 (2.5 gal/yd^2). The prefabricted lining may be on the surface or buried. If buried, it could be covered with a layer of soil or, in some cases, a coating of Allox, which is a stabilized asphalt, is used.[21]

5. *Prefabricated Linings*—Prefabricated asphalt linings consist of a fiber or paper material coated with asphalt. This type of liner has been used exposed and covered with soil. Joints between the material have an asphaltic mastic to seal the joint. When the asphaltic material is covered, it is more effective and durable. When it is exposed it should be coated with aluminized paint every 3–4 years to retard degradation. This is necessary especially above the water line. Joints also have to be maintained when not covered with fine-grained soil. Prefabricated asphalt membrane lining is approximately 0.318 to 0.635 cm (1/8 to 1/4 in.) thick. It may be handled in much the same way as rolled roofing with lapped and cemented joints. Cover for this material is generally earth and gravel, although shot-crete and macadam have been utilized.

Installation procedures for prefabricated asphalt membrane linings and for buried asphalt linings are similar to those stated for buried bentonite linings. The preparation of the subgrade is important, and it should be stable and adequately smooth for the lining.

Best results are obtained with soil cement when the soil mixed with the cement is sandy and well graded to a maximum size of about 1.9 cm (3/4 in.). Soil cement should not be placed in cold weather, and it should be cured for about 7 days after placing. Some variations of the soil cement lining are listed below.

1. Standard soil cement is compacted using a water content of the optimum moisture content of the soil. The mixing process is best accomplished by traveling mixing machines and can be handled satisfactorily in slopes up to four to one. Standard soil cement may be on the surface or buried.

2. Plastic soil cement (surface or buried) is a mixture of soil and cement with a consistency comparable to that of Portland cement concrete. This is accomplished by adding a considerable amount of water. Plastic soil cement contains from three to six sacks of cement per cubic yard and is approximately 3 in. thick.

3. Cement-modified soil contains 2–6% volume of cement. This may be used with plastic fine-grained soils. The treatment stabilizes the soil in sections subject to erosion. The lining is constructed by spreading cement on top of loose soil layers by a fertilizer type spreader. The cement is then mixed with loose soil by a rotary traveling mixer and compacted with a sheeps foot roller. The 7-day curing period is also necessary for a cement-modified soil.

Bentonite linings may be effective if the sodium bentonite used has an adequate amount of exchangeable sodium. Deterioration of the linings has been observed to occur in cases in which magnesium or calcium has replaced sodium as absorbed ions. A layer of bentonite on the soil surface tends to crack if allowed to dry and is, therefore, usually placed as a blanket of bentonite soil mixture with a cover of fine-grained soil on top or as a thicker layer, 6 in. or more, of a soil bentonite material.[22] Surface bentonite cannot be expected to be effective longer than 2–4 years. A buried bentonite blanket may last from 8 to 12 years.

The quality of the bentonite used is a primary consideration in the success of bentonite membranes. Poor-quality bentonite deteriorates rapidly in the presence of hard water, and it also tends to erode in the presence of cur-

FIGURE 5-1

Top anchor detail—alternative 1, all linings (Courtesy of John Wiley & Sons, Inc., New York, New York, Ref. 1).

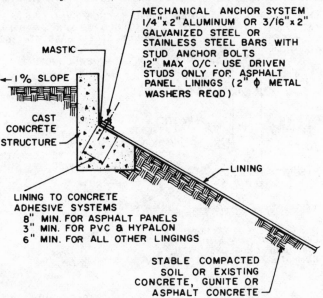

MECHANICAL ANCHOR SYSTEM
1/4" x 2" ALUMINUM OR 3/16" x 2"
GALVANIZED STEEL OR
STAINLESS STEEL BARS WITH
STUD ANCHOR BOLTS
12" MAX O/C. USE DRIVEN
STUDS ONLY FOR ASPHALT
PANEL LININGS (2" φ METAL
WASHERS REQD)

MASTIC

←1% SLOPE

CAST
CONCRETE
STRUCTURE

LINING

LINING TO CONCRETE
ADHESIVE SYSTEMS
8" MIN. FOR ASPHALT PANELS
3" MIN. FOR PVC & HYPALON
6" MIN. FOR ALL OTHER LININGS

STABLE COMPACTED
SOIL OR EXISTING
CONCRETE, GUNITE OR
ASPHALT CONCRETE

N O T E

1. TOP OF CONCRETE SHOULD BE SMOOTH AND FREE OF ALL CURING COMPOUNDS.

2. USE MIN 1/32" x 2" GASKET (MAT'L COMPATIBLE WITH LINING) BETWEEN BAR AND LINING, EXCEPT NO GASKET REQUIRED FOR ASPHALT PANELS OR OTHER LININGS THICKER THAN .040"

rents or waves. Bentonite linings must often be placed by hand, and this is a costly procedure in areas of high labor costs.

Seepage losses through buried bentonite blankets are approximately $0.21-0.26$ m^3/m^2 $(0.7-0.85$ ft^3/ft$^2)$ per day. This figure is for thin blankets and represents about a 60% improvement over ponds with no lining.

Linings of bentonite and asphalt are sometimes unsuitable in areas of high weed growth, since weeds and tree roots puncture the material readily.[23]

Many lining failures occur as a result of rodent and crayfish holes in embankments. Asphalt membrane lining tends to decrease the damage, but in some cases, hard-surface linings are necessary to prevent water loss from embankment failures.

FIGURE 5-2
Top anchor detail—alternative 2, all linings (Courtesy of John Wiley & Sons, Inc., New York, New York, Ref. 1).

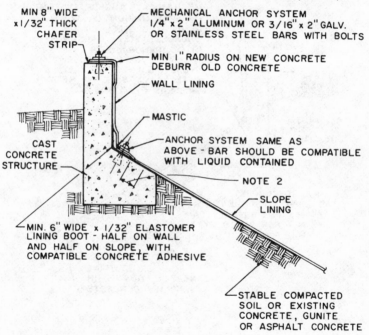

NOTE

1. ALL CONCRETE AT SEALS SHALL BE SMOOTH AND FREE OF ALL CURING COMPOUNDS.

2. USE COMPATIBLE ADHESIVE BETWEEN SLOPE LINING AND ELASTOMER BOOT, AND 3" MIN. WIDTH OF COMPATIBLE ADHESIVE BETWEEN SLOPE LINING AND CONCRETE.

FIGURE 5-3
Top anchor detail—alternative 3, all linings (Courtesy of
John Wiley & Sons, Inc., New York, New York, Ref. 1).

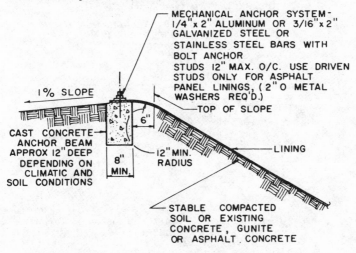

N O T E

1. TOP OF CONCRETE SHOULD BE SMOOTH AND
    FREE OF ALL CURING COMPOUNDS.

2. USE MIN. 1/32"x 2" GASKET (MAT'L COMPATIBLE WITH
    LINING) BETWEEN BAR & LINING EXCEPT NO
    GASKET REQUIRED FOR ASPHALT PANELS OR OTHER
    LININGS THICKER THAN .040"

Linings of hot-applied buried asphalt membrane provide one of the tighest linings available. These linings deteriorate less than other flexible membrane linings.[23]

Asphalt linings composed of prefabricated buried materials are best for small jobs, since the amount of special equipment and labor connected with installation is a minimum. For larger jobs, sprayed asphalt is more economical.

When fibers and fillers are used in asphalt membranes, there is a greater tendency to deteriorate when these fillers are composed of organic materials. Inorganic fibers are, therefore, more useful.[23] Typical volume of seepage through one buried asphalt membrane after 10 years of service was consistently 0.024 m^3/m^2 (0.08 ft^3/ft^2) per day.[20]

Asphalt membrane linings can be constructed at any time of year, and since it is usually convenient in canals and ponds to use the late fall and

FIGURE 5-4
Top anchor detail—alternative 4, all linings except
asphalt panels (Courtesy of John Wiley & Sons,
Inc., New York, New York, Ref. 1).

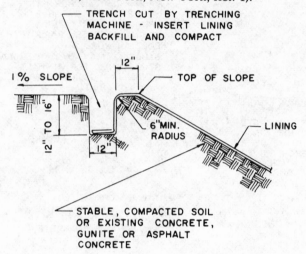

winter seasons for installing lining, this may dictate the buried asphalt
membrane lining as the proper one to use in many cases.[23]

Buried asphalt membranes, in general, perform satisfactorily for more
than 15 years. When these linings fail, it is generally due to one or more of
the following causes:

FIGURE 5-5
Top anchor detail—alternative 5, all linings (Courtesy of
John Wiley & Sons, Inc., New York, New York, Ref. 1).

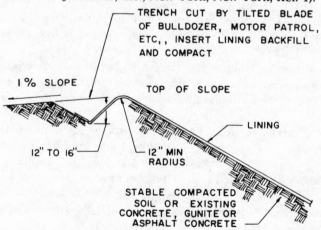

1. Placement of lining on unstable side slopes.
2. Inadequate protection of the membrane.
3. Weed growth.
4. Surface runoff.
5. Type of subgrade material.
6. Cleaning operations.
7. Scour of cover material.
8. Membrane puncture.

Soil cement has been used successfully in some cases in mild climates. When wetting or drying is a factor or if freezing-thawing cycles are present, the lining will deteriorate rapidly.[23]

Thin Membrane Liners. Plastic and elastomeric membranes are popular in applications requiring essentially zero permeability. These

FIGURE 5-6

Seal at pipes through slope, all linings (Courtesy of John Wiley & Sons, Inc., New York, New York, Ref. 1).

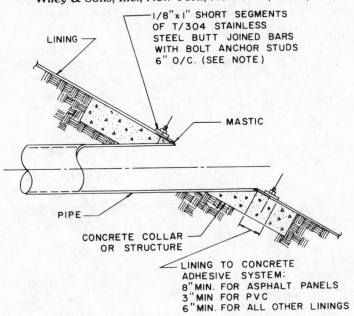

LINING

1/8"x I" SHORT SEGMENTS OF T/304 STAINLESS STEEL BUTT JOINED BARS WITH BOLT ANCHOR STUDS 6" O/C. (SEE NOTE)

MASTIC

PIPE

CONCRETE COLLAR OR STRUCTURE

LINING TO CONCRETE ADHESIVE SYSTEM: 8"MIN. FOR ASPHALT PANELS 3"MIN FOR PVC 6"MIN. FOR ALL OTHER LININGS

NOTE

FOR ASPHALT PANEL LININGS, PERCUSSION DRIVEN STUDS THRU 2" MIN. DIA. x 1/16" THICK GALVANIZED METAL DISCS AT 6" O/C ENCASED IN MASTIC MAY BE SUBSTITUTED FOR ANCHOR SHOWN.

materials are economical, resistant to most chemicals if selected and installed properly, available in large sheets simplifying installation, and essentially impermeable. As discharge standards continue to become more stringent, the application of plastic and elastomeric membranes as lagoon liners will increase because of the need to guarantee protection against seepage. This is particularly true in the sealing of lagoons containing toxic wastewaters or the sealing of landfills containing toxic solids and sludges.

Typical standards being developed to control the application of liners are presented in a report by Middlebrooks et al.[2] A partial listing of the trade names, product description and manufacturer of plastic and elastomer lining materials is presented in Table 5–3.

The most difficult design problem encountered in liner applications involves placing a liner in an existing reservoir (lagoon). Effective design practices are essentially the same as those used in new systems, but additional care must be exercised in the evaluation of the existing structure and the required results. Lining materials must be selected so that compatibility is obtained. For example, a badly cracked concrete lining to be covered with a flexible synthetic material must be properly sealed and placed in such a way that additional movement will not destroy the new liner. Sealing

FIGURE 5-7
Seal at floor columns, asphalt panels (Courtesy of John Wiley & Sons, Inc., New York, New York, Ref. 1).

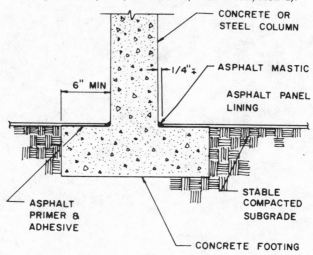

CONCRETE OR STEEL COLUMN

ASPHALT MASTIC

ASPHALT PANEL LINING

6" MIN

1/4"±

STABLE COMPACTED SUBGRADE

ASPHALT PRIMER & ADHESIVE

CONCRETE FOOTING

N O T E
MECHANICAL FASTENERS NOT REQUIRED

FIGURE 5-8

Pipe boot detail, all linings except asphalt panels (Courtesy of John Wiley & Sons, Inc., New York, New York, Ref. 1).

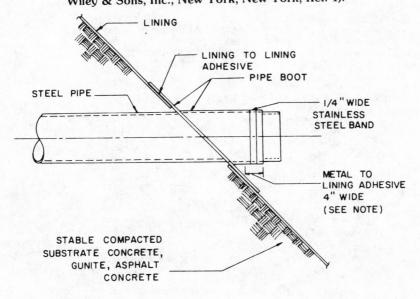

N O T E

CLEAN PIPE THOROUGHLY AT AREA OF ADHESIVE APPLICATION.

around existing columns, footings, and the like are other examples of items to be considered.

The following paragraphs are a condensation of the discussion by Kays[1] of effective design practices which have been summarized in Table 5-4. Emphasis is placed on the details describing the installation of plastic or elastomeric materials.

Formal and informal anchor systems are used at the top of the slope of dikes. Details of three types of formal anchors are presented in Figures 5–1 to 5–3. Recommended informal anchors are shown in Figures 5–4 and 5–5.

When the lining is pierced, seals can be made in two ways. The techniques illustrated in Figures 5–6 and 5–7 are commonly used, and the second technique utilizes a pipe boot that is sealed to the liner and clamped to the entering pipe, as shown in Figure 5–8.

It is recommended that inlet-outlet pipes enter a reservoir through a structure such as that shown in Figure 5–9. A better seal can be produced when the liner is attached to the top of the structure. However, such an ar-

FIGURE 5-9
Seal at inlet-outlet structure, all linings (Courtesy of John Wiley
& Sons, Inc., New York, New York, Ref. 1).

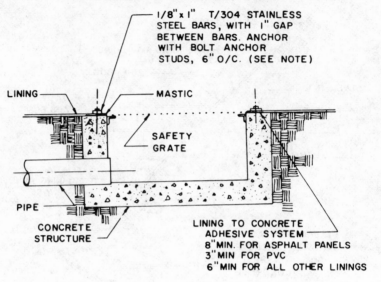

N O T E

WITH ASPHALT PANEL LININGS, PERCUSSION
DRIVEN STUDS THRU I" MIN DIA 1/16" THICK
GALVANIZED METAL DISCS AT 6" O/C, ENCASED
IN MASTIC MAY BE SUBSTITUTED FOR ANCHOR
SHOWN.

rangement can result in solids accumulation, and a direct free entry into a wastewater lagoon is better.

A drain near the outlet can be constructed as shown in Figure 5–10. As mentioned in Table 5–4, large reservoirs containing above 20–30 million gal should be equipped to empty quickly in case of an emergency.

The structure supporting the liner must be smooth enough to prevent damage to the liner. Rocks, sharp protrusions, and other rough surfaces must be controlled. In areas with particularly rough surfaces, it may be necessary to add padding to protect the liner. Cracks can be repaired, as shown in Figures 5–11 and 5–12.

Thin membrane liners may have problems with wind on the leeward slopes. Vents built into the lining control this problem as well as serve as an outlet for gases trapped beneath the liner (Figure 5–13).

Protection of a thin membrane lining is essential, and Kays[1] recommends that the fence be at least 6 ft high and be placed on the outside berm slope, with the top of the fence below the top elevation of the dike.

FIGURE 5-10

Mud drain detail, all linings (Courtesy of John Wiley & Sons, Inc., New York, New York, Ref. 1).

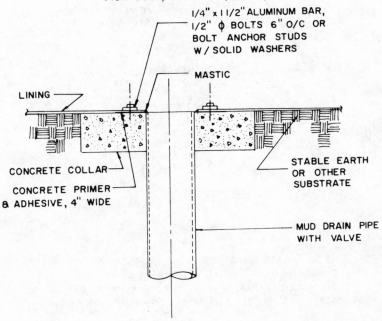

A partial listing of the manufacturers of plastic and elastomeric liners is presented in Table 5-3. In addition to these manufacturers, there are many firms specializing in the installation of lining materials. Most of the installation companies and the manufacturers publish specifications and installation instructions and design details for use by customers and design

FIGURE 5-11

Crack treatment—alternative A (Courtesy of John Wiley & Sons, Inc., New York, New York, Ref. 1).

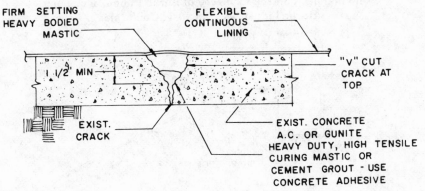

FIGURE 5-12
Crack treatment—alternative B (Courtesy of John Wiley
& Sons, Inc., New York, New York, Ref. 1).

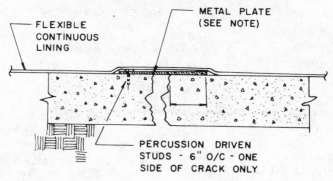

METAL PLATE MUST BE ABLE TO SPAN
CRACK WITHOUT BUCKLING FROM WEIGHT
OF WATER BRIDGING THE CRACK. COPPER &
STAINLESS STEEL ARE MOST COMMON CHOICES.

engineers. Most of the recommendations by the manufacturers and installers are similar, but there are differences worthy of consideration when designing a system requiring a liner. Consult the manufacturers for details.

New products continue to be developed, and with each new material the options available to designers continue to improve. The future should bring even more versatile and effective liners to select for seepage control. If care and common sense are applied to the application of existing and new

FIGURE 5-13
Wind and gas control. Courtesy of Burke Rubber Company,
Burke Industries, San Jose, California.

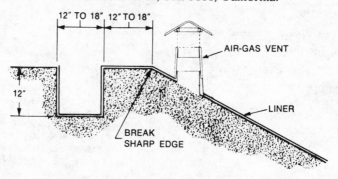

materials, the control of seepage pollution should become a minor problem of the future.

Failure Mechanisms

Kays[1] presented a classification of the principal failure mechanisms observed in cut-and-fill reservoirs (Table 5–6). The list is extensive, and case histories involving all of the categories are available; however, the most frequently observed failure mechanisms have been the lack of integrity in the lining support structure and the abuse of the liner.

Cost of Linings

The following costs of linings for lagoons and reservoirs are approximations and are estimated based on values at specific jobs in terms of 1978 U.S. dollars.

Table 5-6
Classification of the Principal Failure Mechanisms
for Cut-and-Fill Reservoirs.

Supporting structure problems

    The underdrains
    The substrate
        Compaction
        Texture
        Voids
        Subsidence
        Holes and cracks
        Groundwater
        Expansive clays
        Gassing
        Sluffing
        Slope anchor stability
        Mud
        Frozen ground and ice
    The appurtenances

Lining problems

    Mechanical difficulties
        Field seams
        Fish mouths
        Structure seals
        Bridging

    Porosity
    Holes
    Pinholes
    Tear strength
    Tensile strength
    Extrusion and extension
    Rodents, other animals, and birds
    Insects
    Weed growths

Weather
    General weathering
    Wind
    Ozone
    Wave erosion
    Seismic activity

Operating problems

    Cavitation
    Impingement
    Maintenance cleaning
    Reverse hydrostatic uplift
    Vandalism

SOURCE: From Ref. 1, courtesy of John Wiley & Sons, Inc., New York, New York.

Bentonite linings cost approximately $1.20–$2.40/m² ($1–$2/yd²) when applied on the surface. The greater cost will occur for harrowed blankets. Buried blankets cost approximately $2.99/m² ($2.50/yd²).

The average cost of buried asphalt membrane linings with adequate cover is about $4.19/m² ($3.50/yd²).

Prefabricated asphalt materials are generally cheaper than buried asphalt membrane linings if the prefabricated material can be obtained for less than $1.08/m² ($0.90/yd²).

Table 5-7
Cost of Installed Liner

LINER	$/FT²
Bentonite	
2 lb/ft²	0.14
Chemical	
Sodium carbonate	0.02
Sodium silicate	0.02
Sodium pyrophosphate	0.03
Zeogel	0.03
Coherex	0.03
Asphalt	
Asphalt membrane	0.14
Asphalt concrete	0.20
Rubber[a]	
Butyl	
1/16 in.	0.42
3/16 in.	0.36
1/32 in.	0.30
EPDM	
1/16 in.	0.41
3/64 in.	0.35
1/32 in.	0.29
Synthetic membrane	
PVC	
10 mils	0.13
20 mils	0.18
30 mils	0.22
Chlorinated polyethylene (CPE)	
20 mils	0.26
30 mils	0.34
Hypalon	
20 mils	0.26
30 mils	0.34
Fiber glas	
1/8 in.	0.55

SOURCE: From Ref. 24.

[a] Nylon-reinforced rubber costs an additional $0.10/ft.²

Cover material over buried membranes composes the most expensive part of the placing procedure. The cover materials should, therefore, be as thin as possible and still provide adequate protection for the membrane. If a significant current is present in the pond, the depth of coverage should be greater than 25.4 cm (10 in.), and this minimum depth should only be used when the material is erosion resistant and also cohesive. Such a material as a clayey gravel is suitable. If the material is not cohesive or if it is fine grained, a higher amount of cover is needed.[23]

Maintenance costs for different types of linings are difficult to estimate. Maintenance should include repair of holes, cracks, and deterioration; weed control expenses; and animal damages and damages caused by cleaning the pond, if necessary. Climate, type of operation, type of terrain, and surface conditions also influence maintenance costs.

Plastic soil cement containing from three to six sacks of cement per cubic yard and approximately 3 in. thick costs about $3.59/m^2$ ($3.00/yd^2$).

Cost comparisons of various liners (Tables 5–7 and 5–8) indicate that natural and chemical sealants are the most economical sealers. Costs presented in Tables 5–7 and 5–8 can be used only as an approximation of relative costs, because the costs for various materials have not increased consistently. Unfortunately, natural and chemical sealers are dependent on local soil conditions for seal efficiency and never form a complete seal. Asphalt and synthetic liners compete competitively on a cost basis but have different practical applications. Synthetic liners are most practical for zero or minimum seepage regulations, for industrial waste that might degrade concrete or earthen liners, and for extremes in climatic conditions.

Table 5-8
Comparison of Various Installed
Liner Costs, 1962 Cost Figures

Liner Type	Cost ($/ft^2$)
Prefabricated plastic	0.30–0.10
Composite PVC and asphalt	0.09
Butyl rubber membranes	0.40
Bentonite clay	0.60
Prefabricated asphalt	0.11
Spray-type cutback Emulsion asphalt	0.02
Spray-type catalytically blown asphalt	0.08
Asphalt/concrete (hot mix)	0.30
Soil cement	> 0.30

Source: Compiled from text of Ref. 25, Courtesy of *Public Works*, Ridgewood, New Jersey.

References

1. Kays, W. B. 1977. *Construction of Linings for Reservoirs, Tanks, and Pollution Control Facilities.* Wiley–Interscience, New York, New York.

2. Middlebrooks, E. J., C. D. Perman, and I. S. Dunn. 1978. Wastewater Stabilization Pond Linings. *Special Report 78–28.* U. S. Army Cold Regions Research and Engineering Laboratory, Hanover, New Hampshire.

3. Thomas, R. E., W. A. Schwartz, and T. W. Bendixen. 1966. Soil Changes and Infiltration Rate Reduction Under Sewage Spreading. *Soil Sci. Soc. Amer. Proc.*, **30**, 641–646.

4. Bhagat, Surinder K., and Donald E. Proctor. 1969. Treatment of Dairy Manure by Lagooning. *WPCF*, **41**,(5), 785–795.

5. Hills, David J. 1976. Infiltration Characteristics from Anaerobic Lagoons. *JWPCF,* **48**, 4, 695.

6. Chang, A. C., W. R. Olmstead, J. B. Johanson, and G. Yamashita. 1974. The Sealing Mechanism of Wastewater Ponds. *JWPCF*, **46**, 7, 1715–1721.

7. Wilson, L. G., Wayne L. Clark, and Gary G. Small. 1973. Subsurface Quality Transformations During Preinitiation of a New Stabilization Lagoon. *Water Res. Bull.*, **9**, 2, 243–257.

8. Matthew, Floyd L., and Leland L. Harms. 1969. Sodium Adsorption Ratio Influence on Stabilization Pond Sealing. *JWPCF*, **4**, 11, Part 2, R383–R391.

9. Rosene, R. B., and C. F. Parks. 1973. Chemical Method of Preventing Loss of Industrial and Fresh Waters from Ponds, Lakes, and Canals. *Water Res. Bull.* **9**, 4, 717–722.

10. Davis, S., W. Fairbank, and H. Weisbeit. 1973. Dairy Waste Ponds Effectively Self-Sealing. *Am. Soc. Agric. Eng. Trans.* **16**, 69–71.

11. Robinson, F. E. 1973. Changes in Seepage Rate from an Unlined Cattle Waste Digestion Pond. *Trans. Amer. Soc. Agri. Eng.* **16**, 95.

12. Stander, G. J., P. G. J. Meiring, R. J. L. C. Drews, and H. Van Eck. 1970. A Guide to Pond Systems for Wastewater Purification. In *Developments in Water Quality Research.* Edited by H. I. Shuval. Ann Arbor Science Publishers, Ann Arbor, Michigan.

13. Hannaman, M. C., E. J. Johnson, and M. A. Zagar. 1978. *Effects of Wastewater Stabilization Pond Seepage on Groundwater Quality.* Prepared by Eugene A. Hickok and Associates, Wayzata, Minnesota for Minnesota Pollution Control Agency, Roseville, Minnesota.

14. California State Water Pollution Control Board. 1956. *Report on Continued Study of Waste Water Reclamation and Utilization*, Publication No. 15, Sacramento, California.

15. California State Water Pollution Control Board. 1957. *Third Report on the Study of Waste Water Reclamation and Utilization*, Publication No. 18, Sacramento, California.

16. Neal, J. K., and G. J. Hopkins. 1956. Experimental Lagooning of Raw Sewage. *Sewage Ind. Wastes*, **28**, 11, 1326.

17. Voights, D. 1955. Lagooning and Spray Disposal of Neutral Sulphite Semi-chemical Pulp Mill Liquors. *Proceedings of the Tenth Purdue Industrial Waste Conference*, Purdue University Extension Service, West Lafayette, Indiana. No. 89, p. 497.

18. Shaw, V. A. 1962. An Assessment of the Probable Influence of Evaporation and Seepage on Oxidation Pond Design and Construction. Journal 2, Proc. Ins. Sewage Purification, Part 4.

19. Rollins, M. B., and A. S. Dylla. 1970. Bentonite Sealing Methods Compared in the Field. *J. Irr. Dr. Div., ASCE Proc.* 96, IR2, 193.

20. USDI. 1968. *Buried Asphalt Membrane Canal Lining.* Research Report No. 12. U.S Government Printing Office, Washington, D. C.

21. USDA. 1972. *Asphalt Linings for Seepage Control: Evaluation of Effectiveness and Durability of Three Types of Linings.* Tech. Bull. No. 1440.

22. Dedrick, A. R. 1975. *Storage Systems for Harvested Water.* U.S. Department of Agriculture, ARS W–22, p. 175.

23. USDI. 1963. *Linings for Irrigation Canals.* Bureau of Reclamation, U.S. Government Printing Office, Washington, D.C.

24. Clark, Don A., and James E. Moyer. 1974. An Evaluation of Tailings Ponds Sealants. EPA–660/2–74–065. U.S. Environmental Protection Agency, Washington, D.C.

25. Stoltenberg, D. H. 1970. Design, Construction and Maintenance of Waste Stabilization Lagoons. *Public Works.* 101, 9, 103–106.

CHAPTER

6

Disinfection of Lagoon Effluents

Municipal wastewaters contain a variety of infectious microorganisms such as salmonellae, shigellae, enteropathogenic *E. coli*, Pseudomonas aeruginosa, and enteric viruses. Outbreaks of gastroenteritis, typhoid, shigellosis, salmonellosis, ear infections, and infectious hepatitis have been reported for people drinking or swimming in waters mixed with municipal wastewaters. Outbreaks also have occured when people eat raw shellfish harvested from waters contaminated with municipal wastewaters.

All discharges do not necessarily contain all or part of the pathogens mentioned above, and it may be possible to escape contact with these organisms most of the time. However, practical public health practices dictate that constant protection be provided, because it is impossible to detect the presence of pathogens before they are discharged in a wastewater. Consequently, disinfection of wastewater discharges must be practiced continuously.

Experience and judgment have shown that reducing the fecal coliform concentration to 14/100 ml or the total coliform concentration to 70/100 ml will prevent disease outbreaks caused by shellfish.[1] Limited epidemiological data indicates that concentrations of fecal coliform of approximately 200/100 ml in recreational waters reduces the probability of contact to an acceptable level.[1] Although the U.S. Environmental Protection Agency Secondary Effluent Standards no longer contain a standard for fecal coliform, the logic for its inclusion initially was based on the limited epidemiological data referred to above.

Disinfectants

Many chemicals and physical agents are good disinfectants. Heat, sunlight, chlorine, bromine, iodine, potassium permanganate, chlorine dioxide, ozone, and ultraviolet light are effective disinfectants, but experience with most of these materials as a wastewater disinfectant is limited. Because of the extensive experience with chlorine, it is and will likely continue to be the most widely used disinfectant. The principal disadvantage of chlorination is the production of toxic substances and its effect on aquatic life.

Toxic Effects

The production of halogenated organic compounds suspected of being toxic to man has produced concern by public health officials, and efforts are being directed toward developing other methods of disinfection. The implementation of dechlorination to reduce the toxicity of wastewater discharges to natural environments has resulted in concern about the compounds produced by the reactions between the forms of residual chlorine and dechlorinating agents such as sulfur dioxide, sodium bisulfite, sodium sulfite, or activated carbon. This effort will eventually result in a reduction in the use of chlorine, but immediate changes are not likely to occur.

Many of the proposed substitutes may have the same disadvantages associated with chlorination. Ozone is used extensively in Europe to disinfect drinking water, but little is known about its interaction with organic matter in wastewaters. All the disinfectants listed above have disadvantages, and most suffer from high costs, inefficiency in wastewaters with solids, toxic side effects, and no residual. The other halogens will probably have disadvantages similar to those of chlorine.

Basic Principles of Chlorination

To understand the effects of chlorinating stabilization pond effluents, it is necessary to review the basic principles of chlorination. When chlorine gas is used, the gas reacts with water to form hypochlorous acid (HOCl). In a pure water system, the reaction is as follows:

$$Cl_2 + H_2O \rightarrow HOCl + H^+ + Cl^- \qquad (6\text{--}1)$$

The hypochlorous acid then disassociates to form OCl^- and H^+

$$HOCl \rightleftharpoons H + OCl^- \qquad (6\text{--}2)$$

When Ca $(OCl)_2$, for example, is used to chlorinate, OCl^- is formed by the following reaction:

$$Ca(OCl)_2 \rightarrow Ca^{2+} + 2OCl^- \qquad (6-3)$$

The OCl^- is then free to form hypochlorous acid in contact with hydrogen ions. Chlorine in the form of HOCl or OCl^- is referred to as free chlorine. Both forms of free chlorine are powerful disinfectants and react quickly to destroy bacteria and most viruses.

In wastewater, such as stabilization pond effluents, various chemical components react with free chlorine to form compounds that are ineffective as disinfectants. That is, the rates of reactions between chlorine and these components are faster than the rate at which chlorine attacks and kills bacteria and viruses. Fe^{2+}, Mn^{2+}, NO_2^{2-} and S^{2-} are common reducing agents that combine readily with chlorine to prevent it from disinfecting. A typical reaction is as follows:

$$H_2S + 4Cl_2 + 4H_2O \rightarrow H_2SO_4 + 8HCl \qquad (6-4)$$

Free chlorine also reacts with ammonia found in wastewater to form a series of compounds known as chloramines. Although chloramines are less than 5 % as efficient as free chlorine in destroying bacteria and viruses, they do play an important role in disinfection, because they are fairly stable and can continue to provide disinfection for some time after application. The common forms of chloramines (or combined chlorine, as they are referred to) are monochloramine, dichloramine, and nitrogen trichloride. The reactions for their formation are as follows:

$$NH_3 + HOCl \quad \rightarrow NH_2Cl + H_2O \qquad (6-5)$$
$$NH_2Cl + HOCl \quad \rightarrow NHCl_2 + H_2O \qquad (6-6)$$
$$NHCl_2 + HOCl \rightleftharpoons HCl_3 + H_2O \qquad (6-7)$$

In some cases chlorination is used as a treatment step to drive off undesirable ammonia. This is known as breakpoint chlorination. Basically, chlorine is added until all the chlorine has reacted to form chloramines. With the addition of more chlorine, the ammonia is converted to nitrogen gas and driven off. Any additional chlorine added beyond that point is maintained in solution as free chlorine residual. The mechanisms involved are complex, but the overall reaction may be represented as follows:

$$2NH_3 + 3HOCl \rightarrow N_2\uparrow + 3HCl + 3H_2O \qquad (6-8)$$

A comparison of ideal breakpoint chlorination and wastewater breakpoint chlorination is presented in Figure 6–1. Because the chlorine dose necessary to reach the breakpoint in wastewater is much higher than the dose necessary to achieve adequate disinfection, breakpoint chlorination is seldom used in the treatment of wastewater.

FIGURE 6-1
Comparison of ideal and
wastewater stabilization pond
effluent chlorination curves
(Ref. 10).

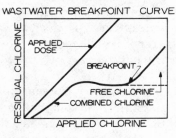

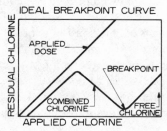

Effects of Chlorinating Lagoon Effluents

Since chlorine, at present, is less expensive and offers more flexibility than other means of disinfection, chlorination, is the most practical method of reducing bacterial populations. However, there is evidence that chlorination of wastewater high in organic nitrogen content, such as stabilization pond effluent, may be accompanied by adverse effects.

White[2] has suggested that chlorine demand is increased by high concentrations of algae commonly found in pond effluents. It was found that to satisfy chlorine demand and to produce enough residual to disinfect algae-laden wastewater within 30–45 min effectively, a chlorine dose of 20–30 mg/l was required. Kott[3] also reported increases in chlorine demand as a result of algae, but he found that a chlorine dose of 8 mg/l was sufficient to produce adequate disinfection within 30 min and that if contact times are kept relatively short, no serious chlorine demand by algae cells is encountered. Of course, the amount of chlorine demand exerted is highly variable. Dinges and Rust[4] found that for pond effluents a chlorine demand of only 2.65–3.0 mg/l was exerted after 20 min of contact. Brinkhead and O'Brien[5] found that at low doses of chlorine, very little increase in chlorine demand is attributable to algae, but at higher doses the destruction of algae cells greatly increases demand. This is because dissolved organic compounds released from destroyed algae cells, as explained by Echelberger et al.,[6] are oxidized by chlorine and thus increase chlorine demand.

Another concern regarding the chlorination of pond effluents is the effects on biochemical oxygen demand (BOD_5) and chemical oxygen demand (COD). Brinkhead and O'Brien[5] and Echelberger et al.[6] found that for higher chlorine doses, increases in BOD_5 due to destruction of algae cells were observed. Echelberger et al.[6] also reported increases in soluble COD. Hom[7] found that when 2.0 m/l chlorine was applied to pond effluent, the BOD_5 measured was 20 mg/l. When 64 mg/l chlorine was applied, the BOD_5 increased to 129 mg/l. However, Zaloum and Murphy[8] observed a 40% reduction of BOD_5 resulting from chlorination. Dinges and Rust[4] also reported reductions of BOD_5. Kott[3] has suggested that increases in BOD_5 may be controlled by using low chlorine doses coupled with long contact periods.

Not all the side effects of chlorinating pond effluents are detrimental. Kott[9] observed reductions of suspended solids (SS) as a result of chlorination. Dinges and Rust[4] reported reductions of volatile suspended solids (VSS) by as much as 52.3% and improved water clarity (turbidity) by 31.8% following chlorination. Echelberger et al.[6] reported that chlorine enhances the flocculation of algae masses by causing algae cells to clump together.

Four systems of identically designed chlorine mixing and contact tanks, each capable of treating 189.3 m³ (50,000 gal) per day, were used by Johnson et al.[10] to study the chlorination of lagoon effluents. Three of the four chlorination systems were used for directly treating pond effluent. The effluent treated in the fourth system was filtered through an intermittent sand filter to remove algae prior to chlorination. The filtered effluent was also used as the solution water for all four chlorination systems.

Following recommendations by Collins et al.,[11] Kothandaraman and Evans,[12,13] and Marske and Boyle,[14] the chlorination systems were constructed to provide rapid initial mixing, followed by chlorine contact in plug flow reactors. A serpentine flow configuration having a length-to-width (L/W) ratio of 25:1, coupled with inlet and outlet baffles, was used to enhance plug flow hydraulic performance. The maximum theoretical detention time for each tank was 60 min, while the maximum actual detention time averaged about 50 min.

The pond effluent was chlorinated at doses ranging from 0.25 to 30.0 mg/l under a variety of contact times, temperatures, and seasonal effluent conditions from August 1975 to August 1976. A variety of chemical, physical, and bacteriological parameters were monitored during this period in evaluating the chlorination of pond effluents. A series of laboratory experiments were also conducted to compliment the field study. Some of the major findings of this study are summarized below.

FIGURE 6-2
Chlorine dose versus residual for initial
sulfide concentrations of 1.0–1.8 mg/l (Ref. 10).

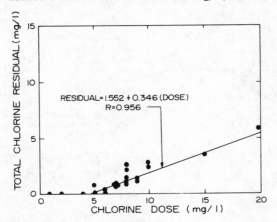

1. Sulfide, produced as a result of anaerobic conditions in the ponds during winter months when the ponds are frozen over, exerts a significant chlorine demand (Figure 6–2). For sulfide concentrations of 1.0–1.8 mg/l, a chlorine dose of 6–7 mg/l was required to produce the same residual as a chlorine dose of about 1 mg/l for conditions of no sulfide.

2. For all concentrations of ammonia encountered, it was found that adequate disinfection could be obtained with combined chlorine residual in 50 min or less of contact time. Therefore, breakpoint chlorination, and the subsequent production of free chlorine residual, was found to be rarely, if ever, necessary in disinfecting pond effluent.

3. It was found that total COD is virtually unaffected by chlorination. Soluble COD was found to increase with increasing concentrations of free chlorine only. This increase was attributed to the oxidation of SS by free chlorine. Increases in soluble COD versus free chlorine residual are shown in Figure 6–3.

4. Some reduction in SS, due to the break down and oxidation of suspended particulates, and resulting increases in turbidity were attributed to chlorination. However, this reduction was found to be of limited importance when compared with reductions of SS resulting from settling. The SS were reduced by 10–50% by settling in the contact tanks.

5. Filtered pond effluent exerted a lower chlorine demand than unfiltered pond effluent, because of the removal of algae (Figure 6–4). The rate of exertion of chlorine demand was determined to be directly related to chlorine dose and total COD.

6. A summary of coliform removal efficiencies as a function of total chlorine residual for filtered and unfiltered effluent is illustrated in Figure

FIGURE 6-3
Changes in soluble COD versus free chlorine residual—unfiltered lagoon effluent (Ref. 10).

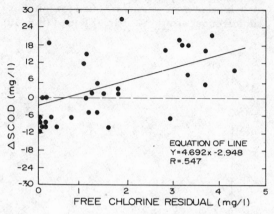

EQUATION OF LINE
Y=4.692x -2.948
R =.547

FIGURE 6-4
Chlorine dose versus total residual—filtered and unfiltered effluent (Ref. 10).

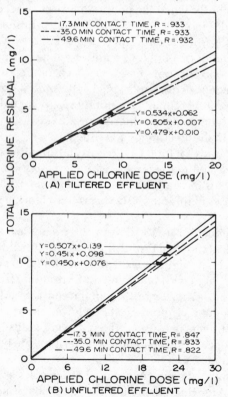

17.3 MIN CONTACT TIME, R = .933
35.0 MIN CONTACT TIME, R =.933
49.6 MIN CONTACT TIME, R =.932

Y=0.534x+0.062
Y=0.505x+0.007
Y=0.479x+0.010

(A) FILTERED EFFLUENT

Y=0.507x+0.139
Y=0.451x+0.098
Y=0.450x+0.076

17.3 MIN CONTACT TIME, R= .847
35.0 MIN CONTACT TIME, R =.833
49.6 MIN CONTACT TIME, R =.822

(B) UNFILTERED EFFLUENT

272

6–5. The rate of disinfection was a function of the chlorine dose and bacterial concentration. Generally, the chlorine demand was found to be about 50% of the applied chlorine dose except during periods of sulfide production.

7. Disinfection efficiency was temperature dependent. At colder temperatures the reduction in the rate of disinfection was partially offset by reductions in the exertion of chlorine demand; however, the net effect was a reduction in the chlorine residual necessary to achieve adequate disinfection with increasing temperature for a specific contact period.

8. In almost all cases, adequate disinfection was obtained with combined chlorine residuals of between 0.5 and 1.0 mg/l after a contact period of appoximately 50 min. This indicated that disinfection can be achieved without discharging excessive concentrations of toxic chlorine residuals into receiving waters. Also, it was found that adequate bacterial removal can be achieved with relatively low doses of applied chlorine during most of the year.

FIGURE 6-5
Coliform removal efficiencies—filtered and unfiltered effluent (Ref. 10).

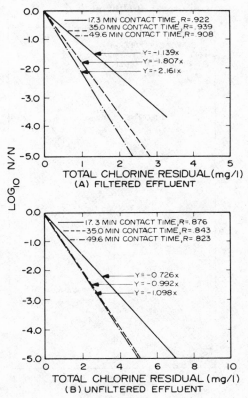

Predicting Required Residuals

Using the data from the study mentioned above, Johnson et al.[10] developed a model to predict the chlorine residual required to obtain a specified bacterial kill.

The model was used to construct a series of design curves for selecting chlorine doses and contact times for achieving desired levels of disinfection. An example may best illustrate how these design curves are applied. Assume that a particular lagoon effluent is characterized as having a fecal coliform (FC) concentration of 10,000/100 ml, 0 mg/l sulfide, 20 mg/l total chemical oxygen demand (TCOD), and a temperature of 5°C. If it is necessary to reduce the FC counts to 100/100 ml, a combined chlorine residual sufficient to produce a 99% bacterial reduction must be obtained. If an existing chlorine contact chamber has an average residence time of 30 min, the required chlorine residual is obtained from Figure 6–6. A 99% percent

FIGURE 6-6
Combined chlorine residual at 5°C for
coliform = 10⁴/100 ml (Ref. 10).

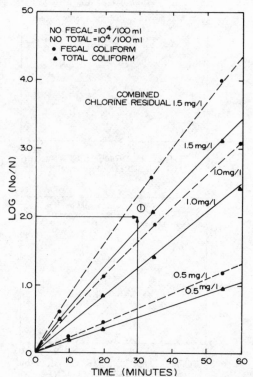

bacterial reduction corresponds to log (N_O/N) equal to 2.0 For a contact period of 30 min, a combined chlorine residual of between 1.0 and 1.5 mg/l is required to produce that level of FC reduction. On interpolation, the actual chlorine residual is determined to be 1.3 mg/l. This is indicated by point 1 in Figure 6–6.

Going to Figure 6–7, it is determined that if a chlorine dose produces a residual of 1.30 mg/l at 5°C, the same dose would produce a residual of 0.95 mg/l at 20°C. This is because of the faster rate of reaction between TCOD and chlorine at the higher temperature. This is indicated by point 2 in Figure 6–7. For an equivalent chlorine residual of 0.95 mg/l at 20°C and 20 mg/l TCOD, it is determined from Figure 6–8 that the same chlorine dose would produce a residual of 0.80 mg/l if the TCOD were 60 mg/l. This is because higher concentrations of TCOD increase the rate of chlorine demand. Point 3 on Figure 6–8 corresponds to this residual. The chlorine dose required to produce an equivalent residual of 0.80 mg/l to 20°C and 60 mg/l TCOD is determined from Figure 6–9. For a chlorine contact period of 30 min, a chlorine dose of 2.15 mg/l is necessary to produce the desired combined residual, as indicated by point 4 on Figure 6–9. This dose will produce a reduction of FC from 10,000/100 ml to 100/100 ml within 30 min at 5°C and with 20 mg/l TCOD.

If, in the previous example, the initial sulfide concentration were 1.0

FIGURE 6-7
Conversion of combined chlorine residual at temperature 1 to equivalent residual at 20°C (Ref. 10).

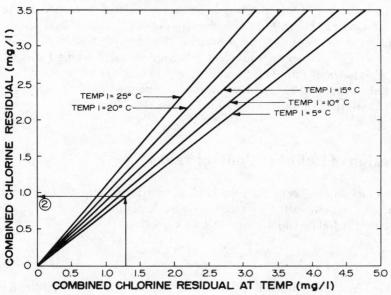

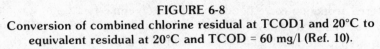

FIGURE 6-8

Conversion of combined chlorine residual at TCOD1 and 20°C to
equivalent residual at 20°C and TCOD = 60 mg/l (Ref. 10).

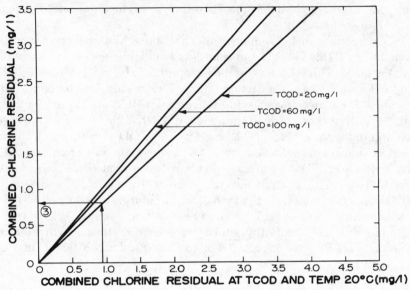

mg/l instead of 0 mg/l, it would be necessary to go directly from Figure 6–6
to Figure 6–10. Here, chlorine residual of 1.30 mg/l at the TCOD of 20 mg/l
and a temperature of 5°C is converted to an equivalent chlorine residual of
1.10 mg/l for a TCOD of 60 mg/l. This is represented by point 5 on Figure
6–10. Going to Figure 6–11, which corresponds to an initial sulfide concentration of 1.0 mg/l, it is determined that a chlorine dose of 6.65 mg/l is
necessary to produce an equivalent chlorine residual of 1.1 mg/l after a contact period of 30 min. Point 6 on Figure 6–11 corresponds to this dose. The
sulfide remaining after chlorination is determined to be 0.44 mg/l from
Figure 6–12, as indicated by point 7.

Design of Chlorine Contact Tanks

Although the degree of bacterial kill is proportional to the concentration of
chlorine dose times the contact time, disinfection of wastewater does not
necessarily follow Chick's law.[11] Chick's law states that,

$$\ln \frac{N}{N_0} = -kt \qquad (6-9)$$

where N is the number of organisms remaining at time t, N_0 is the initial
number of organisms, and k is a constant. The deviation from Chick's law

FIGURE 6-9
Determination of chlorine dose required for equivalent combined
residuals at TCOD = 60 mg/l and temperature = 20°C (Ref. 10).

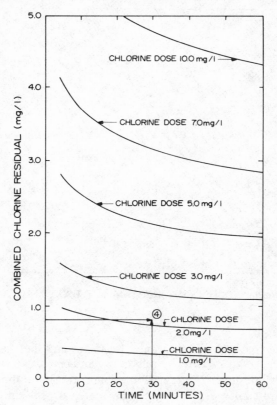

can largely be attributed to the fact that the disinfectant forms of chlorine
in wastewater are mostly chloramines, rather than free chlorine.
Chloramines not only decrease the disinfectant properties of the chlorine
residual but may also result in differences in susceptability of organisms exposed to chloramines. Differences in degree of exposure and increases in
resistance triggered by the exposure of organisms to the disinfectant also affect the way in which chlorine acts to destroy microorganisms. As a result of
deviations from Chick's law, either the time of exposure or the chlorine dose
must be increased to produce the same bacterial kill in wastewater as in
water.

Problems associated with the design of contact tanks stem from the fact
that most designs are based on a theoretical detention time determined by
dividing the tank volume by the flow rate. In practice, actual detention
times may vary between 30 and 80% of the theoretical detention times.[15]
Shorter residence times are caused by dead spaces and short-circuiting and

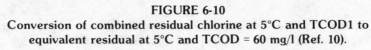

FIGURE 6-10

Conversion of combined residual chlorine at 5°C and TCOD1 to
equivalent residual at 5°C and TCOD = 60 mg/l (Ref. 10).

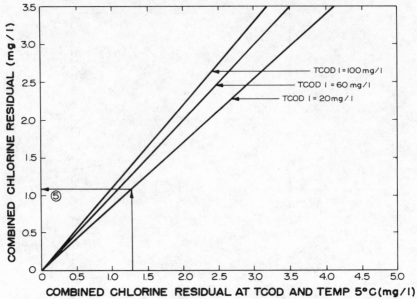

result in decreases in chlorination efficiencies and increases in solids accumulations.[16] With shorter contact times and extra chlorine demands exerted by the buildup of solids, chlorine concentrations must be increased to
produce desired degrees of disinfection. Increasing the chlorine dose often
has serious drawbacks. As well as being an inefficient way to utilize the
disinfectant properties of chlorine, it also increases operational costs. This
approach also increases the concentration of undesirable compounds
discharged into the environment.[17] Increasing the chlorine dose is also hard
on equipment, because of corrosion resulting from the contact of equipment
with high chlorine concentrations.

    Short-circuiting has another effect on adequate operations of chlorine
contact tanks. With short-circuiting, residence times may be continually
changing. This causes difficulty in maintaining prescribed levels of chlorine
residuals. The frequent attention of an operator is required to alter chlorine
doses in maintaining constant chlorine residuals.[18]

    To provide adequate disinfection of wastewater, the basic approach to
good contact tank design should include a thorough investigation of
hydraulic characteristics of various designs and then the selection of design
features that will optimize hydraulic performance. Some important design
considerations include optimization of mixing, contact time, and chlorine
dose.

FIGURE 6-11
Determination of chlorine dose required when $S^{2-} = 1.0$ mg/l,
TCOD = 60 mg/l, and temperature = 5°C (Ref. 10).

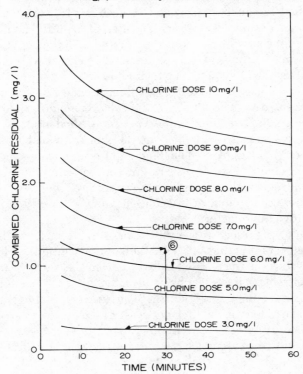

FIGURE 6-12
Sulfide reduction as a function of chlorine dose (Ref. 10).

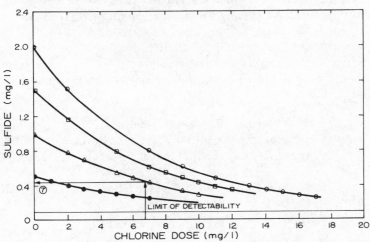

Evaluation of Hydraulic Characteristics. The hydraulic character-istics of a chlorine contact tank are generally determined by conducting tracer studies on flow patterns through the tank. Several possible tracers are available. Salt is a common tracer and has been used to determine detention times in contact tanks.[19] However, it is often difficult to handle the large amounts of salt generally required for such studies. Radioactive tracers are another possibility; however, these are almost never used because of the hazard and regulations controlling their release.

Perhaps the most useful tracers are fluorescent dyes. Most of these are inexpensive and easy to obtain. Two of the dyes commonly used in contact tank tracer studies are Rhodamine WT[17] and Rhodamine B.[15,16] Other fluorescent dyes are also available, and the choice of which dye to use is a matter of personal judgment. The Rhodamine dyes, however, offer the ad-vantages of being detectable at very low concentrations and having low sorption tendencies. Also, turbidity has very little effect on the response of the dye. Fluorescence of the dyes at concentrations as low as 0.01 ppb can be detected with a fluorometer.[20]

In conducting tracer studies the dye or other tracer should be injected into the contact tank at the same point the chlorine solution would enter the tank. If possible, the tracer should also be injected below the water surface to avoid scattering of the tracer by wind on the surface. The most desirable method of conducting tracer studies is to obtain a continuous record of the tracer concentrations at the tank outlet. If fluorescent dyes are used, this may be done by using a continuous flow fluorometer connected to a recorder. This type of approach is more reliable than the collection of grab samples.

The flow characteristics of the contact tank may be determined by evaluating the data obtained from tracer studies in one of several ways. The methods include conventional, statistical, and dynamic analyses.[21] Con-ventional and statistical analyses are the most commonly used. The dynamic approach is basically a mathematical modeling technique and is not discussed.

The conventional method of analysis consists of selecting specific points from the dispersion flow curve as indices to describe the performance characteristics of a tank. The points and indices commonly used are de-scribed as follows:[17,14]

T = V/Q (theoretical detention time)
t_i = time for tracer to initially appear at the tank outlet
t_p = time for tracer at outlet to reach peak concentration
$t_{10}, t_{50},$
t_{90} = time for 10, 50, and 90 % of the tracer to pass at the outlet of the tank
t_g = time to reach the centroid of the effluent curve

FIGURE 6-13
Typical dispersion flow curve.

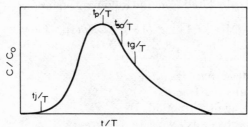

t_i/T = index of short-circuiting
t_p/T = index of modal detention time
t_{50}/T = index of mean detention time
t_g/T = index of average detention time
t_{90}/t_{10} = Morrill dispersion index—indication of degree of mixing

In constructing dispersion flow curves, it is common practice to use dimensionless expressions for tracer concentrations and times. This is done to facilitate comparisons of hydraulic performance between tanks where different tracer concentrations and detention times are involved. The dimensionless dispersion flow curve is obtained by ploting C/C_O against t/T, where C is the tracer concentration at any time t, C_0 is the initial tracer concentration, and T is the theoretical detention time (V/Q). A typical dispersion flow plot is presented in Figure 6–13.

The parameter that is probably the most useful in accurately describing hydraulic performance is the Morrill index (MI).[14] As the MI approaches 1.0, the flow through the tank approaches ideal plug flow. The larger the MI, the more closely the flow in the tank approaches backmix (complete mixed) reaction conditions. The two extreme flow conditions are displayed in Figure 6–14.

There are several different statistical approaches used to evaluate hydraulic performance. One approach, which has gained widespread acceptance, describes the flow regime of a basin in terms of plug flow and perfect mixing.[14,22] It also uses descriptive parameters to define effective

FIGURE 6-14
Comparison of plug and backmix flow.

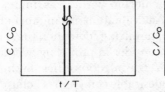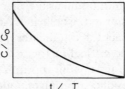

space and dead space. A variation of this approach uses the entire tracer curve to describe hydraulic efficiency in terms of a function of time, $F(t)$.[23] This function is calculated from the following equation:

$$\text{Log}[1 - F(t)] = [-\text{Log } e/(1-p)(1-m)][t/T - p(1-m)] \qquad (6\text{--}10)$$

where

m	=	dead space fraction
$1 - m$	=	effective fraction
p	=	plug flow fraction
$1 - p$	=	perfect mixing fraction
t	=	any time corresponding to time used to get $F(t)$
T	=	theoretical detention time

Probably the most widely used statistical approach is the chemical engineering dispersion index. It is considered to be reliable, since it is calculated using the entire dispersion flow curve. The dispersion index, δ, is calculated from the following equations:[14]

$$\delta = \sigma^2 = \frac{\delta t^2}{t_g^{\,2}} \qquad (6\text{--}11)$$

$$\sigma t^2 = \left(\frac{\Sigma t^2 c}{\Sigma c}\right) - \left(\frac{\Sigma t c}{\Sigma c}\right)^2 \qquad (6\text{--}12)$$

$$t_g = \frac{\Sigma t c}{\Sigma c} \qquad (6\text{--}13)$$

In these equations c is the tracer concentration at any time t, and $\delta = \sigma^2$ is equal to the variance of the flow-through curve.

The dispersion index has the strongest statistical probability of correctly describing the hydraulic performance, because it includes all points on the dispersion flow curve. Conventional parameters only use one point or, at the best, only a portion of the curve. In comparing the dispersion index with conventional parameters, it has been found that the MI is closely correlated with the dispersion index and can be considered as the most reliable conventional parameter in accurately describing the hydraulic performance of a tank. The least reliable indicators of flow characteristics are considered to be the percentage of effective space, t_{50}/T, and t_i/T.[14]

Elements of Contact Tank Design. The primary objective of good chlorine contact tank design is to design for hydraulic performance that will allow for a minimum usage of chlorine with a maximum exposure of microorganisms to the chlorine. An evaluation of a number of wastewater chlorine contact tanks indicates that mixing, detention time, and chlorine dosage are the critical factors to consider in providing adequate disinfection. Good design not only optimizes disinfection efficiency but should also minimize the concentration of undesirable compounds being discharged to the environment and reduce the accumulation of solids in the tank by keeping the flow-through velocity high enough to prevent solids from settling.[17]

Initial mixing of the chlorine solution with wastewater is necessary for providing uniform contact of chlorine with microorganisms and for preventing chlorine stratification in the contact tank. This is especially important because most of the disinfection takes place within the first few minutes of contact. It is also important to note that most of the chlorine demand is exerted during this same period. Since the formation of chloramines in wastewater is extremely rapid, it must be remembered that free chlorine is much more effective as a disinfectant than chloramines. Chloramines are ineffective in killing viruses in comparison with free chlorine. Although the reaction rates involved in the formation of chloramines are more rapid than the rate at which free chlorine reacts with microorganisms, it is important to provide as much exposure as possible of free chlorine to the microorganisms for efficient disinfection. Rapid mixing provides this exposure if there is any free chlorine remaining in solution by the time the chlorine solution is mixed with the wastewater.

Mixing can be accomplished by applying the chlorine solution to the wastewater either in a pressure conduit under turbulent conditions or with a mechanical mixer. A turbulent reactor is generally considered to be the most effective in producing maximum bacterial kill in the shortest contact time. It has been found that a contact time of 0.1–0.3 min is generally sufficient in a turbulent reactor. Slightly longer time might be required when a mechanical mixer is used.[11] If a mechanical mixer is used, the chlorine solution should be added to the wastewater immediately upstream from the mixer. Another form of mixing, which has been found to be effective, is the use of a hydraulic jump in combination with over and under baffles.[19] Both the turbulent reactor and the baffle system of mixing offer the advantage of reducing operation and maintenance costs over those for the mechanical mixer.

Rapid mixing is followed by flow of the chlorinated wastewater into the contact tank. Most approaches to good contact tank design are based on the idea that plug flow is the most desirable hydraulic performance characteristic to achieve in producing efficient disinfection. Plug flow decreases short-circuiting, dead spaces, spiraling, and eddy currents and also closes the gap between theoretical and actual detention times. However, not all designs are based on the objective of achieving plug flow. At least one design suggests the use of a series of backmix reactors to improve chlorination efficiency.[24] In this approach the tank shapes are not important as long as stratification and short-circuiting are eliminated. One advantage to series reactors is the ease with which treatment capacity could be increased by simply adding another reactor. However, high initial and operational costs could offset this advantage.

For the design of tanks in which plug flow is the objective, tank shape is an important consideration. Ideally, plug flow conditions could best be achieved by using a long, narrow, straight contact chamber. A pipe, for example, would be a good contact chamber. However, because of cost and space limitations, this approach is generally not practicable. Circular contact tanks have been used, but generally they do not perform efficiently with respect to hydraulic characteristics.[25] Most tanks are based on a rectangular shape, which is the most practical design.

Conventional design practices can be enhanced by paying particular attention to inflow and outflow structures. They should be designed in such a fashion as to distribute waste flow uniformly across the tank cross section. One of the most effective designs is that of a sharp-crested weir covering the width of the contact tank at the inlet and outlet.[14] This design minimizes the weir overflow rate and greatly enhances hydraulic characteristics through the tank.

A common practice for improving plug flow conditions in a contact tank involves the use of baffles. Longitudinal baffles are generally more effective than cross baffles. In a study of seven different types of chlorine contact tank configurations, it was found that the longitudinally baffled serpentine flow and flow resulting in an annular ring around a secondary clarifier were the best configurations for approaching ideal plug flow. Both have the effect of increasing the ratio of the length to the width (L/W) of the contact tank. The L/W ratio is often considered to be the most important design consideration for chlorine contact tanks. It has been recommended that a minimum L/W ratio of 40:1 be used to obtain maximum plug flow performance;[14] however, Johnson et al.[10] obtained excellent hydraulic characteristics using a L/W ratio of 25:1. Baffles have also been used effectively across the width of contact tanks. Hydraulic performance has been improved by placing baffles near the inlet end of tanks to suppress the kinetic energy of incoming jets.

Often, baffles by themselves are not sufficient to produce desired hydraulic characteristics. Hammerhead shapes at baffle tips have been demonstrated to reduce short-circuiting and flow separation. Corner fillers have been used to eliminate dead spaces and to decrease the buildup of solids in corners. These fillers, however, seem to have little effect on flow characteristics. In some cases, directional vanes around the ends of baffles have been found to produce lower head losses and to produce more uniform flow through the contact tank.[19]

Another approach to improving the effectiveness of chlorine contact tanks has involved aeration. It has been found that mild agitation with compressed air improves hydraulic characteristics and may improve bacterial kill by providing closer contact of microorganisms with residual chlorine.[16] This method also reduces solids accumulation and thus decreases the chlorine demand caused by putrefaction of solids. Using this approach in a field evaluation, it has been found that adequate bacterial kill can be obtained in secondary sewage with a dose of 2–3 mg/l chlorine and a contact time of only 15 min, which should be considered as the minimum actual hydraulic residence time for chlorine contact tanks. If the accumulation of solids is not adequately prevented by aeration, it is recommended that they be removed at least once a day by mechanical or other means to keep chlorine demand as low as possible.

One final design consideration is that of depth. In shallow contact tanks, it is possible for wind to cause short-circuiting. However, this is generally not a problem in tanks designed with standard design depths.

For existing chlorine contact tanks, it is generally not possible to completely redesign the tank. However, improvements can be made in flow characteristics with practical alterations. Gates added to screen and sludge notches have been found to reduce short-circuiting. Spiraling flow patterns have been eliminated by circular baffle plates placed at tank inlets. Additional improvements can be made by using directional vanes to direct flow in a more uniform fashion and by using stop baffles with curved vanes to reduce eddying. In one example, the improvements reduced short-circuiting by 80% in an existing contact tank.[17] Another way to improve hydraulic performance in existing tanks is to use precast baffles. These can be installed with minimum down time. Although it is more efficient to use longitudinal baffles, it may be more economical to use cross baffles. It has been demonstrated that baffles installed in a maze configuration improved performance sufficiently to make economical factors more important in choosing a design than efficiency considerations.[18]

In conclusion, the most important design considerations for efficient disinfection are rapid and complete initial mixing, an adequate L/W ratio to produce near plug flow conditions, and a sufficiently long residence time to produce an optimal amount of disinfection for the chlorine dose applied with a minimum amount of chlorine residual remaining in the effluent. Design considerations are summarized in Table 6–1.

Table 6-1
Summary of Chlorination Design Criteria

Mixing
   I. Rapid initial mixing should be accomplished within 5 sec and before liquid enters contact tank. Design hydraulic residence time ≥ 30 sec for tanks using mechanical mixers.

  II. Methods available
     1. Hydraulic jump in open channels.
     2. Mechanical mixers located immediately below point of chlorine application.
     3. Turbulent flow in a restricted reactor.
     4. Pipe flowing full. Least efficient and should not be used in pipes with diameter > 30 in.

(continued)

Table 6-1 (*continued*)

Contact Chamber

I. Hydraulic residence time
1. $\geqq$ 60 min at average flow rate.
2. $\geqq$ 30 min at peak hourly flow rate.

II. Hydraulic performance
1. Modal value obtained in dye studies $\geqq$ 0.6, $t_p/T \leq$ 0.6 (Figure 6-13).
2. Efficiency of disinfection increases as t_p/T increases.
3. Design for maximum economical t_p/T.

III. Length to width ratio
1. L/W $\geqq$ 25:1.
2. Cross-baffles used to eliminate short circuiting caused by wind.

IV. Solids removal
1. Arrange baffles to remove floating solids.
2. Provide drain to remove solids and liquid for maintenance.
3. Provide duplicate contact chambers.
4. Width between channels should be adequate for easy access to clean and maintain chamber.

V. Storage
1. Provide a minimum of one filled chlorine cylinder for each one in service.
2. Maintain storage area at a temperature $\geqq$ 55°F.
3. Never locate cylinders in direct sunlight or apply direct heat.
4. Limit maximum withdrawal rate from 100- and 150 lb cylinders to 40 lb/day.
5. Limit maximum withdrawal rate from 2000-lb cylinders to 400 lb/day.
6. Provide scales to weigh cylinders.
7. Provide cylinder handling equipment.
8. Install automatic switch-over system.

VI. Piping and valves
1. Use Chlorine Institute approved piping and valves.
2. Supply piping between cylinder and chlorinator; it should be Sc. 80 black seamless steel pipe with 2000-lb forged steel fitting. Unions should be ammonia type with lead gaskets.
3. Chlorine solution lines should be Sc. 80 PVC, rubber-lined steel, saran-lined steel, or fiber cast pipe approved for moist chlorine use. Valves should be PVC, rubber-lined, or PVC lined.
4. Injector line between chlorinator and injector should be Sc. 80 PVC or fiber cast approved for moist chlorine use.

VII. Chlorinators
1. Should be sized to provide dosage $\geqq$ 10 mg/l.
2. Maximum feed rate should be determined from knowledge of local conditions.
3. Direct feed gas chlorinators should be used only in small installations. Check state regulations. Prohibited in certain states.
4. Vacuum feed gas chlorinators are most widely used and are much safer.
5. Hypochlorite solutions should be considered in small installations where safety is major concern.

(continued)

Table 6-1 (*continued*)

VIII. Safety equipment and training
1. Install an exhaust fan near floor level with switch actuated when door is opened.
2. Exhaust fan should be capable of one air exchange per minute.
3. Gas mask should be located outside chlorination room.
4. Emergency chlorine container repair kits should be supplied.
5. Chlorine leak detector should be installed.
6. Alarms should be installed to alert operator when deficiencies or hazardous conditions exist.
7. Operator should receive detailed, hands-on training with all emergency equipment.

IX. Diffusers
1. Minimum velocity through diffuser holes $\geq$ 10-12 ft/sec.
2. Diffusers should be installed for convenient removal, cleaning, and replacement.

References

1. Environmental Protection Agency. 1976. *Disinfection of Wastewater: Task Force Report.* EPA–430/9–75–012. Washington, D.C.

2. White, G. Clifford. 1973. Disinfection Practices in the San Francisco Bay Area. *JWPCF*, **46**, 1, 89–101.

3. Kott, Y. 1971. Chlorination Dynamics in Wastewater Effluents. *J. Sanit. Eng. Div. ASCE*, 97(SA5):647–659.

4. Dinges, R, and A. Rust. 1969. Experimental Chlorination of Stabilization Pond Effluent. *Public Works*, **100**, 3, 98–101.

5. Brinkhead, C. E., and W. J. O'Brien. 1973. Lagoons and Oxidation Ponds. *JWPCF*, **45**, 10, 1054–1059.

6. Echelberger, W. F., J. L. Pavoni, P. C. Singer, and M. W. Tenney. 1971. Disinfection of Algal Laden Waters. *J. Sanit. Eng. Div. ASCE*, **97**, SA5, 721–730.

7. Hom, W. 1972. Kinetics of Chlorine Disinfection in an Ecosystem. *J. Sanit. Eng. Div. ASCE*, **98**, SA1, 183–194.

8. Zaloum, R., and K. L. Murphy. 1974. Reduction of Oxygen Demand of Treated Wastewater by Chlorination. *JWPCF*, **46**, 12, 2770–2777.

9. Kott, Yehuda. 1973. Hazards Associated with the Use of Chlorinated Oxidation Pond Effluents for Irrigation. *Water Research*, **7**, 853–862.

10. Johnson, B. A., J. L. Wight, E. J. Middlebrooks, J. H. Reynolds, and A. D. Venosa. 1979. *Waste Stabilization Lagoon Microorganisms Removal Efficiency and Effluent Disinfection with Chlorine.* EPA–600/2–79–018. Municipal Environmental Research Laboratory, Cincinnati, Ohio.

11. Collins, H. F., R. E. Selleck, and G. C. White. 1971. Problems in Obtaining Adequate Sewage Disinfection. *J. Sanit. Eng. Div. ASCE*, **97**, SA5, 549–562.

12. Kothandaraman,V., and R. L. Evans. 1972. Hydraulic Model Studies of Chlorine Contact Tanks. *JWPCF*, **44**, 4, 626–633.

13. Kothandaraman, V., and R. L. Evans. 1974. *Design and Performance of Chlorine Contact Tanks*. Circular 119, Illinois State Water Survey, Urbana, Illinois.

14. Marske, D. M., and J. D. Boyle. 1973. Chlorine Contact Chamber Design—A Field Evaluation. *Water and Sewage Works*, **120**, 1, 70–77.

15. Deaner, D. G. Undated. *A Procedure for Conducting Dye Tracer Studies in Chlorine Contact Chambers to Determine Detention Times and Flow Characteristics*. Technical Paper printed by G. K. Turner Associates, Palo Alto, California.

16. Kothandaraman, V., and R. L. Evans. 1974. A Case Study of Chlorine Contact Tank Inadequacies. *Public Works*, **105**, 1, 59–62.

17. Hart, F. L., R. Allen, J. DiAlesio, and J. Dzialo. 1975. Modifications Improve Chlorine Contact Chamber Performance, Parts I and II. *Water Sewage Works*, **122**, 9, 73–75 and **122**, 10, 88–90.

18. Stephenson, R. L., and J. R. Lauderbaugh. 1971. Baffling Chlorine Contact Tanks. *Water Sewage Works*, Ref. 1971, R–100–103.

19. Louie, D. S., and M. S. Fohrman. 1968. Hydraulic Model Studies of Chlorine Mixing and Contact Chambers. *JWPCF*, **40**, 2, Part 1, 174–184.

20. Deaner, D. G. 1970. *Chlorine Contact Chamber Study at Redding Sewage Treatment Plant*. Technical Paper, State of California, Department of Public Health.

21. Sawyer, C. M. 1967. Tracer Studies on a Model Chlorine Contact Tank. M. S. thesis, Virginia Polytechnic Institute Library, Blacksburg, Virginia.

22. Wolf, D., and W. Resnick. 1963. Residence Time Distribution in Real Systems. *Ind. Eng. Chem. Fund.*, **2**, 4, 287–293.

23. Rebhun, M., and Y. Argaman. 1965. Evaluations of the Hydraulic Efficiency of Sedimentation Basins. *Proc. ASCE Sanit. Eng. Div.*, **91**, SA5, 37–45.

24. Kokoropoulos, P. 1973. Designing Post-Chlorination by Chemical Reactor Approach. *JWPCF*, **45**, 10, 2155–2165.

25. Warwick, J. W. 1968. Tracer Studies on a Circular Chlorine Contact Tank. B.S. degree thesis. Virginia Military Institute, Lexington, Virginia.

CHAPTER
7

Energy Requirements

ENERGY CONSUMPTION is a major factor in the operation of wastewater treatment facilities. Many of the plans for water pollution management in the United States were developed before the cost of energy and the limitations of energy resources became serious concerns for the nation. As wastewater treatment facilities are built or updated to incorporate current treatment technology and to meet regulatory performance standards, energy must be a major consideration in designing and planning the facilities. Information on energy requirements for various systems must be made available to planners and designers so that a treatment system may be developed that incorporates the most efficient use of energy for each particular wastewater problem.

Energy requirements for several alternatives currently available for the treatment of small flow rates (0.05–5 mgd) of wastewater are presented. Energy requirements for various treatment combinations are compared and presented in tabular form for the most suitable alternatives.

Energy Equations

Equations of the lines of best fit for the energy requirements of the unit operations and processes used in wastewater treatment based on the graphs reported by Wesner et al.[1] were used to develop the tables and graphs. Details about the conditions imposed on the equations can be obtained from the Wesner et al.[1] report.

Effluent Quality and Energy Requirements

Table 7–1 shows the expected effluent quality and the energy requirements for various combinations of operations and processes. Energy requirements and effluent quality are not directly related. By utilizing facultative lagoons and land application techniques, it is possible to obtain an excellent quality effluent and expend small quantities of energy. Although one system may be more energy efficient, the selection of a wastewater treatment facility must be based on a complete economic analysis. However, with rising energy costs, energy requirements are assuming a greater proportion of the annual cost of operating a wastewater treatment facility, and it is likely that energy costs will become the predominant factor in the selection of small flow treatment systems. Operation and maintenance requirements, and consequently costs, are frequently kept to a minimum at small installations because of the limited resources and operator skills normally available. This favors the selection of systems employing units with low energy requirements. It is likely that all future wastewater treatment systems at small installations in isolated areas will be designed for low-energy-consuming units and simple operation and maintenance. The only exceptions to this will be in areas with limited space or construction materials, or where surplus energy is available.

The effluent quality expected with each of the treatment systems and the energy requirements shown in Table 7–1 are presented in the order of decreasing BOD_5 concentration in the effluent. The other parameters (SS, total P, and total N) do not necessarily decrease in the same manner, because most treatment facilities are designed to remove BOD_5; in general, however, there is a trend in overall improvement in effluent quality as one reads down the table. As shown in Table 7–1, there are many systems available to produce an effluent that will satisfy EPA secondary or advanced effluent standards; however, energy requirements for the various systems are varied and can differ by a factor of greater than 20 to produce the same quality effluent.

For purposes of comparison the total energy (electricity plus fuel) for a typical 1-mgd system has been extracted from Table 7–1 and listed in Table 7–2 in order of increasing energy requirements. It is apparent from Table 7–2 that increasing energy expenditures do not necessarily produce increasing water-quality benefits. The four systems at the top of the list, requiring the least energy, produce effluents comparable to the bottom four that require the most. Three of the top four are facultative lagoon-land treatment systems, and their adoption depends on local site conditions. The facultative lagoon followed by intermittent sand filter and surface discharge to receiving waters is less constrained by local soil and groundwater conditions.

Conventional versus Land Treatment

A comparison of the energy requirements for a conventional wastewater treatment system (consisting of a trickling filter system followed by nitrogen removal, granular media filtration, and disinfection) with a facultative lagoon followed by overland flow and disinfection is shown in Figure 7–1. This comparison is made because of the approximately equivalent quality effluents produced by the two systems (Table 7–1). The relationships in Figure 7–1 clearly show that there are significant electricity and fuel savings with the facultative lagoon-land application system. Similar comparisons for modifications of the two systems can be made by referring to the report by Middlebrooks and Middlebrooks[2] and selecting combinations to produce equivalent effluents.

Figure 7–2 shows a comparison of the energy requirements for an activated sludge plant producing a nitrified effluent, followed by granular media filtration and disinfection; a facultative lagoon followed by rapid infiltration land treatment; and primary treatment followed by rapid infiltration land treatment. The facultative lagoon system followed by rapid infiltration land treatment is the most energy-efficient wastewater treatment system, but it is closely followed in energy efficiency by the primary treatment and rapid infiltration system. The energy requirements for both the rapid infiltration land treatment alternatives are less than 15% of the energy required for the activated sludge system.

In Figure 7–3 energy requirements for slow rate land application systems using ridge and furrow and center pivot systems to distribute facultative lagoon effluent are compared with the energy requirements for an activated sludge plant practicing nitrogen and phosphorus removal, granular media filtration of the effluent, and disinfection prior to discharge. Both the activated sludge and advanced treatment system and the facultative lagoon and slow rate systems produce approximately equivalent quality effluents. The ridge and furrow flooding technique of land treatment requires less than 5% of the energy required by the advanced treatment scheme. Utilizing a center pivot mechanism to distribute the facultative lagoon effluent increases the energy requirements by a factor of five compared with the ridge and furrow flooding technique, but the energy requirements for the center pivot system are less than 11% of the energy requirements for the advanced treatment system.

In an energy-conscious environment, the lagoon-land application techniques of treating wastewater have a distinct advantage over the more conventional wastewater treatment systems. When land is available at a reasonable cost, the lower energy requirements for lagoon-land application systems usually result in a more cost-effective as well as a more energy-effective system of wastewater treatment.

Table 7-1
Expected Effluent Quality and Total Energy Requirements for
in the Intermountain

| TREATMENT SYSTEM | EFFLUENT QUALITY, MG/L | | | | TOTAL ENERGY | | | |
| | | | | | 0.05 mgd | | 0.1 mgd | |
	BOD$_5$	SS	Total Phos. as P	Total Nitrogen as N	Electricity (kwh/yr)	Fuel (million Btu/yr)	Electricity (kwh/yr)	Fuel (million Btu/yr)
Trickling filter with anaerobic digestion	30	30	—	—	18,300	225	29,600	335
Rotating biological contactor with anaerobic digestion	30	30	—	—	18,200	225	29,700	335
Facultative lagoon + microscreens 23μ	30	30	—	15	11,300	148	20,300	181
Physical-chemical advanced secondary treatment	30	10	1	—	62,300	801	91,300	1,490
Activated sludge with anaerobic digestion	20	20	—	—	38,600	214	58,700	313
Activated sludge with sludge incineration	20	20	—	—	52,800	305	73,400	493
Extended aeration with sludge-drying beds	20	20	—	—	40,000	161	64,100	208
Trickling filter + granular media gravity filtration	20	10	—	—	19,400	225	31,800	335
Trickling filter + N-removal (ion exchange) + Gran. media filter	20	10	—	5	21,000	225	32,900	335
Facultative lagoon + intermittent sand filter	15	15	—	10	5,840	150	10,920	186
Aerated lagoon + intermittent sand filter	15	15	—	20	20,800	151	39,500	186
Extended aeration + intermittent sand filter	15	15	—	—	40,600	164	65,300	213
Activated sludge (A.D.) + Granular media gravity filter	15	10	—	—	39,700	214	60,900	313
Activated sludge + nitrification + Granular media gravity filter	15	10	—	—	46,700	214	74,900	313
Overland flow-facultative lagoon flooding	5	5	5	3	5,700	148	10,700	181
Rapid infiltration-facultative lagoon flooding	5	1	2	10	1,540	148	2,810	181
Slow rate (irrigation) fac. lagoon-ridge and furrow flooding	1	1	0.1	3	2,800	149	5,300	183
Activated sludge + advanced treatment	< 10	5	< 1	< 1	131,690	565	244,560	900

SOURCE: From Ref. 2.

Various Sizes and Types of Wastewater Treatment Plants Located Area of the USA

REQUIREMENTS AT VARIOUS FLOW RATES

0.5 mgd		1.0 mgd		3.0 mgd		5.0 mgd	
Elec-tricity (kwh/yr)	Fuel (million Btu/yr)	Elec-tricity (kwh/yr)	Fuel (million Btu/yr)	Elec-tricity (kwh/yr)	Fuel (million Btu/yr)	Elec-tricity (kwh/yr)	Fuel (million Btu/yr)
106,500	1,100	196,600	2,000	542,900	5,490	880,200	8,910[a]
111,000	1,100	208,300	2,000	589,700	5,490	966,900	8,910[a]
83,100	320	154,600	433	419,800	745	670,900	988
307,900	6,760	567,200	13,300	1,583,000	39,300	2,585,000	65,344
203,100	975	376,000	1,750	1,044,000	4,680	1,703,000	7,540[a]
222,100	1,880	397,200	3,560	1,065,000	10,000	1,711,000	16,600[b]
248,500	457	476,300	707	1,382,000	1,570	2,283,000	2,360
117,500	1,100	218,600	2,000	608,900	5,480	990,200	8,910
134,000	1,100	251,700	2,000	708,300	5,480	1,155,800	8,910
50,540	345	99,270	483	291,800	896	482,200	1,240
184,800	345	364,500	483	1,079,100	896	1,790,900	1,240
253,500	482	486,000	757	1,408,700	1,720	2,326,200	2,610
214,100	975	398,000	1,750	1,109,900	4,680	1,813,300	7,540
284,100	975	538,000	1,750	1,529,900	4,680	2,513,300	7,540
50,070	320	98,810	433	292,600	745	485,080	988
12,140	320	23,050	433	64,300	745	103,900	988
24,700	330	48,050	453	139,100	805	228,400	1,090
1,029,400	3,320	1,981,000	6,240	5,701,200	17,600	9,374,700	28,900

[a] No energy recovery.
[b] Theoretically could recover enough heat to generate all needed electricity.

Table 7-2
Total Annual Energy for Typical 1 mgd System
(electrical plus fuel, expressed as 1000 kwh/yr)

Treatment System	Effluent Quality				Energy (1000 kwh/yr)
	BOD	SS	P	N	
Rapid infiltration (facultative lagoon)	5	1	2	10	150
Slow rate, ridge + furrow (facultative lagoon)	1	1	0.1	3	181
Overland flow (facultative lagoon)	5	5	5	3	226
Facultative lagoon + intermittent sand filter	15	15	—	10	241
Facultative lagoon + microscreens	30	30	—	15	281
Aerated lagoon + intermittent sand filter	15	15	—	20	506
Extended aeration + sludge drying	20	20	—	—	683
Extended aeration + intermittent sand filter	15	15	—	—	708
Trickling filter + anaerobic digestion	30	30	—	—	783
RBC + anaerobic digestion	30	30	—	—	794
Trickling filter + gravity filtration	20	10	—	—	805
Trickling filter + N removal + filter	20	10	—	5	838
Activated sludge + anaerobic digestion	20	20	—	—	889
Activated sludge + anaerobic digestion + filter	15	10	—	—	911
Activated sludge + nitrification + filter	15	10	—	—	1,051
Activated sludge + sludge incineration	20	20	—	—	1,440
Activated sludge + AWT	< 10	5	< 1	< 1	3,809
Physical chemical advanced secondary	30	10	1	—	4,464

Source: From Ref. 2.

Summary

1. With increasing energy costs, energy consumption is assuming a greater proportion of the annual cost of operating wastewater treatment facilities of all sizes, and because of this trend, it is likely that energy costs will become the predominant factor in the selection of cost-effective small-flow wastewater treatment systems.

2. Small-flow wastewater treatment systems are frequently designed to minimize operation and maintenance, and as energy costs increase, design engineers will tend to select low-energy-consuming systems.

3. Low-energy-consuming wastewater treatment systems are generally easier to operate and maintain than energy-intensive systems, making the low-energy-consuming systems even more advantageous, because it is usually difficult to employ highly skilled operators at small facilities.

4. When suitable land and groundwater conditions exist, a facultative lagoon followed by rapid infiltration is the most energy-efficient system described in this chapter.

FIGURE 7-1
Comparison of energy requirements for trickling filter effluent treated for nitrogen removal and filtered versus facultative pond effluent followed by overland flow treatment (Ref. 2).

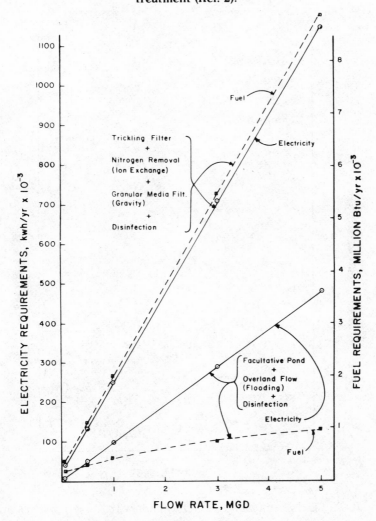

5. When surface discharge is necessary and impermeable soils exist, a facultative lagoon followed by overland flow is the second-most energy-efficient system described in this chapter.

6. Facultative lagoons, followed by slow or intermittent sand filters, are the fourth-most energy-efficient systems discussed, and are not limited by local soil or groundwater conditions.

FIGURE 7-2
Comparison of energy requirements for activated sludge, nitrification, filtration, and disinfection versus facultative pond effluent followed by rapid infiltration and primary treatment followed by rapid infiltration (Ref. 2).

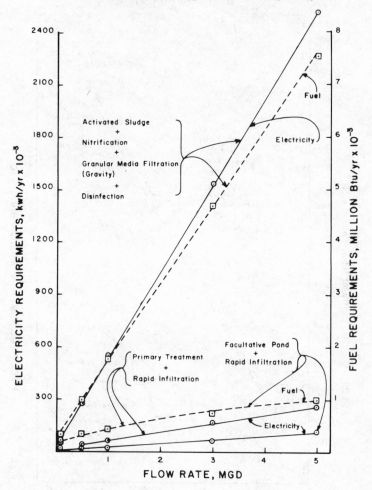

7. Physical-chemical advanced secondary treatment systems utilize the most energy of the conventional methods of producing an effluent meeting the federal secondary effluent standard of 30 mg/l of BOD_5 and suspended solids.

8. Slow rate land application systems following facultative lagoons are more energy efficient than most forms of mechanical secondary treatment systems, while also providing benefits of nutrient removal, recovery, and reuse.

FIGURE 7-3
Comparison of energy requirements for secondary
treatment followed by advanced treatment versus
facultative pond effluent followed by slow rate land
treatment (Ref. 2).

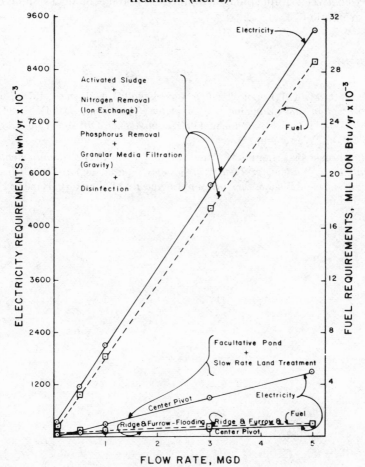

9. Advanced physical-chemical treatment following conventional secondary treatment consumes approximately 34 times as much electrical energy and 13 times as much fuel as slow rate land treatment to produce an equivalent effluent.

10. Land application wastewater treatment systems following storage ponds (aerated or facultative), preliminary treatment (bar screens, comminutors, and grit removal), or primary treatment are by far the most energy-efficient systems capable of producing secondary effluent quality or better.

11. Although the energy requirements for production of all materials consumed in the treatment process were not considered, it is not believed that inclusion of such factors would significantly change the relative ranking of the systems discussed. Such inclusion would rather make the differences between simple biological processes and mechanical systems even more dramatic.

References

1. Wesner, G.M., L.J. Ewing, Jr., T. S. Lineck, and D. J. Hinricks. 1978. *Energy Conservation in Municipal Wastewater Treatment.* MCD–32. EPA–430/9–77–011. U. S. Environmental Protection Agency, Office of Water Program Operations, Washington, D.C.
2. Middlebrooks, E.J., and C.H. Middlebrooks. 1979. *Energy Requirements for Small Flow Wastewater Treatment Systems.* Special Report 79–7, Corps of Engineers, Cold Regions Research and Engineering Laboratory, Hanover, New Hampshire.

CHAPTER 8

Nutrient Removal and Treatment with Water Hyacinth Systems

THE USE OF AQUATIC PLANTS in wastewater treatment systems to reduce the concentration of contaminants to levels equivalent to secondary, advanced secondary, and tertiary effluents has been proposed and tried on small-scale operations beginning about 1970.[1] These systems have been advocated as a means to overcome the disadvantages of high costs and energy-intensive processes associated with conventional wastewater treatment systems. The water hyacinth system appears to have potential for application to small community wastewater treatment. A summary of the results obtained with water hyacinth wastewater treatment systems and an evaluation of the systems from an engineering design point of view follow.

A summary of the known water hyacinth systems that were constructed or modified to treat wastewater are summarized in Table 8–1. There are few consistencies in the design criteria used or developed during the evaluation of these systems. Water hyacinth wastewater treatment systems are used to treat raw wastewater as well as effluents from various stages of treatment (Figure 8–1). The most common system incorporates a stabilization pond followed by series-type water hyacinth culturing tanks. The design characteristics and performance of hyacinth systems are discussed in the following paragraphs.

Table 8-1
Summary of Design and Performance Characteristics

Location	Type of Pre-treatment	Surface Area of Hydraulic Pond (ha)	Depth of Hyacinth Pond (m)	Hydraulic Loading Rate (m^3/ha)	Hydraulic Residence Time Days	Organic Loading Rate (kg/ha/day)
Nat. Space Tech. Lab Before Hyacinths After Hyacinths	None	2	1.22	240	54	26
Lucedale, Miss. Before Hyacinths After Hyacinths	None	3.6	1.73	260^a	$\sim 67^a$	44
Orange Grove, Miss. Without Hyacinths With Hyacinths	2-Aerated 3-Facul. in parallel Exp conducted in parallel systems	0.28	1.83	3,570	6.8	179
Cedar Lake, Miss. (Duckweed)	1-Aerated and Facultative	0.07	1.5	700	22	31
Williamson Creek, Texas Phase I	Plant A Aeration Basin, Clarifier, 3 Lagoons in series	0.0585	1.0	1,860	5.3	43
Phase II	Plant B 2-Aeration Basins in Parallel, 3-lagoons		0.85	1,860	4.5	89
Full-scale	Plant A' Aeration Basin, Clarifier, two lagoons (1.2 ha each) in series	1.2	0.7-1.3	1,100-1,400	6-9	56-72

Physical Characteristics

Location

All the water hyacinth systems currently treating wastewater are located in tropical or warm-temperature climates. The water hyacinth is very sensitive to temperature and does not grow in water with a temperature of 10°C or lower. The optimum temperature for water hyacinth growth ranges between 21 and 30°C. If a water hyacinth system were to be used in a colder climate, it would be necessary to house the system in a greenhouse.

f Aquatic Weed Wastewater Treatment Systems

PERCENT HYACINTH COVER	BOD (MG/L)		SS (MG/L)		TOTAL NITROGEN (MG/L)		TOTAL PHOSPHORUS (MG/L)		REF.
	Influent	Effluent	Influent	Effluent	Influent	Effluent	Influent	Effluent	
	91	17	70	49	9.8	5.2	2.9	2.1	2, 3
100	110	7	97	10	12.0	3.4	3.7	1.6	
									3
	127	57	140	77					
100	161	23	125	6					
									3
	50	37	49	53					
100	50	14	49	15					
duckweed 100	44	18	188	11					3
100	23	5	43	7	8.2	2.5			4, 5
100	46	6	117	8	9.9	3.6			4, 5
	42	12	40	9					4, 5

(continued)

and maintain the temperature in the range of the optimum. There are possibilities of utilizing methane produced from harvesting the plant to produce heat to control the temperature partially and carbon dioxide to enrich the environment above the plants. The benefits of such a system must be investigated on a large scale to establish the economics as well as the operational problems.

Based on the limited data available, it appears that it would be uneconomical to attempt to develop a water hyacinth wastewater treatment system in cold regions. Even if the system were selected to operate only

Table 8-1 (continued

Location	Type of Pre- treatment	Surface Area of Hydraulic Pond (HA)	Depth of Hyacinth Pond (M)	Hydraulic Loading Rate (M³/HA)	Hy- draulic Residence Time Days	Organic Loading Rate (KG/HA·DAYᵃ)
Austin-Hornsby Bend, Texas	Excess Act. Sludge Lagoons Overflow to Hyacinth Pond	1.4	1.23	430	~3	—
San Juan, Texas	Aerated Lagoon, Stabilization Pond	1.0 (2 hyacinth of 1.0 ha, only 1 in operation	0.61 0.91 0.91 1.37	870 850 1,550 1,860	~7 ~11 ~6 ~7	—
Alamo, Texas Before hyacinths After hyacinths	Imhoff Tank, Trickl. Filter Aeration Basin Two 4.04 ha Stab. Ponds in Parallel	Two Basins 1.35 1.05				
San Benito, Texas Before hyacinths After hyacinths	Stab. Pond, 4 Ponds in Series	3 in series, 0.8, 0.8, & 2.0		3,100;3,100, & 1,200 (overall system 0.068) 680		
Rio Hondo, Texas	None	3 in series 0.41,0.41, & 0.41		1,110; 1,110 1,110		Basin 197
Lakeland, Florida	Secondary Treatment	3 in series, 0.40,0.40, & 0.40				
Walt Disney World, Florida		3 exp. basins 8.84 × 109.8 m	Var. 0.38, 0.61 & 0.91 m	650 to 7,850		
Coral Springs, Florida	Activated Sludge Effluent	5 in series, 0.18,0.08, 0.08,0.08, & 0.08	0.38	870	6 2, 1, 1, 1, 1	Basin 31

Source: From Ref. 1.
ᵃ Based on effluent flow rate.

during the warmer months of the year in the cold region, it would be necessary to provide a culturing unit to maintain a hyacinth crop for introduction into the system in the spring. It is likely that hauling a culture of hyacinths to a relatively large system would be prohibitive, and the cost of maintaining a culture remains to be determined. Introduction of the hyacinth plant to areas in which it does not currently grow must be avoided. The damages from an infestation of hyacinth plants would far exceed the benefits derived in wastewater treatment.

PERCENT HYACINTH COVER	BOD (MG/L)		SS (MG/L)		TOTAL NITROGEN (MG/L)		TOTAL PHOSPHORUS (MG/L)		REF.
	Influent	Effluent	Influent	Effluent	Influent	Effluent	Influent	Effluent	
100									5
100	—	20		30					5
		9		11					
		35		30					
		31		32					
		33		86					
<25		38-50		68-82					5
1st-0		17		35					5
nd-50		17		20					
rd-100									
		16		24					5
	Only partial cover and limited preliminary data								6
	Experimental units not fully operational								7
100	13	3		3	22.4	1.0	11.0	3.6	8

Wastewater Characteristics

Many domestic wastewaters have been applied to water hyacinth systems, and hyacinths thrive in wastewaters because of the high nutrient content normally available. The hyacinth has also been grown in mixtures of industrial waste and municipal wastewaters. The growth of hyacinths has been very good in these mixtures. There is limited experience with projects using hyacinths to treat just industrial wastes. Hyacinth systems have the capability of removing heavy metals and other difficult-to-remove organics. This ability to remove the materials might be a significant disad-

FIGURE 8-1
Water hyacinth covered wastewater stabilization lagoon. Courtesy of
B.C. Wolverton, National Space Technologies Laboratories, Bay St.
Louis, Mississippi.

vantage. The presence of toxic substances would make the disposal of the
solid material difficult and expensive and would prohibit the use of the
plant material as a feed supplement. High accumulations of heavy metals
might also interfere with the anaerobic digestion of the solid materials to
produce methane as a source of energy.

Water hyacinths thrive in municipal wastewater and can survive in
relatively high concentrations of heavy metal contaminants. The impact of
heavy metals and toxic organics has not been investigated to any significant
degree.

Size

All the water hyacinth wastewater treatment systems presently operating
are less than 4 ha in surface area, and the majority of the systems are less
than 1 ha in surface area. The majority of the systems currently in operation
are experimental systems, and even the ones that are serving a community
are still classified as being in the experimental stage.

The recommendation by Dinges[5] that individual water hyacinth wastewater treatment systems be kept to a surface area of 0.4 hectare appears reasonable, and this size selection is based on the convenience of harvesting the water hyacinths and cleaning the basins periodically. However, long rectangular basins would not necessarily be limited by this constraint.

The depth of the hyacinth pond varies from location to location. Depths vary from 0.38 to 1.83 m, with the majority of the investigators recommending a depth of 0.91 m or less. The critical concern is to provide adequate depth for the root system to penetrate through the majority of the liquid flowing through the hyacinth pond. Systems that have been designed for nutrient removal have been designed at a depth of approximately 0.4m to ensure complete contact of the wastewater with the root system.

Hydraulic Loading Rates

The hydraulic loading rates applied to water hyacinth facilities have varied from 240 m³/ha·day up to 3570 m³/ha·day when treating domestic wastewaters. Higher hydraulic loading rates have been applied to the Austin–Hornsby Bend, Texas hyacinth system treating overflow from a lagoon receiving excess activated sludge, but this treatment process was ineffectual because of high organic and hydraulic loading rates. The Disney World, Florida system was designed to process hydraulic loading rates between 650 and 780 m³/ha·day, and once these experiments are completed, a better set of design criteria can be presented for design of hyacinth systems to be used principally as a polishing device for secondary treatment processes.

Based on the results currently available, it appears that an hydraulic loading rate of 2000 m³/ha·day when treating secondary effluent will produce an effluent quality that would satisfy advanced secondary standards ($BOD_5 \leq 10$ mg/l; $SS \leq 10$ mg/l; $TKN \leq 5$ mg/l; and $TP \leq 5$ mg/l.) Hydraulic loading rates applied to three water hyacinth systems treating raw wastewater have ranged from 240 to 680 m³/ha·day. All three of the systems operated effectively, but the lower hydraulic loading rates appear to produce a higher-quality effluent measured in terms of BOD_5 and SS concentrations.

A reasonable design for a hydraulic loading rate for a hyacinth system receiving raw wastewater appears to be approximately 200 m³/ha·day. There are few data supporting this decision, but an analysis of the available data would support this lower hydraulic loading rate for systems treating raw wastewater. Hyacinth systems processing a secondary effluent could be designed to process approximately 2000 m³/ha·day if the principal objective was the control of BOD_5 and suspended solids in the effluent. With nutrient removal as the principal objective, little data exist as to what might be the best hydraulic loading rate. A nutrient removal hyacinth system probably would be used in conjunction with a wastewater stabilization pond, or another secondary effluent would be applied. A shallow pond (0.4 m) and a hydraulic loading rate of approximately 500 m³/ha·day should produce good nitrogen removals (< 2 mg/l). Approximately 50% reduction in the total phosphorus concentration could be expected.

Number of Units in System

The majority of the water hyacinth systems have been designed to operate with three cells in series. Single-cell stabilization ponds with water hyacinths have been employed successfully, but the majority of the systems currently being evaluated are considering the nutrient removal aspects of the hyacinth systems, and the three-cells in series system appears to be preferred. If the objective is the control of algae in the effluent from a wastewater stabilization pond, it is likely that the single unit would work just as effectively as the series configuration. It appears that the control of the algae in wastewater stabilization pond effluents is principally a physical process of shading sunlight.

Active Components

In a water hyacinth system, during the active growth phase hyacinths are capable of sorbing organics, heavy metals, pesticides, and other organic contaminants. The root system of the water hyacinth also supports a very active mass of organisms that assist in breaking down and removing the pollutants in wastewaters. As mentioned above, the control of algae in wastewater stabilization pond effluent by the introduction of water hyacinths apears to be a physical process, limiting the light available to the algae. Nutrient removal apparently is a result of hyacinth growth, physiochemical reactions, and accumulation by other organisms growing in the ecosystem.

Organic Loading Rates

Water hyacinth wastewater treatment systems processing raw wastewater in a stabilization pond appear to be able to process wastewater organics at approximately the same loading rates used in lightly loaded wastewater stabilization ponds. The system operating at the National Space Technology Laboratories (NSTL) was loaded at 26 kg of BOD_5/ha·day and operated without significant odors; in contrast, the system also processing raw wastewater at the Lucedale, Mississippi location was loaded at 44 kg/ha·day, and odors developed at night. These results indicate that organic loading rates of less than 30 kg/ha·day would provide satisfactory results when processing a raw wastewater. Only three systems are known to be processing raw wastewater, and operational data from one of these (Rio Hondo, Texas) were extremely limited.

Water hyacinth wastewater treatment systems receiving secondary effluents or wastewater stabilization pond effluents are more numerous, and a much wider range of organic loading rates have been employed with these systems. Organic loading rates applied to the first basin in hyacinth systems have ranged from 197 kg/ha·day to 31 kg/ha·day. All the systems receiving organic loading rates within this range have produced an effluent that would satisfy the secondary standards of 30 mg/l of BOD_5 and suspended solids. In addition, significant reductions in the total nitrogen concentrations entering the hyacinth system have also been reported. However, the data are limited except for the Williamson Creek, Texas, NSTL, and the Coral Springs, Florida experiments. These studies show significant reductions in total nitrogen as well as total phosphorus. Unfortunately, the phosphorus concentrations were not reduced to the desired level of less than 1 mg/l at the Coral Springs, Florida operation, and total phosphorus concentrations were not measured at the Williamson Creek, Texas experiments. Considerable experimentation remains to be done before phosphorus control with hyacinth systems can be fully evaluated.

Hydraulic Detention Time

With the exception of the Williamson Creek, Texas phase 1 experiment, all the other studies with water hyacinth systems reporting hydraulic retention times are based on theoretical calculations. The degree to which the actual hydraulic residence time approaches the theoretical depends on the care with which the original design was carried out. Systems consisting of long, narrow, rectangular channels probably approach a ratio of actual to theoretical hydraulic detention time of 0.75 as a rough approximation. The circular or free-form ponds and systems adapted to water hyacinths probably have a ratio of actual to theoretical hydraulic detention time of 0.5 or less. All experiments that are currently being conducted should definitely incorporate a dye study to evaluate the actual hydraulic residence time in the hyacinth system.

Engineering Criteria

As mentioned previously, the application of water hyacinth systems to treat wastewater is limited to tropical and warm-temperate climates. It is unlikely that such a system can be economically adapted to cold regions successfully. Greenhouses and plant digestion to produce methane for partial heating and carbon dioxide enrichment are theoretical possibilities, but with the absence of experience in this area, it is impossible to recommend such a system for cold regions. A large-scale research project in a cold climate would be necessary to answer the majority of the questions involving the use of plant systems in cold regions. Many suggestions have been made that a more cold-tolerant plant such as duckweed be considered for cold climates. However, duckweed would not survive the low temperatures and ice cover in the northern United States. Winter protection or only warm-weather use of the plants would be necessary. Duckweeds, in theory, offer a greater geographical range and longer operational season when compared with hyacinths. It is possible that such a system would work, but again there are no data available to prove that the system will operate in cold climates or on which to base engineering design criteria.

In areas with warm-temperate climates, the application of water hyacinth wastewater treatment technology appears to be feasible. The system is based on essentially the same criteria utilized in design of wastewater stabilization ponds. Frequently, a water hyacinth system is installed in an existing wastewater stabilization pond.

The role of hyacinths in algae control appears to be that of a light screening function that controls algae growth. Wolverton[3] has presented results supporting the sorption of nutrients and pollutants by hyacinths, but significant nitrogen and phosphorus reductions occur in lagoons without hyacinths. Numerous reports summarize nitrogen and phosphorus removals by lagoon systems with total nitrogen removals frequently exceeding 70% and total phosphorus removals exceeding 50% without hyacinths. Nutrient reductions in hyacinth systems is far more complicated than plant uptake alone.

If the water hyacinth system is used to remove nutrients, it is necessary to maintain the hyacinth culture in an active growth phase, which means that harvesting must be conducted frequently. There is still need for definition as to what the proper harvesting schedule should be. With intensive harvesting it is necessary to construct the hyacinth ponds so that harvesting can be easily accomplished. This has a tendency to increase the cost of the hyacinth system, and it also develops the problem of disposing of the excess material. Most of the cost data associated with the harvesting and processing of hyacinth plants are based on small-scale experiments.[9] These small-scale experiments indicate that the cost for harvesting and processing will be expensive, but perhaps not prohibitive. In systems such as those recommended by Dinges[5] for use in Texas, where harvesting is recommended only once each year, the cost would be far more attractive.

Sludge accumulation information is very limited for hyacinth systems, but the experimental systems and the full-scale system utilized at Williamson, Texas indicate that a sizable mass of sludge accumulates in the course of a year. With multiple cell hyacinth systems it is likely that one pond could be drained and cleaned while the other ponds assume the total loading. It is unlikely that much of an upset would occur with this type operation.

Therefore, it would be possible to drain the hyacinth ponds completely and allow the materials to dry in place before removing the materials. Whether this would be the most satisfactory method of cleaning out ponds depends on the degree of sophistication an engineer may choose to design into the system. There are numerous harvesting opportunities described in the literature, and as mentioned above, there is too little data at this time to select an optimum harvesting and utilization technique. Basing calculations on one cleaning and harvesting per year, it is very unlikely that the cost associated with this would be prohibitive, and when nutrient control is not a consideration, this is probably the best approach to disposing of the accumulated sludge and plants.

When a hyacinth system is combined with wastewater stabilization pond technology in warm climates, it is an attractive system for the production of an advanced secondary effluent. The system can be efficient and economical, and it requires very little energy for operation. When properly designed and operated, the system apparently does not have an odor problem and can be aesthetically attractive. During the active growing season, the evapotranspiration losses from hyacinth systems can approach half the flow entering the system. The rate of evapotranspiration varies widely and is directly related to the rate of growth of the water hyacinth. In a water-short area such as Arizona and parts of California, this evapotranspiration could be significant and may make the process unattractive because of the loss of water.

In summary, the water hyacinth wastewater treatment process appears to be applicable in warm temperate and tropical climates, and adequate data appear to be available to assist in the design of a system capable of producing an advanced secondary effluent. The recommended design criteria for such a system are summarized in Table 8–2. These design data are based on the work of the individuals referred to in Table 8–1. Similar design criteria developed by Dinges[5] for the State of Texas also appear reasonable.

By-Product Recovery

The literature on water hyacinths as a wastewater treatment process contains considerable speculation on the use of the water hyacinth on harvesting. Composting, anaerobic digestion for the production of methane, and the fermentation of the sugars into alcohol are techniques proposed as a means to cover the costs of wastewater treatment.[10] All these techniques may have application in limited areas; however, it is very unlikely that a production system will be developed in the near future that could even approach paying for the treatment of the wastewater.[11] One cannot deny the possibility of reclaiming a product, but at this stage of

Table 8-2
Design Criteria for Water Hyacinth Wastewater Treatment Systems
Based on Best Available Data and to be Operated in Warm Climates

Parameter	Design Value		Expected Effluent Quality
	Metric	*English*	
A. *Raw Wastewater System* (Algae Control)			$BOD_5 \leq 30$ mg/l
Hydraulic residence time	> 50 days	> 50 days	$SS \leq 30$ mg/l
Hydraulic loading rate	200m³/ha·day	0.0214 mgad	
Depth, maximum	≤ 1.5 m	≤ 5 ft	
Area of individual basins	0.4 ha	1 acre	
Organic loading rate	≤ 30kg BOD_5/ ha·day	≤ 26.7 lbs BOD_5/ac·day	
L/W ratio of hyacinth basin	> 3:1	> 3:1	
Water temperature	> 10°C	> 50°F	
Mosquito control	Essential	Essential	
Diffuser at inlet	Essential	Essential	
Dual systems, each designed to treat total flow	Essential	Essential	
B. *Secondary Effluent System* (Nitrogen removal and algae control)			$BOD_5 \leq 10$ mg/
Hydraulic residence time	> 6 days	> 6 days	$SS \leq 10$ mg/l
Hydraulic loading rate	800 m³/ha·day	0.0855 mgad	$TP \leq 5$ mg/l
Depth, Maximum	0.91 m	3 ft	$TN \leq 5$ mg/l
Area of individual basins	0.4 ha	1 acre	
Organic loading rate	≤ 50 kg BOD_5/ ha·day	≤ 44.5 lb BOD_5/ac·day	
L/W ratio of hyacinth basin	> 3:1	> 3:1	
Water temperature	> 20°C	> 68°F	
Mosquito control	Essential	Essential	
Diffuser at inlet	Essential	Essential	
Dual systems, each designed to treat total flow	Essential	Essential	
Nitrogen loading rate	≤ 15 kg TKN/ ha·day	≤ 13.4 lbs TKN/ac·day	

SOURCE: From Ref. 1.

development, it is very unlikely that the recovery of useful products from water hyacinth wastewater treatment will be economically feasible.

Removal of Pollution

The greatest difficulty in interpreting the data presented by the various papers describing the work with water hyacinth systems is the infrequency of sampling and the lack of 24-hr composite samples. Although many of the studies include relatively large numbers of samples, most are grab samples collected twice each week. Even with large numbers of samples, it is still possible to make sizable errors in predicting the performance of a wastewater treatment system. Only the data for the Coral Springs, Florida system are based on 24-hr or 48-hr composite samples. All others are grab samples collected at various frequencies. The performance of typical water hyacinth systems is summarized in Table 8–1.

The most complete nutrient-removal data were collected at the Williamson Creek, Texas phase 1 and phase 2 experiments and at the Coral Springs, Florida, water hyacinth treatment facility. The organic loading rates, nutrient loading rates, and removals obtained during these three studies are summarized in Table 8–3. The lowest total nitrogen loading rate occurred at the NSTL experimental water hyacinth facility, but the effluent quality at the NSTL facility was no better than that experienced at the Williamson Creek facility. A higher percentage of phosphorus removal was experienced at the Coral Springs, Florida facility than at the NSTL facility. The total phosphorus effluent concentration at the NSTL was lower than that at the Coral Springs, Florida effluent. However, the influent total phosphorus concentration at Coral Springs, Florida was approximately three times greater than that at the NSTL facility. These differences are possibly due to the influence of the low concentrations being applied at the NSTL facility. In general, higher percentage removals are experienced with higher concentrations. In addition, harvesting at the NSTL facility was not conducted at a frequency to optimize nutrient removal.

Sludge Accumulation

Very little data are presented in the water hyacinth studies showing the quantities of sludge that accumulate during the rapid growth of plants. The only measurements of sludge accumulation reported were for the pilot-plant and full-scale studies at Williamson Creek, Texas. In the pilot studies the sludge accumulation was measured after the material dried, and in the full-scale operation the sludge was measured while wet. The area covered by the sludge was not reported in either case, and only the depth of the sludge was apparently measured. However, the dimensions of both the pilot and the full-scale facility were given, and with reasonable assumptions, the quantities of sludge that accumulated were estimated to be between 1.5 and 8×10^{-4} m³ of sludge/m³ of wastewater treated. This compares with 1.8×10^{-3} m³ of sludge/m³ of wastewater treated for conventional primary stabilization ponds.[12] The quantities of sludge accumulated per cubic meter of wastewater treated in the pilot plant were approximately five times less than that estimated in the full-scale unit. However, because of the lack of accurate

Table 8-3
Summary of Nutrient Loading Rates Applied to Water Hyacinths Wastewater Treatment Systems

Location	Organic Loading Rate (kg BOD₅/ha·day)	Nutrient Loading Rate To First Unit		Nutrient Removal (%)		Comments
		(kg TN/ha·day)	(kg TP/ha·day)	TN	TP	
Williamson Creek, Texas						
Phase I (109 m³/d)	43	15.3	—	70	—	Single basin, surface area = 0.0585 ha
Phase II (109 m³/d)	89	18.5	—	64	—	Single basin, surface area = 0.0585 ha
Coral Springs, Florida	31	19.5	4.8	96	67	Five basins in series, total surface area = 0.52 ha
NSTL	26	2.9	0.9	72	57	Single basin receiving raw wastewater, surface area = 2 ha

SOURCE: From Ref 1.

measurement for the quantity of sludge accumulated, these estimates are the best available. Regardless of which figure is used to estimate sludge production, it is apparent that the rate of sludge accumulation in a hyacinth growth basin is relatively slow, and cleaning the systems once each season, as recommended by Dinges[5] would probably be adequate to prevent the passing of solids out of the system. Compared with the accumulation of plants in the system, the mass of sludge would be relatively insignificant and could easily be disposed of along with the harvested hyacinths.

Hydraulics of Triangular Basins

Dinges[5] has recommended that rectangular basins with a length-to-width ratio of 3 to 1 be constructed and then divided into two triangles to improve the hydraulic characteristics of the hyacinth basin. Such a design would result in an increase in cross-sectional velocity as the wastewater flowed toward the apex of the triangle. A preliminary hydraulic analysis of the triangular basin concept indicates that the use of such a hydraulic design should be approached with caution, since small organic particles near the overflow weir may be washed out of the basin. Before installing such a system, a more detailed hydraulic analysis should be conducted.

Mosquito control

Various experiences with mosquito problems at water hyacinth wastewater treatment systems are reported by the investigators listed in Table 8–1. Although some investigators did not encounter a mosquito problem, all recommended that some means of mosquito control be incorporated into the design of such a facility. Most investigators recommended that natural control measures be employed such as the mosquito fish (*Gambusia*). In quiescent bodies of water, the growth of mosquito larvae is encouraged; therefore, it appears imperative that control measures be incorporated into hyacinth wastewater treatment systems.

Recommendations

1. The water hyacinth wastewater treatment process appears to be applicable in warm temperature and tropical climates, and adequate data are available to assist in the design of a system capable of producing an advanced secondary effluent.

2. Water hyacinths thrive in municipal wastewaters and appear to do well in mixtures of municipal and industrial wastewaters.

3. A hydraulic loading rate of 2000 m³/ha·day to a hyacinth system appears reasonable when treating secondary wastewater treatment plant effluent if nutrient control is not an objective. When treating raw wastewater in a hyacinth system, a hydraulic loading rate of 200 m³/ha·day appears reasonable if nutrient control is not an objective.

4. A shallow hyacinth pond (≤ 0.4 m) and a hydraulic loading rate of approximately 500 m³ of stabilization pond effluent per hectare per day should produce an effluent containing a total nitrogen concentration of less than 2 mg/l.

5. Total phosphorus removals of approximaetly 50% are normal with a hyacinth system.

6. Considerable experimentation remains to be performed before phosphorus control with hyacinth systems can be accomplished.

7. Dye studies should be conducted to determine the actual hydraulic residence times in hyacinth systems.

8. Algae growth appears to be controlled in hyacinth systems by simple shading by the plant.

9. Nutrient removal in hyacinth systems is more complex than plant uptake alone. Excellent nitrogen and phosphorus reductions occur in wastewater stabilization ponds without water hyacinths.

10. Sludge accumulation in hyacinth systems does not appear to be a significant problem.

11. Harvesting and utilizing the water hyacinth after harvesting requires considerable investigation to develop satisfactory methods and procedures.

12. The use of more cold-tolerant plants such as duckweed should be investigated more extensively.

13. More extensive investigations should be conducted on the range of organic and hydraulic loading rates that the hyacinth system is capable of treating, particularly with systems processing raw wastewater.

14. Mosquito control is essential in hyacinth wastewater treatment systems.

15. Water hyacinths must not be introduced into areas in which it does not currently grow.

References

1. Middlebrooks, E.J. 1981. Aquatic Plant Processes Assessment. In *Aquaculture Systems for Wastewater Treatment*. Edited by S. C. Reed and R. K. Bastian. EPA–430/9–80–007. Office of Water Program Operations, Municipal Construction Division, Washington, D. C.

2. Wolverton, B.C., and R.C. McDonald. 1979. Upgrading Facultative Wastewater Lagoons with Vascular Aquatic Plants. *JWPCF*, **51**, 2, 305–313.

3. Wolverton, B.C. 1979. Engineering Design Data for Small Vascular Aquatic Plant Wastewater Treatment Systems. Presented at the Seminar on Aquaculture Systems for Wastewater Treatment, September 11–12, 1979, University of California, Davis.

4. Dinges, R. 1978. Upgrading Stabilization Pond Effluent by Water Hyacinth Culture. *JWPCF*, **50**, 5, 833–845.

5. Dinges, R. 1979. Development of Hyacinth Wastewater Treatment Systems in Texas. Presented at the Seminar on Aquaculture Systems for Wastewater Treatment, September 11–12, 1979, University of California, Davis.

6. Stewart, E.A., III. 1979. Utilization of Water Hyacinths for Control of Nutrients in Domestic Wastewater—Lakeland, Florida. Presented at the Seminar on Aquaculture Systems for Wastewater Treatment, September 11–12, 1979, University of California, Davis.

7. Kruzic, A.P. 1979. Water Hyacinth Wastewater Treatment System at Walt Disney World. Presented at the Seminar on Aquaculture Systems for Wastewater Treatment, September 11–12, 1979, University of California, Davis.

8. Swett, D. 1979. A Water Hyacinth Advanced Wastewater Treatment System. Presented at the Seminar on Aquaculture Systems for Wastewater Treatment, September 11–12, 1979, University of California, Davis.

9. Bagnall, L.O. 1979. Resource Recovery from Wastewater Aquaculture. Presented at the Seminar on Aquaculture Systems for Wastewater Treatment, September 11–12, 1979, University of California, Davis.

10. Benemann, J.R. 1979. Energy from Wastewater Aquaculture Systems. Presented at the Seminar on Aquaculture Systems for Wastewater Treatment, September 11–12, 1979, University of California, Davis.

11. Crites, R. W. 1979. Economics of Aquatic Treatment Systems. Presented at the Seminar on Aquaculture Systems for Wastewater Treatment, September 11–12, 1979, University of California, Davis.

12. Middlebrooks, E. J., A. J. Panagiotou, and H. K. Williford. 1965. Sludge Accumulation in Municipal Sewage Lagoons. *Water Sewage Works*, **112**, 2, 62.

CHAPTER
9

Sludge Accumulation
in Lagoons

MANY GROUPS HAVE MADE theoretical calculations that predict sludge accumulation in municipal sewage lagoons at the rate of 1 ft every 135 years in a 4.05-ha (10-acre) lagoon serving a population of 1000 people.[1-4] These calculations have been based on an average wastewater flow of 379 lpcd (100 gpcd) and a yearly sludge contribution of 90.9 m³ (3212 ft³) from 1000 people. These calculations consider only organic materials and neglect the effect of silt and other inorganic substances that may originate in the sewer system or be washed from the dikes.

A limited investigation of sludge accumulation on experimental lagoons receiving municipal wastes from the City of Fayette, Missouri pointed out that the most significant factor in sludge accumulation would probably be the silt and other inorganic material entering the lagoon.[5] The study concluded that a loss of capacity by sedimentation of organic matter would be insignificant for a period of 100 years or more. Silting would probably result in a loss of 1 acre foot per acre of surface area in approximately 25 years.

Two small municipal lagoons were drained in preparation for construction of experimental lagoons that were used to evaluate the effects of recirculation.[6] The drained lagoons were 1.22 m (4 ft) deep and had been receiving wastes from the City of Concord, California for about 20 years. About 46 cm (18 in.) of sludge was found at the inlet end and tapered down to about 1 in. near the outlet end. In the deeper layers of the sludge methane fermentation was in progress, and the sludge was found to have gelatinous characteristics and poor dewatering properties. The history of the drained

316

lagoons was not discussed. Study of their experimental lagoons indicated that the recirculation rate influenced the rate and profile of sludge deposition. Sludge deposition near the inlet was found to decrease with an increase in the recirculation rate. Where masses of sludge accumulated, anaerobic conditions developed and buoyed masses of sludge to the lagoon surface. These masses tended to accumulate in the corners of the lagoons.

At several facultative lagoon sites in Canada and Alaska, sludge accumulation rates varied from 0.25 to 0.40 m³ (8.8 to 14.0 ft³) per 1000 people per day, with the highest accumulation rates occurring at the more northerly sites.[7]

Facultative Lagoons

Mississippi Study

Field data from 15 facultative lagoons located in Mississippi were collected to predict the accumulation of sludge. The location of the lagoons and the data are shown in Table 9–1.[8]

Sludge samples were collected in proportion to the area of the lagoon, and a minimum number of 18 samples were collected in each lagoon. Sampling directions were determined with reference to the direction of the prevailing winds so that wind effects on sludge depth and distribution could be detected. The directions of the sampling lines were determined with a lensatic compass. The core samples were collected in a sampling tube, which allowed the sample to be removed from the lagoon bottom essentially undisturbed.

Silt and Inorganic Substances. Because others had found silting to be significant, the samples were composited and analyzed for volatile solids content. These data are presented in Table 9–1. The composited sample total solids are given in percent by weight of the original sample of sludge, and the volatile solids are shown as percent of the total solids. Where the percent of total solids are high, the volatile portion is generally low, indicating that a large portion of the sludge must be silt and other inorganic matter. These findings tend to support the conclusions reached by Clare, Neel, and Monday that silting is the most significant factor in sludge accumulation.[5] This is also supported by the statistical analysis discussed below.

Wind Action. The prevailing winds for Mississippi vary from the south to southwest. Sampling directions were selected parallel and perpendicular to the prevailing winds to determine the effects of wind direction on the sludge depth and distribution. The mean sludge depth for the parallel and perpendicular directions showed that eight lagoons had a greater mean sludge depth in the direction parallel to the prevailing winds, and seven had

Table 9-1

Summary of the Data for the Mississippi Facultative Lagoon Studies

LOCATION OF LAGOON	SURFACE AREA [HA]	AGE OF LAGOON AT SAMPLING DATE [MONTHS]	BOD APPLIED [KG/HA·DAY]	COMPOSITED SAMPLE TOTAL SOLIDS [%]	COMPOSITED SAMPLE VOLATILE SOLIDS [%]	AVERAGE SLUDGE DEPTH [CM]
Brandon-W	2.02	82.0	39.3	3.48	34.9	17.4
Brandon-E	2.02	82.0	20.2	10.30	25.5	10.0
Fayette-N	1.70	65.0	25.8	17.40	16.10	12.4
Fayette-SW	1.25	65.0	25.8	19.20	11.10	10.4
Newton	1.21	54.0	31.4	4.92	37.20	11.9
Rolling Fork-E	2.08	54.0	28.1	5.58	28.40	12.3
Rolling Fork-W	2.05	54.0	28.1	5.62	36.0	12.3
Ellisville	1.42	44.0	32.0	11.30	15.90	6.6
Canton Club	1.21	31.0	78.6	11.90	29.60	14.1
Canton	2.02	26.0	16.8	7.04	20.20	5.9
Utica	2.43	25.0	8.4	13.40	9.84	8.4
Decatur	1.21	25.0	11.8	4.34	30.30	5.8
Walnut Grove	0.81	19.0	28.1	3.66	31.30	9.0
Ridgeland-N	1.21	5.0	9.4	9.70	21.90	2.6
Philadelphia	1.21	5.0	8.4	10.90	22.60	0.4
Mean				9.25	24.70	

SOURCE: From Ref. 8, courtesy of *Water Engineering and Management*, Des Plaines, Illinois.

a greater mean sludge depth in the direction perpendicular to the prevailing winds. The slightly larger number of lagoons having greater sludge depths in the direction of the prevailing winds indicates there was no significant effect due to the prevailing winds.

There was a greater accumulation of sludge in the corners. This accumulation was probably due to excessive sludge accumulation around the inlet. The sludge becomes anaerobic and is buoyed up by the gaseous products of anaerobic decomposition. These floating masses are then blown into the corners by the wind. Again, there was little difference in the sludge depth in the directions parallel and perpendicular to the prevailing winds.

Sludge Accumulation about Inlets. Sludge concentrations were found at all inlet locations. Indications were that little effect was produced by the prevailing winds. Heavy sludge accumulation was observed to extend to a distance of 24.4–30.5 m (80–100 ft) for gravity flow lines and 15.2–22.9 m (50–75 ft) for force mains terminating with a 90-degree bend which directed the flow upward.

Statistical Analysis of Data. The rate of sludge accumulation as observed in the investigated lagoons is described in terms of an equation based on the least-squares method for curve fitting. The mean sludge depth in the lagoons was described in terms of the following parameters:

1. The age of the lagoon in months.
2. The applied BOD loading rates in kilograms per hectare per day.
3. The average percentage of total and volatile solids of all sludge samples from each lagoon.

The general equation that follows was found to give the best correlation coefficient of all the combinations of the data that were analyzed.

$$Y = 0.861 + 0.0404X_1 - 0.000175X_2 - 0.00289X_3 + 0.0059X_4 \quad (9\text{-}1)$$

where

Y = the average sludge depth in cm

X_1 = the age of the lagoon in months

X_2 = the age of the lagoon in months multiplied by the average total solids in percent

X_3 = the BOD loading rates multiplied by the average total solids in percent

X_4 = the average total solids in percent multiplied by the average volatile solids in percent.

Equation 9-1 has an R^2 value of 0.851, which indicates that 85.1% of the variation is accounted for by the regression sum of squares.

The largest source of variation was explained by the age factor which accounted for 70% of the variation; however, the interaction between the total and volatile solids was the second largest source of variation. This indicates that the interaction between these variables had a considerable effect on sludge depth, as would be expected. Extrapolation of Equation 9–1 indicates that the rate of sludge accumulation was 0.3 m (1 ft) per 27.6 years. This value agrees with expected accumulations reported by Clare, Neel, and Monday.[5]

Utah Study

Field data were collected from two facultative lagoons located in northern Utah.[9] A summary of the information available for the two lagoons is presented in Table 9–2. The data were collected as part of a study to determine the affect of temperature on sludge accumulation in cold climates. At this time (1981) the study is still in progress.

Aerated Lagoons

Sludge accumulation in the Northway, Alaska aerated lagoon varied from 8.26 cm (3.25 in.) to 16.5 cm (6.5 in.) in the primary cell and from 2.54 cm (1 in.) to 3.18 cm (1.25 in.) in the secondary cell after approximately 21 months of operation. After another 33 months, the primary cell sludge varied from 11.4 cm (4.5 in.) to 16.5 cm (6.5 in.) in depth, and the secondary cell sludge increased to 9.5 cm (3.75 in.) and 11.4 cm (4.5 in.). This indicates an accumulation rate of over 2.54 cm/year (1 in./year), or about 0.25 m³/1000 people per day (9 ft³/1000 people per day) in 4.5 years.[7] The lagoon at Eilson A.F.B., Alaska experienced a 1.3 cm (0.5 in.) per year sludge accumulation in the first one-third of the pond and 0.64 cm to 1.27 cm (0.25 in. to 0.5 in.) per year in the final two-thirds of the pond.[10]

The aerated lagoon at Fort Greely, Alaska had 51 cm (20 in.) of accumulated sludge near the inlet, tapering down to 3.2 cm (1.25 in.) at about 15.2 m (50 ft) from the inlet. The secondary cell was about 50% covered with 3.2–5.1 cm (1.25–2 in.) of sludge.[11]

Influent to the aerated lagoon was discontinued for 4 months at a Winnipeg lagoon. Only slight evidence of bottom sludge reduction was observed during this period, indicating a stable sludge mass.[12] Schneiter[9] reported sludge accumulation in three aerated lagoons, as shown in Table 9–3. The data in Table 9–3 were collected as part of the study on temperature effects on lagoon sludge accumulation referred to above.

Anaerobic Lagoons

Clark et al.[7] compiled the sludge accumulation data for anaerobic lagoons shown in Table 9–4. All the lagoons were located in cold climates and

Table 9-2

Sludge Accumulation in Two Facultative Lagoons

Location	Surface Area (HA)	Months of Continuous Operation	Influent BOD$_5$ (MG/L)	Sludge Total Organic Carbon Concentration (G/L)	Sludge Total Solids. (%)	Sludge Volatile Solids. (%)	Mean Sludge Depth (CM)
Logan, Utah	38.0	168	75	6.75	5.0	67.5	8.9
Corinne, Utah	1.49	120		6.30	7.8	82.9	10.2

Source: From Ref. 9.

Table 9-3
Sludge Accumulation in Three Aerated Lagoons

Location	Surface Area (ha)	Months of Continuous Operation	Influent BOD₅ (mg/L)	Sludge Total Organic Carbon Concentration	Sludge Total Solids, (%)	Sludge Volatile Solids, (%)	Mean Sludge Depth (cm)
Palmer, Alaska	0.81	60	180	13.16	8.6	69.1	33.5
Galena, Alaska	0.30	72		3.08	1.0	47.8	27.7
Elmendorf A.F.B., Alaska	0.085	40 48 54	84	—	—	—	33.5 45.7 48.8

SOURCE: From Ref. 9.

322

Table 9-4

Sludge Accumulation in Anaerobic Lagoons

LOCATION	DETENTION TIME (DAYS)	DEPTH (M)	ORGANIC LOADING (KG BOD$_5$/HA·DAY)	SLUDGE ACCUMULATION (M^3/1000 POP.·DAY)
Yellowknife, N.W.T., Canada	0.6	2.44	5807-6930	0.37
Stettler, Alberta, Canada	5.94	3.05	1610	0.25
Sutherland, Saskatchewan, Canada	2.75-1.45	5.18	1460-2920	0.40
Camrose, Alberta, Canada	6.2	3.05	908	0.27-0.29

SOURCE: From Ref. 7.

Table 9-5
Characteristics of Lagoon and Conventional Processes Sludges

Sludge Type	Facility Location	Total Solids (%)	Volatile Solids (%)	pH	Fecal Coliform (MPN/G)	COD	Color	Odor	Ref
Aerated lagoon	Northway, Alaska	12-20	29-50	7.1-7.6	2.4×10^3 4.9×10^3	46-88 mg/g	Black	Musty	7, 11
Aerated lagoon	Winnipeg, Manitoba	8-10	45-55						13
Aerated lagoon	Eilson A.F.B., Alaska	3.6-9.1	55-93				Gray-brownish black	Fresh sewage-humus	14
Facultative lagoon	Columbia, Missouri					800 mg/in.^{2}	Black-gray		15
Anaerobic lagoon	Melbourne, Australia	14.9	38.3	7.5					16
Anaerobic lagoon	Werribee, Australia	1.2-7.0	41-60						17
Untreated primary		2-8	60-80	5.0-8.0	1.1×10^4				18, 19

Source: From Ref. 9.

324

should represent the maximum accumulation likely to occur in an anaerobic lagoon treating domestic wastes.

Lagoon Sludge Characteristics

Lagoon sludge characteristics are as varied as lagoon sites; however, some general sludge characteristics have been defined. Table 9–5 presents sludge characteristics from several sources and provides a comparison of various types of sludges.

The chemical composition of numerous sludge types (anaerobic, aerobic, activated, lagoon, etc.) collected in Wisconsin, Michigan, New Hampshire, New Jersey, Illinois, Minnesota, and Ohio are listed in Table 9–6.[20]

The sludge nitrogen and phosphorus content varies with the waste origin and treatment process. Typical values may range from 2 to 6% for nitrogen and from 1 to 7% for phosphorus.[21]

Table 9-6
Chemical Composition of Wastewater Sludges

Component	Units	Number of Samples	Range	Median	Mean
otal N	Percent	191	0.1-17.6	3.3	3.9
H$_4$ – N	Percent	103	0.1-6.8	0.1	0.7
O$_3$ – N	Percent	45	0.1-0.5	0.1	0.1
	Percent	189	0.1-14.3	2.3	2.5
	Percent	192	0.1-2.6	0.3	0.4
a	Percent	193	0.1-25.0	3.9	4.9
g	Percent	189	0.1-2.0	0.5	0.5
	Percent	165	0.1-15.3	1.1	1.3
n	mg/kg	143	18-7,100	260	380
	mg/kg	109	4-760	33	77
	mg/kg	78	0.5-10,600	5	733
	mg/kg	205	84-10,400	850	1210
	mg/kg	208	101-27,800	1740	2790
	mg/kg	165	2-3,520	82	320
	mg/kg	189	13-19,700	500	1360
	mg/kg	189	3-3,410	16	110

SOURCE: From Ref. 20.

Lagoon Sludge Removal

Information describing lagoon sludge removal is limited. Common practice for removing sludge from unlined or earthen-lined lagoons usually involves draining the lagoon and excavating the sludge with a scraper or a front-end loader.

In Moose Jaw, Alberta accumulated sludge was removed from an aerated lagoon by first removing the aeration tubing during the warm months. With the arrival of winter, the sludge froze and was subsequently removed by front-end loaders.[22]

Successful application of dredges requires that the accumulated sludge be loosened for pumping to holding tanks before final disposal; however, clogging of the pump suctions on a sludge dredging operation at Red Deer, Alberta required the use of cutting heads on the suction pipes.[22] Sludge pumps are available to meet various application requirements.[20]

References

1. Heuveilen, W.V., and J.H. Svore. 1954. Sewage Lagoons in North Dakota. *Sewage Ind. Waste J.* **26**, 775.

2. U.S. Public Health Service, Washington, D.C., *Waste Stabilization Lagoons*, Public Health Service Publications No. 872. August 1961.

3. Williamson, J., Jr., 1956. Sewage Lagoons Are the Answer. The *Consul. Eng.* **8**, 45.

4. Fair, G.M., and K. Imhoff. 1960. *Sewage Treatment.* Wiley, New York, New York, pp. 180–188.

5. Clare, H.C., J.K. Neel, and C.A. Monday, Jr. 1961. *Waste Stabilization Lagoons.* Public Health Service Publication No. 872, pp. 31–34.

6. Oswald, W.J., C.G. Golueke, and H. B. Gotaas. 1959. *Experiments in Algal Culture in a Field Scale Oxidation Pond.* Issue No. 10, I.E.R. Series 44:19.

7. Clark, S.E., H.J. Coutts, and C. Christianson. 1970. *Biological Waste Treatment in the Far North.* Federal Water Quality Administration, Northwest Region, Department of the Interior, Alaska Water Laboratory, College, Alaska.

8. Middlebrooks, E.J., A.J. Panagiotou, and H.K. Williford. 1965. Sludge Accumulation in Municipal Sewage Lagoons. *Water Sewage Works*, **112**, 2, 62.

9. Schneiter, R. W. 1981. Cold Region Wastewater Lagoon Sludge: Accumulation, Characterization and Digestion. Ph.D. dissertation, Utah State University, Logan, Utah. Also to appear as Special Report, U.S. Army Cold Regions Research and Engineering Laboratory, Hanover, New Hampshire.

10. Reid, L.C., Jr. 1965. The Aerated Sewage Lagoon in Arctic Alaska. Presented at the 17th Annual Convention of the Western Canada Water and Sewage Conference, Edmonton, Alberta, Canada, September 15–17.

11. Christianson, C.D., and H. J. Coutts. 1979. *Performance of Aerated Lagoons in Northern Climates.* EPA–600/3-7-003, U.S. Environmental Protection Agency, Washington, D.C.

12. Girling, R.M., et al. 1974. Further Field Investigation on Aerated Lagoons in the City of Winnipeg. Symposium on Wastewater Treatment in Cold Climates, EPS 3–WP–74–3, Environment Canada.

13. Pick, A.R. 1970. Evaluation of Aerated Lagoons as a Sewage Treatment Facility in the Canadian Praire Provinces. *International Symposium on Water Pollution Control in Cold Climates.* Edited by R. S. Murphy and D. Nyquist, University of Alaska, Fairbanks, Alaska, July 22–24.

14. Reid, L.C. 1970. *Design and Operation Considerations for Aerated Lagoons in the Arctic and Subarctic.* Report No. 102, Environmental Engineering Section, Arctic Health Resource Center, Environmental Health Service, U.S.P.H.S., Department of Health, Education, and Welfare, College, Alaska.

15. Howard, D.E., and L.D. Ray. 1966. Deposition of Solids in a Facultative Lagoon. Presented at 3rd Sanitary Engineering Conference, University of Missouri, Columbia, Missouri.

16. Gloyna, E.F. 1971. *Waste Stabilization Ponds.* World Health Organization Monograph Series No. 60, Geneva, Switzerland.

17. Parker, C.D., and G.P. Skerry. 1968. Function of Solids in Anaerobic Lagoon Treatment of Wastewater. *JWPCF*, **40**, 2, Part 1, 192–204.

18. Metcalf and Eddy, Inc. 1979. *Wastewater Engineering: Treatment, Disposal, and Reuse.* 2nd Edition, McGraw-Hill, New York, New York.

19. Farrell, J.B. 1974. *Overview of Sludge Handling and Disposal.* National Environmental Research Center, U. S. Environmental Protection Agency, Cincinnati, Ohio.

20. Environmental Protection Agency. 1978. *Sludge Treatment and Disposal: Sludge Treatment.* Volume 1, EPA–625/4–78–012.

21. Anderson, M.S. 1959. Fertilizing Characteristics of Sewage Sludge. *Sewage Ind. Wastes*, **31**, 6, 678–682.

22. Lawson, P.D. 1977. Sludge Handling and Disposal Problems in the Praire Provinces. Presented at the Technology Transfer Seminar on Sludge Handling and Disposal, Calgary, Alberta, February 16–18.

CHAPTER 10

Cost and Performance Comparisons of Various Upgrading Alternatives and Lagoons

COST AND PERFORMANCE VALUES shown in Table 10–1 represent the best available information for all of the processes listed. In several cases the costs are based on estimates derived from pilot-plant studies or engineering estimates. Where actual bid prices are available, the location of the facility is given. All costs are site specific and can be expected to vary widely. Costs are reported as shown in the literature and are not corrected for changes in value of the dollar. This was done to allow the reader to use the system appropriate for his or her area to adjust the costs to a current base. Corrections were made to all capital costs to reflect a 7 % interest rate and a 20-year life except for systems known to have shorter operating periods. The exceptions are identified in Table 10–1.

The selection of the cost-effective alternative must be made, based on good engineering judgment and local economic conditions. Cost variations in one item, such as filter sand or land, can change the relative position of a process dramatically. In brief, Table 10–1 cannot be substituted for good engineering.

All the processes listed in Table 10–1 are capable of meeting secondary standards, and several are capable of producing an effluent superior to 30 mg/l of BOD_5 and suspended solids. Variations in design and operation also alter the quality of the effluent dramatically in most of the processes. A

Comparative Costs and Performance of Various Upgrading Alternatives and Lagoon Systems

Process or System and Location	Design Flow Rate (MGD)	Design Loading	Annual Costs[a] $/1000 Gallons			Cost Base	Ref.	Effluent Concentration (mg/L)	
			Capital	O&M	Total			BOD$_5$	SS
Overland flow[b]									
EPA Estimate	0.3	2 in./wk	0.27	0.14	0.41	1973	1	<10	<10
EPA Estimate	0.3	8 in./wk	0.19	0.10	0.29	1973	1		
Davis, California	5.0	8 in./wk	0.10	0.05	0.15	1976	2		
Surface irrigation[b]									
EPA Estimate	0.3	2 in./wk	0.20	0.19	0.39	1973	1	<10	<10
EPA Estimate	0.3	4 in./wk	0.17	0.15	0.32	1973	1		
Spray irrigation center pivot[b]									
EPA Estimate	0.3	2 in./wk	0.19	0.18	0.37	1973	1	<10	<10
EPA Estimate	0.3	4 in./wk	0.16	0.13	0.29	1973	1		
Spray irrigation-solid set[b]									
EPA Estimate	0.3	2 in./wk	0.26	0.15	0.41	1973	1	<10	<10
EPA Estimate	0.3	4 in./wk	0.19	0.12	0.31	1973	1		
Rapid infiltration									
EPA Estimate	0.3	8 in./wk	0.17	0.10	0.27	1973	1	<10	<10
EPA Estimate	0.3	24 in./wk	0.13	0.08	0.21	1973	1		
Intermittent sand filtration									
Metcalf and Eddy capital cost estimate & European O&M[c]	0.3	—	0.16	0.16	0.32	1975	3 4	<15	<15
Huntington, Utah	0.3	0.3 MGAD	0.18			1975	5		
Kennedy, Alabama	0.084	0.1 MGAD	0.36			1975	6		
Ailey, Georgia	0.08	0.6 MGAD	0.20			1975	7		
Moriarty, N.M.	0.2	0.3 MGAD	0.12			1975	8		
White Bird, Idaho	0.03	0.4 MGAD	0.18			1978	9		

(continued)

329

Table 10-1 (cont.)

Comparative Costs and Performance of Various Upgrading Alternatives and Lagoon Systems

Process or System and Location	Design Flow Rate (MGD)	Design Loading	Annual Costs[a] $/1000 Gallons			Cost Base	Ref.	Effluent Concentration (mg/L)	
			Capital	O&M	Total			BOD_5	SS
Microscreens ENVIREX	1.7-2.25		0.11-0.14[e]	0.05[e]	0.16-0.19	1978	10	<30	<30
Dissolved air flotation Snider	0.8	—	0.14	0.06	0.20[d]	1975	11	<30	<20
Coagulation-flocculation sedimentation-filtration Los Angeles Co., Calif.	0.5	—	0.13	0.30	0.43	Cap 1970 O&M 1973 1974	12	<10	<10
Rock filters									
Wardell, Mo.	0.08	1 yd³/pop. eq.	0.04			1974	13	<30	<30
Delta, Mo.	0.08	1 yd³/pop. eq.	0.05			1974	13		
California, Mo.	0.36	3 gal/day-ft³	0.04			1974	14		
Luxemburg, Wis.	0.4		0.03[e]			1976	15		
Veneta, Oregon	0.22	2 gal/day-ft³	0.05			1975	16		
Intermittent discharge-chemical addition Canadian Experience	0.3	Alum: 150 mg/l Det. time: 120 days	0.04[f]	0.08	0.12	1976[h]		<30	<10

Location							
Total containment lagoons Huntington, Utah	0.3		0.36[g]	1975	5	No discharge	No discharge
Wellsville, Utah	0.285		0.37	1974	5		
Total containment lagoons Southshore, Utah	0.00075		0.30	1978	11		
Facultative lagoons Huntington, Utah	0.3		0.33	1975	17 / 5	Varies with design & time of year	Varies with design & time of year
Wardell, Mo.	0.08	Primary cell 34 lb BOD_5/acre-day 2nd cell 0.3 (primary cell surface area) 3rd cell 0.1 (Primary cell surface area)	0.29	1974	13		
Delta, Mo.	0.08	Same as Wardell	0.41	1974	13		
Idaho	1.7	20 lb/acre day	0.31	1978	17		

(continued)

331

Table 10-1 (cont.)

Comparative Costs and Performance of Various Upgrading Alternatives and Lagoon Systems

Process or System and Location	Design Flow Rate (MGD)	Design Loading	Annual Costs[a] $/1000 Gallons			Cost Base	Ref.	Effluent Concentration (MG/L)	
			Capital	O&M	Total			BOD$_5$	SS
Aerated lagoons									
Luxemburg, Wis.	0.40	Det. time, = 35 days	0.31[e]			1977	15	Varies with design & time of year	Varies with design & time of year
Sugarbush, Va.	0.163		0.48			1974	18		
Paw Paw, Mich.	0.40		0.26			1974[i]			
Blacksburg, Va.	0.04	Det. time = 30 days				1972	19		
Luxemburg, Wis.	0.40	807 lbs BOD$_5$/day	0.45	0.32	0.77	1978	20		
White Bird, Idaho	0.03	50 lbs BOD$_5$/day	1.40			1978	9		

[a] Costs amortized at 7% and a 20-year life.
[b] Values can vary by 50% and prices do not include land costs.
[c] Includes land costs with no credit for salvage value.
[d] Excludes sludge disposal costs.
[e] Engineer's estimate.
[f] Amortized at 7% and a 10-year life.
[g] Bid but not constructed.
[h] Authors' estimate.
[i] Anonymous.

careful study of all alternatives must be made before selecting a system. The literature referenced herein will provide all details needed, but engineers should remain aware of current developments and use other alternatives as more information becomes available.

References

1. Environmental Protection Agency. 1975. *Cost of Wastewater Treatment by Land Application.* Office of Water Program Operations. EPA–430/9–75–003, Washington, D.C., June 1975.

2. Brown and Caldwell. 1976. Draft Project Report, City of Davis—Algae Removal Facilities. Walnut Creek, California, November 1976.

3. Metcalf & Eddy, Inc. 1975. *Draft Report to National Commission on Water Quality on Assessment of Technologies and Costs for Publicly Owned Treatment Works Under Public Law 92–500.* April.

4. Huisman, L., and W.E. Wood. 1974. *Slow Sand Filtration.* World Health Organization, Geneva, Switzerland.

5. Valley Engineering. 1976. Personal communication, Logan, Utah, December 1976.

6. Gilbreath, Foster & Brooks, Inc. 1976. Personal communication, Tuscaloosa, Alabama, November 29, 1976.

7. McCrary Engineering Corporation. 1976. Personal communication, Atlanta, Georgia, November 29, 1976.

8. Molzen-Corbin & Associates. 1976. Personal communication, Albuquerque, New Mexico.

9. Hamilton and Voeller, Inc. 1978. Personal communication. Moscow, Idaho. December 1978.

10. Kormanik, R.A., and J.B. Cravens. 1979. Microscreening and Other Physical-Chemical Techniques for Algae Removal. In: *Performance and Upgrading of Wastewater Stabilization Ponds,* EPA–600/9–79–011, Municipal Environmental Research Laboratory, Cincinnati, Ohio.

11. Snider, E.F., Jr. 1976. Algae Removal by Air Flotation. In: *Ponds as a Wastewater Treatment Alternative,* edited by E.F. Gloyna, J.F. Malina, Jr., and E.M. Davis, Center for Research in Water Resources, College of Engineering, The University of Texas at Austin.

12. Parker, D.S. 1976. Performance of Alternative Algae Removal Systems. In *Ponds as a Wastewater Treatment Alternative,* edited by E.F. Gloyna, J.F. Malina, Jr., and E.M. Davis, Center for Research in Water Resources, College of Engineering, The University of Texas at Austin.

13. Gaines, G.F. 1977. Personal communication, C. R. Trotter and Associates, Dexter, Missouri.

14. Kays, F. 1977. Personal communication, Lane-Riddle Associates, Kansas City, Missouri.

15. Miller, D.L. 1977. Personal communication, Robert E. Lee and Associates, Inc., Green Bay, Wisconsin.

16. Williamson, K.J., and G.R. Swanson. 1979. Field Evaluation of Rock Filters for Removal of Algae from Lagoon Effluent. In *Performance and Upgrading of Wastewater Stabilization Ponds*, EPA–600/9–79–011, Municipal Environmental Research Laboratory, Cincinnati, Ohio.

17. Nielsen, Maxwell & Wangsgard. 1981. Personal communication, Salt Lake City, Utah, February 1981.

18. Jupka, J. 1977. Personal communication, Lane-Riddle Associates, Kansas City, Missouri.

19. Gearheart, R.M. 1972. Aeration Process Efficiency. *Water Wastes Eng.*, 9, 6, 35–36.

20. Robert E. Lee & Associates. 1980. Personal communication, Green Bay, Wisconsin, November 1980.

APPENDIX
A

Conversion Factors

Instructions on Use:
To convert, multiply in
direction shown by arrows.

	SI Units	↑	↓	U. S. Units
Length	centimeter	0.3937	2.5400	inch
	centimeter	0.032808	30.480	foot
	meter	39.3701	2.540×10^{-2}	inch
	meter	3.2808	0.30480	foot
	meter	1.0936	0.91441	yard
	kilometer	3,280.833	3.0480×10^{-4}	foot
	kilometer	0.6214	1.6093	mile
Area	centimeter^2	0.1550	6.4516	inch^2
	meter^2	10.7639	9.2903×10^{-2}	foot^2
	meter^2	1.1960	0.83612	yard^2
	meter^2	2.4711×10^{-4}	4046.78	acre
	kilometer^2	3.8610×10^{-7}	2.5900×10^6	mile^2
	kilometer^2	1.0764×10^7	9.29023×10^{-8}	foot^2
	kilometer^2	247.1044	4.0469×10^{-3}	acre
	hectare	0.3861006	2.59000	mile^2
	hectare	107,638.7	9.290339×10^{-6}	foot^2
	hectare	2.47104	0.40468	acre
Volume	centimeter^3	0.06102	16.3934	inch^3
	centimeter^3	3.5314×10^{-5}	2.8317×10^4	foot^3
	centimeter^3	2.6417×10^{-4}	3.7854×10^3	gallon
	meter^3	61,023.38	1.638716×10^{-5}	inch^3
	meter^3	35.3147	2.83168×10^{-2}	foot^3
	meter^3	1.3079	0.76458	yard^3
	meter^3	264.1720	3.7854×10^{-3}	gallon
	meter^3	8.3865	0.11924	barrel
	meter^3	8.1071×10^{-4}	1,233.487	acre-foot

	liter	33.8143	0.0295733	ounce
	liter	1.05668	0.946360	quart
	liter	0.2642	3.7853	gallon
	liter	61.025	0.016387	inch^3
	liter	0.0353	28.329	foot^3
Mass	milligram	0.015432	64.8004	grain
	milligram	3.5274×10^{-5}	28,349.49	ounce
	milligram	2.2046×10^{-6}	4.536×10^5	pound
	gram	0.035274	28.34949	ounce
	gram	0.002205	453.6	pound
	kilogram	2.2046	0.4536	pound
	kilogram	0.0011023	907.194	ton
Velocity	meters/second	3.2808	0.304804	feet/second
	kilometers/sec	2.2369	0.44705	miles/hour
Acceleration	meters/second^2	3.2808	0.30480	feet/second^2
	meters/second^2	39.3701	2.5400×10^{-2}	inches/second^2
Temperature	Celsius (°C)	1.8(°C) + 32	$\dfrac{(°F) - 32}{1.8}$	Fahrenheit (°F)
	Kelvin (°K)	1.8(°K) − 459.67	$\dfrac{(°F) + 459.67}{1.8}$	Fahrenheit (°F)
Flow Rate	liters/second	15.8508	0.063088	gallons/minute
	liters/second	22,824.5	4.38126×10^{-5}	gallons/day
	liters/second	0.0228245	43.8126	million gallons/day
	liters/second	0.035316	28.3158	feet^3/second
	meters^3/second	15,850.3	6.3088×10^{-5}	gallons/minute
	meters^3/second	2.28245×10^7	4.38126×10^{-8}	gallons/day
	meters^3/second	22.8245	4.38126×10^{-2}	million gallons/day
	meters^3/second	35.316	0.028316	feet^3/second
Energy	joule	0.9478	1.0551	British thermal unit

(continued)

Instructions on Use:
To convert, multiply in
direction shown by arrows. (cont)

SI Units	→	←	U. S. Units
joule	2.778×10^{-7}	3.600×10^{6}	kilowatt-hour
joule	0.7376	1.3557	foot-pound (force)
joule	1.000	1.0000	watt-second
joule	0.2388	4.1876	calorie
joule	2.778×10^{-4}	3,599.71	watt-hour
Power			
watt	0.7376	1.35575	foot-pounds (force)/second
watt	0.001341	745.7	horsepower
watt	9.478×10^{-4}	1,055.1	British thermal units/second
watt	0.014333	69.7691	calories/minute
Pressure			
pascal	1.4504×10^{-4}	6,894.65	pounds (force)/inch^2
pascal	2.0885×10^{-2}	47.88125	pounds (force)/foot^2
pascal	2.9613×10^{-4}	3,376.895	inches of mercury (60°F)
pascal	4.0187×10^{-3}	248.8367	inches of water (60°F)
pascal	9.8687×10^{-6}	101,330	atmosphere

Source: Courtesy of Garland STPM Press, New York, New York.

APPENDIX
B

Ten State Standards
for Lagoons

General

This Appendix deals with generally used variations of treatment ponds to achieve secondary treatment including controlled-discharge pond systems, flow-through pond systems and aerated pond systems. Ponds utilized for equalization, percolation, evaporation and sludge storage will not be discussed in this Appendix.

Supplement to Engineer's Report

The engineer's report shall contain pertinent information on location, geology, soil conditions, area for expansion and any other factors that will affect the feasibility and acceptability of the proposed project. The following information must be submitted in addition to that required in Chapter 10.

Supplementary Field Survey Data

Location of Nearby Facilities. The location and direction of all residences, commercial developments, parks, recreational areas, and water

Source: *Recommended Standards for Sewage Works.* 1978. A report of Committee of the Great Lakes-Upper Mississippi River Board of State Sanitary Engineers. Published by Health Education Service, Inc., P.O. Box 7126, Albany, New York 12224. Courtesy of Health Education Service Inc.

supplies (including a log of each well unless waived by the reviewing authority) within one mile (*1.6 km*) of the proposed pond shall be included in the engineer's report.

Land Use Zoning.　Land use zoning adjacent to the proposed pond site shall be included.

Site Description.　A description, including maps showing elevations and contours of the site and adjacent area shall be provided. Due consideration shall be given to additional treatment units and/or increased waste loadings in determining land requirements. Current U.S. Geological Survey and Soil Conservation Service maps may be considered adequate for preliminary evaluation of the proposed site.

Location of Field Tile.　The location, depth, and discharge point of any field tile in the immediate area of the proposed site shall be identified.

Soil Borings.　Data from soil borings conducted by an independent soil testing laboratory to determine subsurface soil characteristics and groundwater characteristics (including elevation and flow) of the proposed site and their effect on the construction and operation of a pond shall also be provided. At least one boring shall be a minimum of 25 feet (*7.6 m*) in depth or into bedrock, whichever is shallower. If bedrock is encountered, rock type, structure and corresponding geological formation data should be provided. The boring shall be filled and sealed. The permeability characteristics of the pond bottom and pond seal materials shall also be studied. (See Section 104.2)

Sulfate Content of Water Supply.　Sulfate content of the basic water supply shall be determined.

Location

Distance from Habitation

A pond site should be located as far as practicable, with a minimum of 1/4 mile (*0.4 km*), from habitation or any area which may be built up within a reasonable future period. Consideration should be given to site specifics such as topography, prevailing winds, forests, etc.

Prevailing Winds

If practicable, ponds should be located so that local prevailing winds will be in the direction of uninhabited areas.

Surface Runoff

Location of ponds in watersheds receiving significant amounts of storm-water runoff is discouraged. Adequate provision must be made to divert stormwater runoff around the ponds and protect pond embankments from erosion.

Hydrology

Construction of ponds in close proximity to water supplies and other facilities subject to contamination should be avoided. A minimum separation of 4 feet (*1.2 m*) between the bottom of the pond and the maximum groundwater elevation should be maintained.

Geology

Ponds shall not be located in areas which may be subjected to karstification, (i.e., sink holes or underground streams generally occurring in areas underlain by limestone or dolomite).

A minimum separation of 10 feet (*3.0 m*) between the pond bottom and any bedrock formation is recommended.

Basis of Design

Area and Loadings for Controlled-Discharge Stabilization Ponds

Pond design for BOD_5 loading may range from 15 to 35 pounds per acre per day (*17–45 kg/ha · d*) at the mean operating depth in the primary cells and at least 180 days detention time between the 2 foot (*0.6 m*) and the maximum operating depth of the entire pond system. The detention time and organic loading rate shall depend on climatic or stream conditions.

Area and Loadings for Flow-Through Stabilization Ponds

Pond design for BOD loading may vary from 15 to 35 pounds per acre per day (*17–45 kg/ha · d*) for the primary pond(s). The major design considerations for BOD loading must be directly related to the climatic conditions.

Design variables such as pond depth, multiple units, detention time, and additional treatment units must be considered with respect to applicable standards for BOD_5, total suspended solids (TSS), fecal coliforms, dissolved oxygen (DO), and pH.

A detention time of 90–120 days should be provided; however, this must be properly related to other design considerations. It should be noted that the major factor in the design is the duration of the cold weather period (water temperature less than 5°C.).

Aerated Pond Systems

For the development of final design parameters it is recommended that actual experimental data be developed; however, the aerated pond system design for minimum detention time may be estimated using the following formula:

$$t = \frac{E}{2.3\,K_1 \times (100 - E)}$$

t   =   detention time, days

E   =   percent of BOD_5 to be removed in an aerated pond

K_1 =   reaction coefficient, aerated lagoon, base 10. For normal domestic sewage, the K_1 value may be assumed to be 0.12/day at 20°C and 0.06/day at 1°C.

The reaction rate coefficient for domestic sewage which includes some industrial wastes, other wastes and partially treated sewage must be determined experimentally for various conditions which might be encountered in the aerated ponds. Conversion of the reaction rate coefficient at other temperatures shall be made based on experimental data.

Raw sewage strength should also consider the effect of any return sludge. Also, additional storage volume should be considered for sludge, and in northern climates, ice cover.

Oxygen requirements generally will depend on the BOD loading, the degree of treatment, and the concentration of suspended solids to be maintained. Aeration equipment shall be capable of maintaining a minimum dissolved oxygen level of 2 mg/l in the ponds at all times. Suitable protection from weather shall be provided for electrical controls.

See Chapter 80 for details on aeration.

Industrial Wastes

Consideration shall be given to the type and effects of industrial wastes on the treatment process. In some cases it may be necessary to pretreat industrial or other discharges.

Industrial wastes shall not be discharged to ponds without assessment of the effects such substances may have upon the treatment process or discharge requirements in accordance with state and federal laws.

Multiple Units

At a minimum, a pond system should consist of 3 cells designed to facilitate both series and parallel operations. The maximum size of a pond cell should be 40 acres (*16 ha*). Two-cell systems may be utilized in very small installations.

All systems should be designed with piping flexibility to permit isolation of any cell without affecting the transfer and discharge capabilities of the total system. In addition, the ability to discharge the influent waste load to a minimum of 2 cells and/or all primary cells in the system should be provided.

Controlled-Discharge Stabilization Ponds. For controlled-discharge systems the area specified as the primary ponds should be equally divided into two cells with the third or secondary cell volume a minimum of 1/3 the total volume of the entire system.

In addition, design should permit for adequate elevation difference between primary and secondary ponds to permit gravity filling of the secondary from the primary. Where this is not feasible, pumping facilities shall be provided.

Flow-Through Pond Systems. At a minimum, primary cells shall provide adequate detention time to maximize BOD removal. Secondary cells should then be provided for additional detention time with depths to eight feet (*2.4 m*) to facilitate solids reduction. Design should also consider recirculation within the system.

Aerated Pond Systems. For a total aerated system, a minimum of 3 cells employing a tapered mode of aeration is recommended.

Pond Shape

The shape of all cells should be such that there are no narrow or elongated portions. Round, square or rectangular ponds with a length not exceeding three times the width are considered most desirable. No islands, peninsulas or coves shall be permitted. Dikes should be rounded at corners to minimize accumulations of floating materials. Common-wall dike construction, wherever possible, is strongly encouraged.

Additional Treatment

Consideration should be given in the design stage to the utilization of additional treatment units as may be necessary to meet applicable discharge standards. See Section 101.13.

Pond Construction Details

Embankments and Dikes

Material. Dikes shall be constructed of relatively impervious material and compacted to at least 90 percent Standard Proctor Density to form a stable structure. Vegetation and other unsuitable materials shall be removed from the area where the embankment is to be placed.

Top Width.   The minimum dike width shall be 8 feet (*2.4 m*) to permit access of maintenance vehicles.

Maximum Slopes.   Inner and outer dike slopes shall not be steeper than 1 vertical to 3 horizontal (1:3).

Minimum Slopes.   Inner slopes should not be flatter than 1 vertical to 4 horizontal (1:4). Flatter slopes can be specified for larger installations because of wave action but have the disadvantage of added shallow areas being conducive to emergent vegetation. Outer slopes shall be sufficient to prevent surface runoff from entering the ponds.

Freeboard.   Minimum freeboard shall be 3 feet (*1.0 m*). For very small systems, 2 feet (*0.6 m*) may be acceptable.

Design Depth.   The minimum operating depth should be sufficient to prevent growth of aquatic plants and damage to the dikes, bottom, control structures, aeration equipment and other appurtenances. In no case should pond depths be less than 2 feet (*0.6 m*).

Controlled-Discharge Stabilization Ponds

The maximum water depth shall be 6 feet (*1.8 m*) in primary cells. Greater depths in subsequent cells are permissible although supplemental aeration or mixing may be necessary.

Flow-Through Stabilization Ponds: Same as Controlled-Discharge Stabilization Ponds

Aerated Pond Systems

The design water depth should be 10–15 feet (*3–4.5 m*). This depth limitation may be altered depending on the aeration equipment, waste strength and climatic conditions.

Erosion Control.   A justification and detailed discussion of the method of erosion control which encompasses all relative factors such as pond location and size, seal material, topography, prevailing winds, cost breakdown, application procedures, etc., shall be provided.

Seeding

The dikes shall have a cover layer of at least 4 inches (*10 cm*), of fertile topsoil to promote establishment of an adequate vegetative cover wherever riprap is not utilized. Prior to prefilling (in accordance with 104.24), adequate vegetation shall be established on dikes from the outside toe to 2 feet (*0.6 m*) above the pond bottom on the interior as measured on the slope. Perennial-type, low-growing, spreading grasses that minimize erosion and can be mowed are most satisfactory for seeding on dikes. In general, alfalfa and other long-rooted crops should not be used for seeding since the roots of this type are apt to impair the water holding efficiency of the dikes.

Additional Erosion Protection

Riprap or some other acceptable method of erosion control is required as a minimum around all piping entrances and exits. For aerated cells the design should ensure erosion protection on the slopes and bottoms in the areas where turbulence will occur. Additional erosion control may also be necessary on the exterior dike slope to protect the embankment from erosion due to severe flooding of a watercourse.

Alternate Erosion Protection

Alternate erosion control on the interior dike slopes may be necessary for ponds which are subject to severe wave action. In these cases riprap or acceptable equal shall be placed from one foot (0.3 m) above the high water mark to two feet (0.6 m) below the low water mark (measured on the vertical).

Pond Bottom

Soil. Soil used in constructing the pond bottom (not including seal) and dike cores shall be relatively incompressible and tight and compacted at or up to 4 percent above the optimum water content to at least 90 percent Standard Proctor Density.

Seal. Ponds shall be sealed such that seepage loss through the seal is as low as practicably possible. Seals consisting of soils, bentonite, or synthetic liners may be considered provided the permeability, durability, and integrity of the proposed material can be satisfactorily demonstrated for anticipated conditions. Results of a testing program which substantiates the adequacy of the proposed seal must be incorporated into and/or accompany the engineering report. Standard ASTM procedures or acceptable similar methods shall be used for all tests.

To achieve an adequate seal in systems using soil, bentonite, or other seal materials, the coefficient of permeability (K) in centimeters per second specified for the seal shall not exceed the value derived from the following expression:

$$K = 3.0 \times 10^{-9} \cdot L$$

where L equals the thickness of the seal in centimeters. The "K" obtained by the above expression corresponds to a percolation rate of pond water of less than 500 gallons per day per acre (4.68 m^3/ha·d) at a water depth of six feet (2 m).

For a seal consisting of a synthetic liner, seepage loss through the liner shall not exceed the quantity equivalent to seepage loss through an adequate soil seal.

Uniformity.   The pond bottom shall be as level as possible at all points. Finished elevations shall not be more than 3 inches (7.5 cm) from the average elevation of the bottom.

Prefilling.   Prefilling the pond should be considered in order to protect the liner, to prevent weed growth, to reduce odor, and to maintain moisture content of the seal. However, the dikes must be completely prepared as described in Sections 104.171 and 104.172 before the introduction of water.

Influent Lines

Material.   Generally accepted material for underground sewer construction will be given consideration for the influent line to the pond. Unlined corrugated metal pipe should be avoided, however, due to corrosion problems. In material selection, consideration must be given to the quality of the wastes, exceptionally heavy external loadings, abrasion, soft foundations, and similar problems.

Manhole.   A manhole or vented cleanout wye shall be installed prior to entrance of the influent line into the primary cell and shall be located as close to the dike as topography permits. Its invert shall be at least 6 inches (15 cm) above the maximum operating level of the pond and provide sufficient hydraulic head without surcharging the manhole.

Flow Distribution.   Flow distribution structures shall be designed to effectively split hydraulic and organic loads equally to primary cells.

Placement.   Influent lines shall be located along the bottom of the pond so that the top of the pipe is just below the average elevation of the pond seal; however, the pipe shall have adequate seal below it.

Point of Discharge.   All primary cells shall have individual influent lines which terminate at approximately the center of the cell so as to minimize short-circuiting. Consideration should be given to multi-influent discharge points for primary cells of 20 acres (8 ha) or larger to enhance the distribution of waste load in the cell.

All aerated cells shall have influent lines which distribute the load within the mixing zone of the aeration equipment. Consideration of multiple inlets should be closely evaluated for any diffused aeration system.

Influent Discharge Apron.   The influent line shall discharge horizontally into a shallow, saucer-shaped, depression.

The end of the discharge line shall rest on a suitable concrete apron large enough to prevent the terminal influent velocity at the end of the apron from causing soil erosion. A minimum size apron of 2 feet (0.6 m) square shall be provided.

Control Structures and Interconnecting Piping

Structure.   Where possible, facilities design shall consider the use of multi-purpose control structures to facilitate normal operational functions such as drawdown and flow distribution, flow and depth measurement, sampling, pumps for recirculation, chemical additions and mixing, and minimization of the number of construction sites within the dikes.

As a minimum, control structures shall be (a) accessible for maintenance and adjustment of controls; (b) adequately ventilated for safety and to minimize corrosion; (c) locked to discourage vandalism; (d) contain controls to permit water level and flow rate control, complete shutoff, and complete draining; (e) constructed of non-corrodible materials (metal-on-metal contact in controls should be of similar alloys to discourage electrochemical reactions); and (f) located to minimize short-circuiting within the cell and avoid freezing and ice damage.

Recommended devices to regulate water level are valves, slide tubes or dual slide gates. Regulators should be designed so that they can be preset to stop flows at any pond elevation.

Piping. All piping shall be of cast iron or other acceptable material. The piping shall not be located within or below the seal. Pipes should be anchored with adequate erosion control.

Drawdown Structure Piping

a. *Submerged Takeoffs*—For ponds designed for shallow or variable depth operations, submerged takeoffs are recommended. Intakes shall be located a minimum of 10 feet (*3.0 m*) from the toe of the dike and 2 feet (*0.6 m*) from the top of the seal, and shall employ vertical withdrawal.

b. *Multi-level Takeoffs*—For ponds that are designed deep enough to permit stratification of pond content, multiple takeoffs are recommended. There shall be a minimum of 3 withdrawal pipes at different elevations. The bottom pipe shall conform to a submerged takeoff. The others should utilize horizontal entrance. Adequate structural support shall be provided.

c. *Surface Takeoffs*—For use under constant discharge conditions and/or relatively shallow ponds under warm weather conditions, surface overflow-type withdrawal is recommended. Design should evaluate floating weir box or slide tube entrance with baffles for scum control.

d. *Maintenance Drawdown*—All ponds shall have a pond drain to allow complete emptying, either by gravity or pumping, for maintenance. These should be incorporated into the above-described structures.

e. *Emergency Overflow*—To prevent overtopping of dikes, emergency overflow should be provided.

Hydraulic Capacity

The hydraulic capacity for continuous discharge structures and piping shall allow for a minimum of 250 percent of the design flow of the system.

The hydraulic capacity for controlled-discharge systems shall permit transfer of water at a minimum rate of six inches (*15 cm*) of pond water depth per day at the available head.

Miscellaneous

Fencing

The pond area shall be enclosed with an adequate fence to prevent entering of livestock and discourage trespassing. Fencing should not obstruct vehicle traffic on top of the dike. A vehicle access gate of sufficient width to accommodate mowing equipment shall be provided. All access gates shall be provided with locks.

Access

An all-weather access road shall be provided to the pond site to allow year-round maintenance of the facility.

Warning Signs

Appropriate permanent signs shall be provided along the fence around the pond to designate the nature of the facility and advise against trespassing. At least one sign shall be provided on each side of the site and one for every 500 feet (*150 m*) of its perimeter.

Flow Measurement

Flow measurement requirements are presented in Section 46.6. Effective weather protection shall be provided for the recording equipment.

Groundwater Monitoring

An approved system of wells or lysimeters may be required around the perimeter of the pond site to facilitate groundwater monitoring. The need for such monitoring will be determined on a case-by-case basis.

Laboratory Equipment

For laboratory equipment refer to Chapter 40.

Pond Level Gauges

Pond level gauges shall be provided.

Service Building

Consideration in design should be given to a service building for laboratory and maintenance equipment.

Index